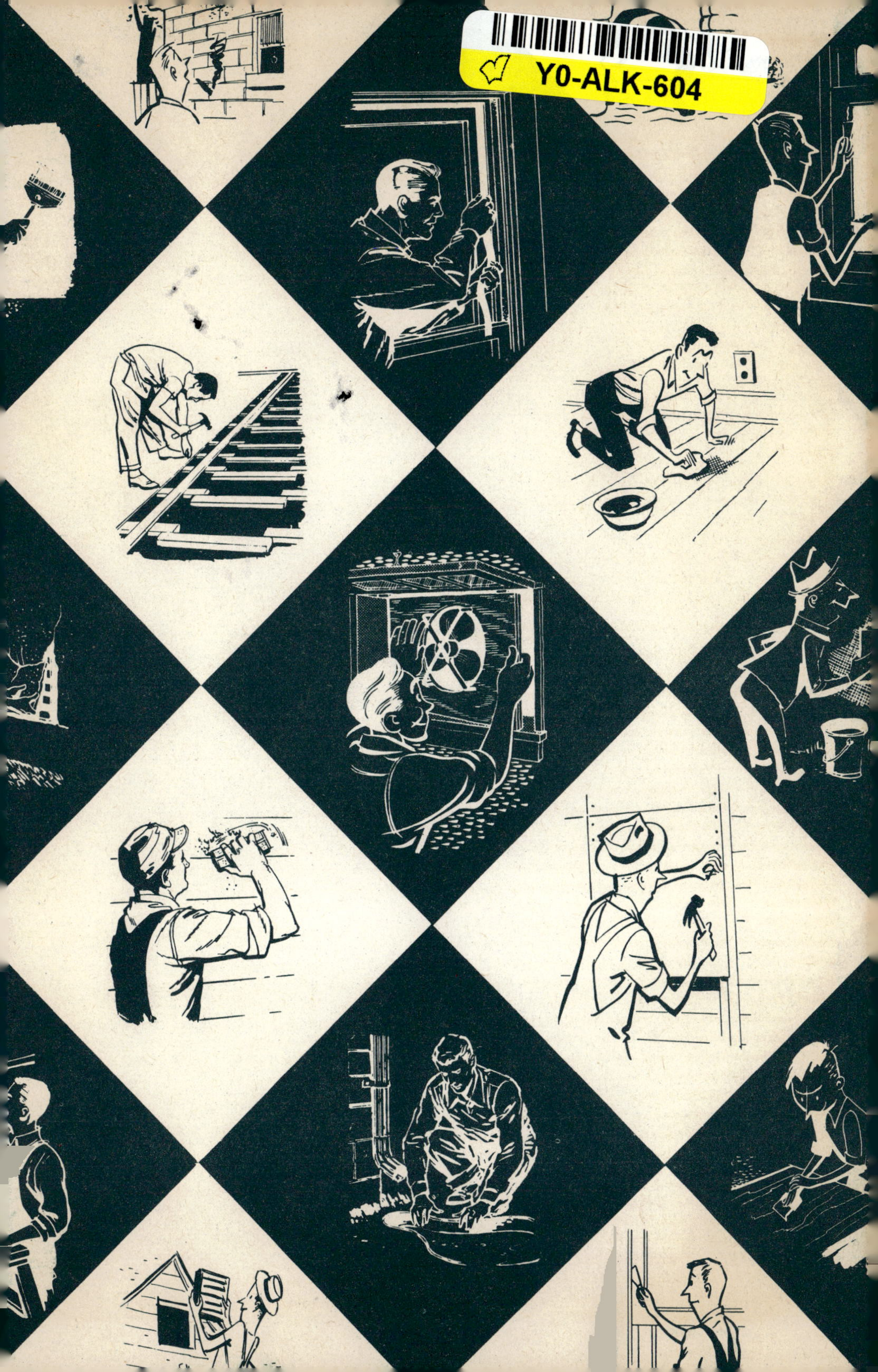

POPULAR MECHANICS DO-IT-YOURSELF ENCYCLOPEDIA

POPULAR MECHANICS DO-IT-YOURSELF ENCYCLOPEDIA

FOR

HOME OWNER, CRAFTSMAN

AND HOBBYIST

IN TWELVE VOLUMES

Volume VI

Complete Index in Volume XII

J. J. LITTLE & IVES Co., INC. • NEW YORK

COPYRIGHT 1955 BY POPULAR MECHANICS COMPANY

DESIGNED AND PRODUCED BY
BOOKSERVICE AMERICA, NEW YORK

NO PART OF THE TEXT OR
ILLUSTRATIONS IN THIS BOOK
MAY BE USED WITHOUT
FORMAL PERMISSION FROM
POPULAR MECHANICS COMPANY

ALL RIGHTS RESERVED UNDER
PAN-AMERICAN COPYRIGHT CONVENTION
AND THE INTERNATIONAL COPYRIGHT CONVENTION

FIRST PRINTING 1955
PRINTED IN THE UNITED STATES OF AMERICA

KITCHEN FOOD-CHOPPER BRACKET

HERE'S THE ANSWER to the question of where to clamp the food chopper—a simple two-piece wooden bracket that clamps firmly to the side rail of a kitchen counter or table. The bracket prevents marring of linoleum or plastic work surfaces and is attached or removed by means of a single thumbscrew. The lower member of the bracket, which serves as a support, is made from 3/4-in. maple or birch, and its design and size will depend to some extent on where it is placed, either on the side of a table or cabinet. In any case, be sure that you have the measurements correct before cutting stock. Make the curved cuts with a jigsaw or coping saw. Then cut the top piece to over-all size and cut the two rabbets on the front and back edges. Join the two parts with screws and waterproof glue. Then solder a 1/8 or 1/4-in. nut over a hole drilled in a small plate cut from heavy sheet metal. Drill two holes through the plate near opposite ends for small screws and screw the plate to the inside face of a kitchen cabinet or table rail as shown in the sectional view. Locate and drill the hole for the thumbscrew through the lower member of the bracket and also through the cabinet rail as shown. The food chopper is clamped to the top member of the bracket.

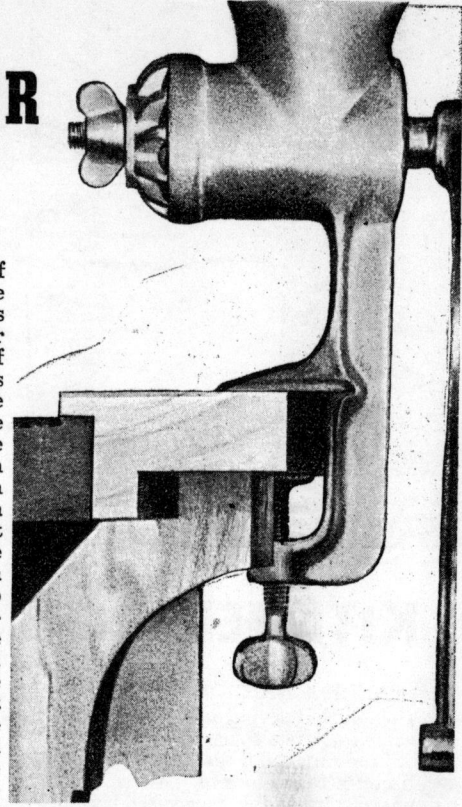

There's no danger of marring the finish of a table or cabinet top with this food-chopper bracket, as it attaches to the rail with a single thumbscrew. The chopper clamps to the top member of the bracket and is held firmly in position while being operated

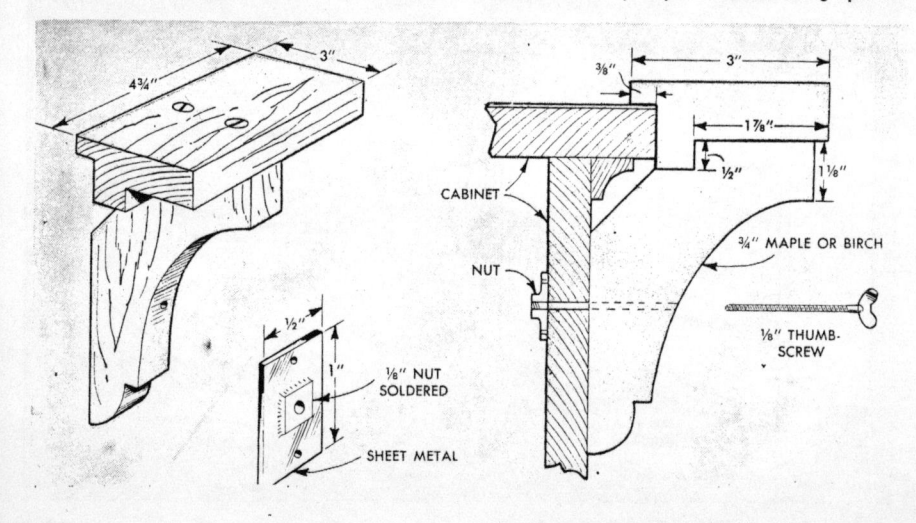

KITCHEN ROLL-OUT CABINET

IS THERE A NARROW SPACE in your kitchen that seems destined to be wasted? Such a waste space often is found between the range and the sink or between a wall and a range or sink. Too small for a standard cabinet, the space often becomes a bothersome cleaning problem and adds nothing to the convenience or appearance of the kitchen.

One solution that puts this waste space to work is a roll-out shelf cabinet that is simple to build yet adds greatly to the accessible storage space. Made to fit the opening, the cabinet consists of three shelves that pull out like a drawer, rolling on rubber-wheeled casters. The casters are non-pivoting and should be placed as wide apart as possible to increase stability.

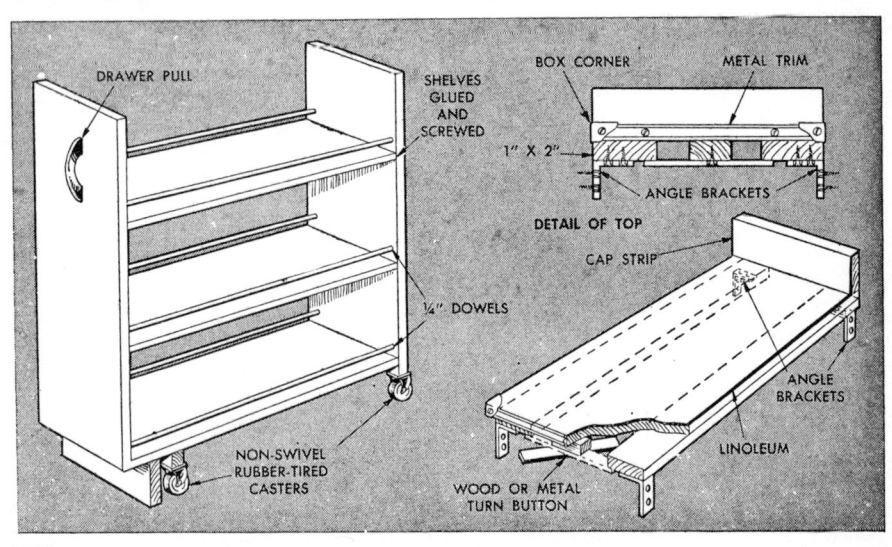

For the cabinet on the facing page, 8-in. pine boards were used throughout. The simple butt joints were glued and screwed for strength. The drawing shows the simple design of the cabinet. To prevent the stored items from sliding off the edges of the shelves, quarter-inch dowels are run along each side. The dowels are forced into holes drilled in the front and back uprights.

The top shelf, which is separate from the pull-out cabinet, rests on 1 x 2s attached to the sink and range with angle irons. If you prefer, the shelf can be supported at the rear by an angle iron screwed into the wall and at the front by a sheet-metal strap that slides into the joints between the top and sides of the range and sink. The shelf should be built so it can be easily removed for cleaning, as crumbs frequently lodge along its edges. Linoleum covering and metal trim add to its neatness and serviceability.

KITCHEN SHELVES

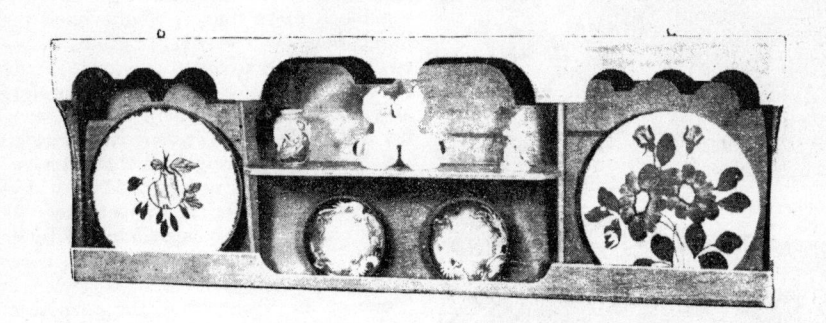

USUALLY a small hanging shelf is the answer to the decorative problem posed by a bare wall in the kitchen. Although pictured above as a china and knickknack shelf, this one, with its scrolled cornice, also serves equally well as a storage space for spices and condiments. The back is cut from ¾-in. plywood and all other parts are of ⅛ and ¼-in. plywood as indicated. Note especially the method of fitting the scrolled parts A, B and C. Part C is notched to fit inside the open end of the shelf while part B is an overlay. After sanding, join all parts with glue and brads and finish in the natural color of the wood with shellac, or in color with two coats of enamel.

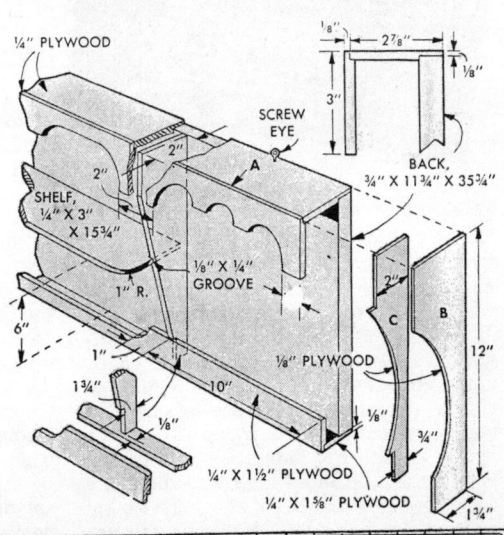

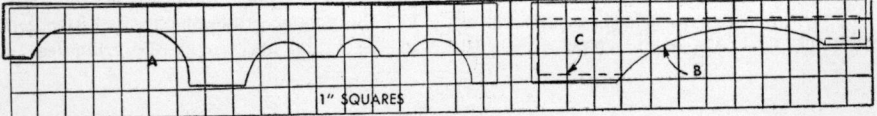

How to make

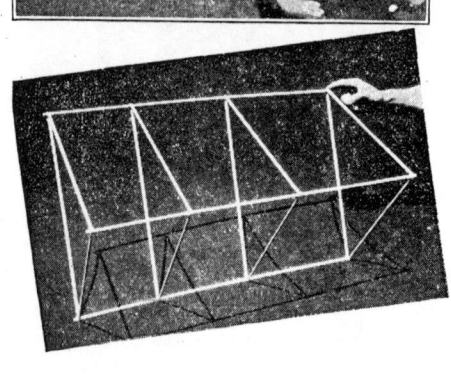

B OX kites varying in size from midgets to mammoths, to be used for either sport or utility, can be built by anyone at small expense. Such variable factors as material, thickness of the frame, weight of the twine, and other information on construction, are all given in a handy chart. Photo enthusiasts may use this method to obtain interesting aerial views by suspending the camera from a frame tied to the flying twine. For advertising purposes, banners can be lifted high into the air, a 6-ft. kite being capable of keeping a weight of from 5 to 8 lb. aloft. However, this capacity varies and is, of course, mainly dependent on the wind. To get a 6-ft. kite into the air and keep it there, requires a breeze that will give a pulling strain of about 18 lb. on the twine.

For best results when carrying banners, the bellybands are adjusted so that the kite will fly almost vertically above you, and the load will swing directly beneath it. If the kite were flown at a 35° or 40° angle, the actual pulling power would be a little greater, but the weight of the banner would then cause the twine to sag much more than it would when the kite is flown in a vertical position. Material used for an advertising banner should be the lightest goods obtainable, preferably about 4 oz. per square yard. A banner 3 ft. wide and 24 ft. long would weigh 2 lb. plus the weight of the paint required for illustrating it. The widest part of the banner should be tacked to a ⅜-in. stick which is tied to the flying twine and is also fitted with two cords to carry an 8-oz. weight as shown in one of the illustrations, the purpose of the weight being to hold the stick in as much of a vertical position as possible, which in turn keeps the banner floating at an even keel.

In constructing a kite, the first step is to get the proper material and prepare the triangles, four in number, around the vertical members. The parts used for the triangles should fit snugly against the vertical members by cutting small angular pieces off the ends. Refer to the top view for the relation of vertical and triangular members. The four braces comprising the back of the kite should be cemented at each joint with a good grade of wood cement and temporarily held in place with pins until all of the triangular sections have been attached and securely tied in position. After the triangles and the vertical sections have been completely assembled, they should be set aside and allowed to dry. As soon as the triangular framework is dry, this portion of the kite should be covered. Select the proper grade of paper as specified in the table, and cover the triangles in

KITES

one piece, overlapping about 2 in. On the inside of the triangles, cement a strip of paper about 2½ in. wide, over and against the vertical members, in order to reinforce and hold the paper rigidly at these points. On the 5 and 6-ft. kites, short vertical members are placed between the horizontal triangular members at three points, meeting the small triangle, which is placed inside of the large triangle, as shown in the top view. After the two triangles have been covered, the wings are cemented in place. The wing is cemented to the front of both triangles for a distance of 1½ in. and at the open section the paper folds around the string at the outline of the wing. As there is considerable strain on the front vertical member of the kite, it is important that the flying twine be attached around the four triangular sections to embrace the vertical pieces of the frame. This will prevent the kite from being pulled apart in a breeze, which incidentally occurred during experiments to determine the strength of these kites and the best method of bracing them. Flying large kites is usually no one-man job, the 6-ft. model requiring the combined strength of three huskies to hold it in a good wind.

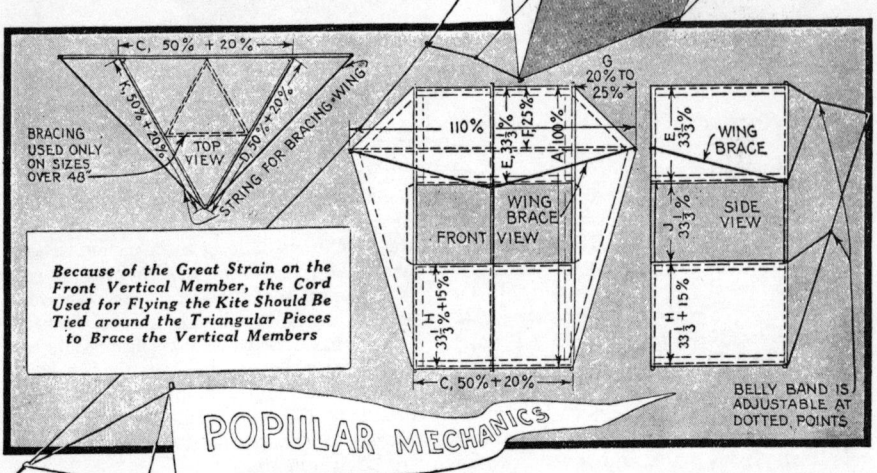

Great sport can be had by using a runner and parachute dropper. The runner is arranged so that it cannot fall off the string. However, it may be removed by opening the axle clips holding the axle which in turn slides through a brace on each of the wheels. The length of the runner is 8 in. and the main body is ½ in. square. The slot running lengthwise is ¼ in. square. Four side braces may be constructed from thin plywood, aluminum or light tin. Note that a part of the body flange is cut away to allow the wheel to set down evenly with the bottom of the slot through which the twine travels. This prevents the wheel from catching and binding. Two small stubs of thin wood are attached on each side of the body about midway between the two front guides. The stubs hold a cardboard wing which catches wind that drives the runner up to the kite. Three small wood guides are cemented to the bottom of the body. The holes in the guides should all be drilled alike. A piece of piano wire, .032 in. in diameter, bent to the required shape for the trip, is passed through the guides. The cardboard wing and parachute are pierced by the lower piece of trip wire, which is then pulled through the lower hole in the front guide and is ready to be sent on its way. A simple parachute can be made from an 18 or 24-in. square of tissue paper. Attach four threads, one and one-half times as long as one side, to each corner and tie them together as shown. To prevent the threads from becoming tangled, a small cork is placed at the knot. Attach this cork by cutting four slots and slipping the threads into them. If a small weight is attached below the cork, the threads will untwist immediately upon the release of the parachute, and the chute will open quickly.

Table of Measurements, Stock Thickness, and Weight of Twine for Various Size Kites
(See drawing for reference of letters and proportions)

Height A 100%	Width B 100 to 110%	Triangles C-D-K, 50% plus 20%	Wing G 20 to 25%	Height of triangle (upper) E 33⅓%	Height of space (middle) J 33⅓%	Height of triangle (lower) H 33⅓%+	Size of spruce or pine frame	Size twine
12	12 to 13	7.2	3	4	4	4	3/32 sq.	No. 60 thread
18	18 to 20	10.8	4 to 5	6	6	6	⅛ sq.	Light store string
24	24 to 26	14.4	5 to 6	8	7 to 8	8 to 9	5/32 sq.	Store string
30	30 to 33	18.	6 to 7	10	9 to 10	10 to 12	3/16 sq.	No. 18, 42-lb. twine
36	36 to 39	21.6	7 to 9	12	10 to 12	12 to 14	¼ sq.	No. 24, 56-lb. twine
48	48 to 53	28.8	10 to 12	16	13 to 16	16 to 19	5/16 sq.	No. 36, 86-lb. twine
60	60 to 66	36.	12 to 15	20	16 to 20	20 to 24	⅜ sq.	No. 48, 110-lb. twine
72	72 to 79	43.2	15 to 18	24	19 to 24	24 to 29	½ sq.	No. 60, 150-lb. twine

Note: Allow 15% extra for height of triangle H on kites under 4-ft. size. Thickness of triangle bracing is only 60% of the thickness of regular frame size. Use polished cotton twine where twine is referred to. Numerals in chart indicate size in inches. Covering material: Tissue paper for kites from 12 to 24 in. in height. 30 and 36-in. sizes have No. 30 Kraft paper triangles and tissue-paper wings. 48 to 72-in. sizes have No. 40 Kraft paper triangles and closely woven cloth wings.

It's KITE TIME *again!*

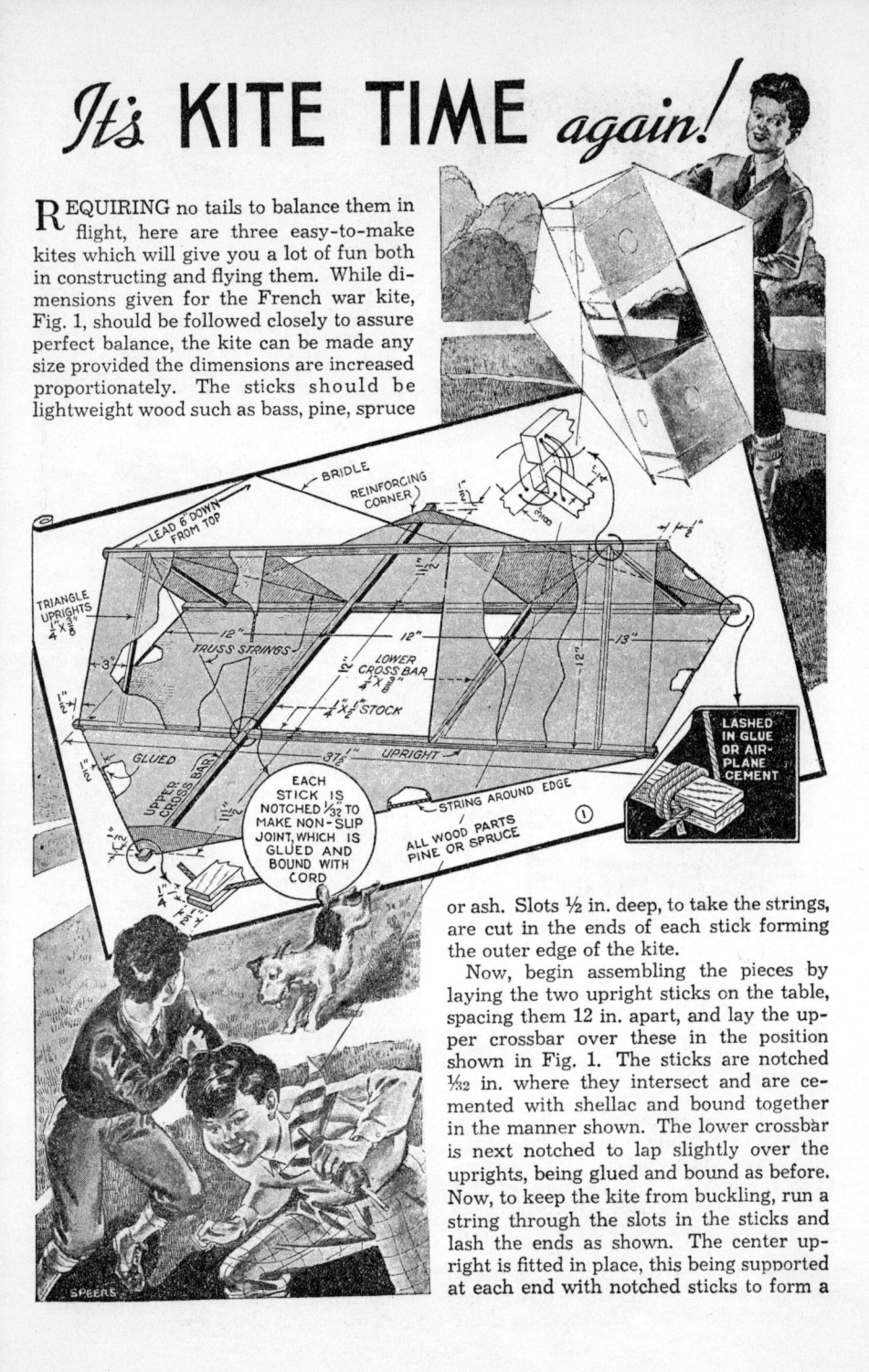

REQUIRING no tails to balance them in flight, here are three easy-to-make kites which will give you a lot of fun both in constructing and flying them. While dimensions given for the French war kite, Fig. 1, should be followed closely to assure perfect balance, the kite can be made any size provided the dimensions are increased proportionately. The sticks should be lightweight wood such as bass, pine, spruce or ash. Slots ½ in. deep, to take the strings, are cut in the ends of each stick forming the outer edge of the kite.

Now, begin assembling the pieces by laying the two upright sticks on the table, spacing them 12 in. apart, and lay the upper crossbar over these in the position shown in Fig. 1. The sticks are notched 1/32 in. where they intersect and are cemented with shellac and bound together in the manner shown. The lower crossbar is next notched to lap slightly over the uprights, being glued and bound as before. Now, to keep the kite from buckling, run a string through the slots in the sticks and lash the ends as shown. The center upright is fitted in place, this being supported at each end with notched sticks to form a

lower triangle is made ½ in. wider than that on the upper one. This is done to correct balancing. The capacity of the wing surface should be smaller than the surface covering of the triangle. This prevents the kite from diving. The ends and center, you will notice, are left open. The wings are covered next, turning and gluing the edges over the strings as before. Do not pull the covering tight, but allow it to bag slightly to catch all the air currents. Protect the corners of the wings by reinforcing with an extra covering of paper. The bridle should be attached to the top and run to a point two-thirds the way down. The towing point should be 4 to 6 in. from the top.

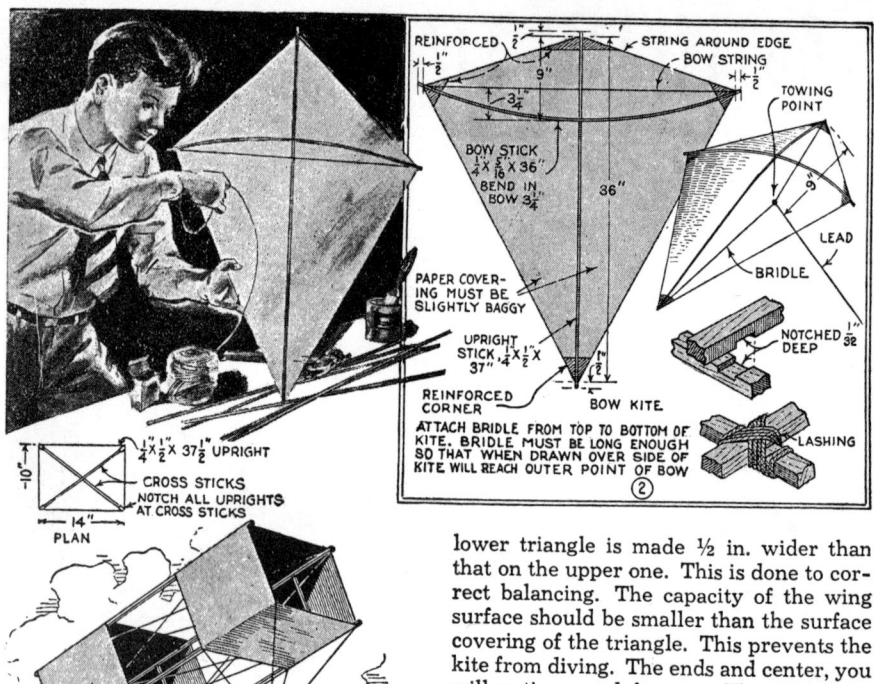

The simplicity of the bow kite, shown in Fig. 2, makes it very easy to construct. Here a cross strip, bowed 3¼ in. with a cord stretched through slots made in the ends, is notched 1/32 in. at midpoint as shown, to fit a similar notch cut in a center upright. When these two are joined a string is run around the outside of the kite and then the covering is applied as before.

To make the box kite in Fig. 3, four slotted uprights of identical size are assembled into two pairs of corresponding units by notched cross sticks. The units are joined then, into box-shape, and truss strings are stretched from the corner slots to keep it rigid and straight. The paper covering is made ½ in. wider on the lower box than on the upper. The towing point should equal the length of the kite.

triangular shaped assembly. When dry, run a string around the inner edges of the triangles to support the covering and, then brace them with several cross strings stretched on each side. Parchment or heavy-grade kite paper is the best covering material. Cellophane can be used also.

Cover the three sides of the two triangles and fold and glue the edges of the paper over the string. The covering of the

KNITTING TABLE

ATTRACTIVE in appearance and adaptable to a wide variety of uses, this handy knitting and sewing table may also be utilized to hold books and newspapers, candies or floral decorations. It may be constructed easily from 3/4-in. white pine and smartly finished in paint or varnish. With the exception of the bottom section, which measures 9 1/2 x 14 3/4 in., lay out all members on 3/4 x 5 5/8-in. stock. A paper pattern is helpful in tracing the outline of the legs and divider section onto wood. Cut the separate pieces just outside the guide lines, using a compass saw on the curved portions. Now assemble ends, sides, divider and bottom sections with glue and finishing nails. After fastening two legs to each leg brace with glue and flatheaded screws, rule diagonal lines from corner to corner of the bottom section. Then center the tops of legs over the lines and attach them to the underside of section. For accurate fitting of parts, glue and nail the pieces together before drilling pilot, shank and countersunk holes and driving in screws. Use a wood rasp to round all edges and sandpaper to provide a smooth surface before application of varnish or paint.

A driftwood finish may be obtained by using yellow pine, sanding down the soft wood around its pronounced grain and applying a light-gray stain—turpentine, 3 parts, to 1 part clear varnish—to which has been added a little dry pigment or oil colors. When dry, rub with No. 3 steel wool to bring out the grain and leave gray remaining in the low surrounding areas. Tack 1/2-in. furniture glides onto sides to give appearance of dowel pegs and paint with gray enamel. Apply two coats of dull-finish clear varnish to complete the job.

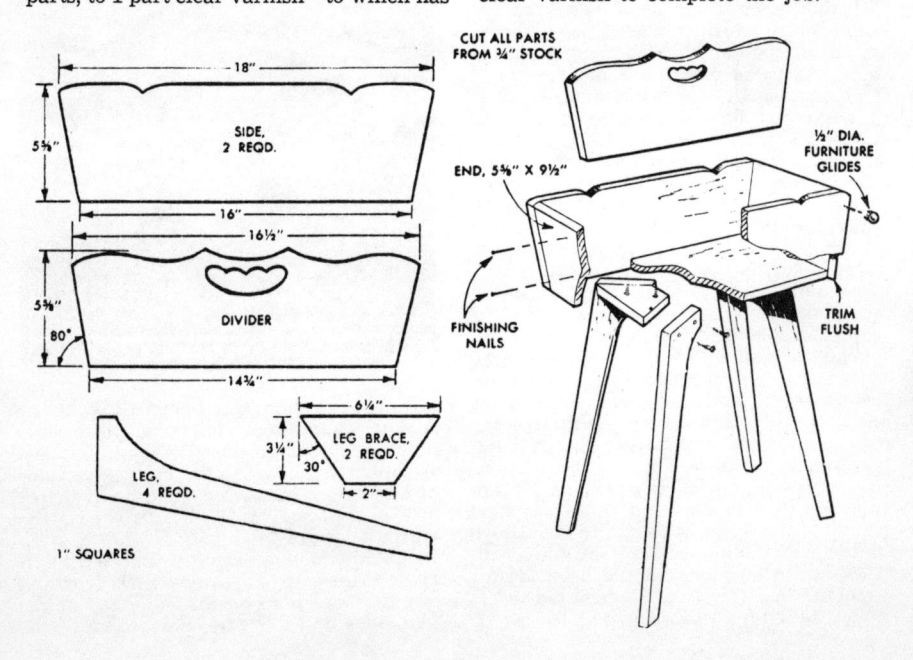

KNIVES FROM HACKSAW BLADES

STANDARD 1" POWER-HACKSAW BLADE

A B C
DEGREE OF BEVEL NEEDED FOR VARIOUS TYPES OF CUT

③ ROUGH-GRIND TAPER ON SIDE OF FINE WHEEL

WELL tempered, razor-sharp knives that keep their keen edge can be ground from worn or broken power-hacksaw blades, obtainable at many machine shops. As these blades are made of hard steel alloy such as molybdenum, which is so tough that a file will not cut it, shaping the blades must be done on a wet grindstone or very lightly on an emery wheel. Heavy pressure results in overheating and withdrawing the temper from the steel, which renders the blade useless. When grinding, a blade should be dipped in warm water every half minute or so, but if it is hot enough to sizzle, allow it to cool in the air. Tempered steel should never be immersed in cold water while the metal is hot.

When grinding knives from hacksaw blades, the best procedure is to work on two or three blades at once. As soon as one becomes too warm for the fingers, lay it aside and work on the next. The carving-knife blade shown in Fig. 1, utilizes an entire hacksaw blade. Broken blades are used for shorter knives. Before you start grinding, the shape of the knife is outlined on the blade with a wax crayon, and a full-size pattern on paper should be made for checking. A fairly coarse wheel is used to blank the knife to shape.

Rough-grinding the tempered sides to remove excess material, which is a slow process, is done on the side of a fine wheel as shown in Fig. 3. As the blank gets thinner it tends to heat more rapidly and therefore it is necessary to take light cuts the entire length of the blade. Be sure to keep the blade moving rapidly, not letting it come to rest for an instant. First work on one side and then on the other to cut the two sides down evenly. Smoothing of the

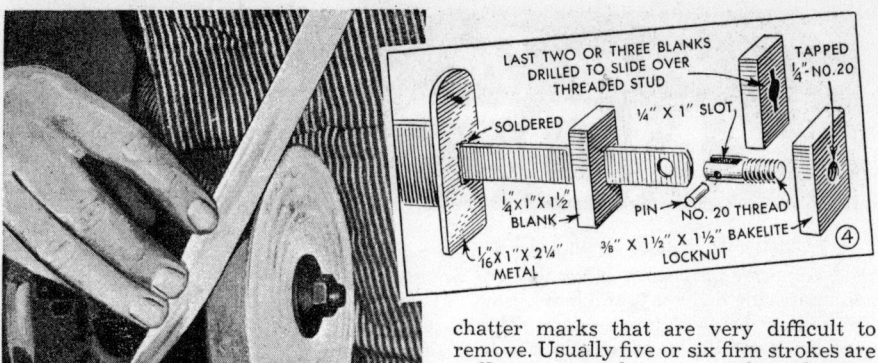

⑤ USE "GREASE WHEEL" TO REMOVE GRINDING MARKS

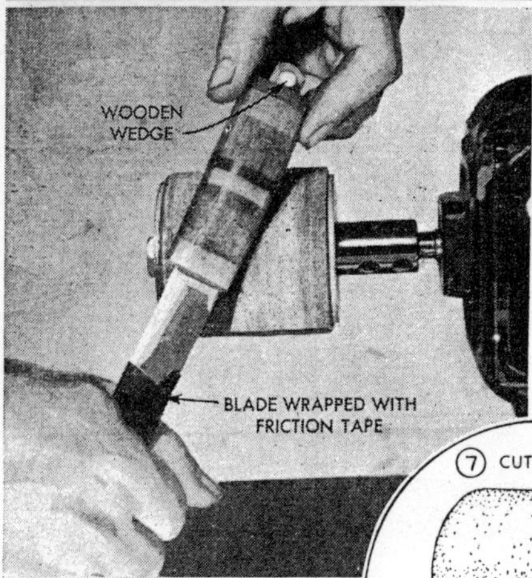

④ LAST TWO OR THREE BLANKS DRILLED TO SLIDE OVER THREADED STUD — SOLDERED — ¼" X 1" SLOT — TAPPED ¼"-NO. 20 — ¼" X 1" X 1½" BLANK — PIN — NO. 20 THREAD — ⅜" X 1½" X 1½" BAKELITE LOCKNUT — 1/16" X 1" X 2¼" METAL

⑥ WORK HANDLE TO SHAPE ON SANDING DRUM
— WOODEN WEDGE
— BLADE WRAPPED WITH FRICTION TAPE

⑦ CUTLERY BOB
1½" X 2½" HARD-MAPLE WHEEL COATED WITH EMERY GRITS

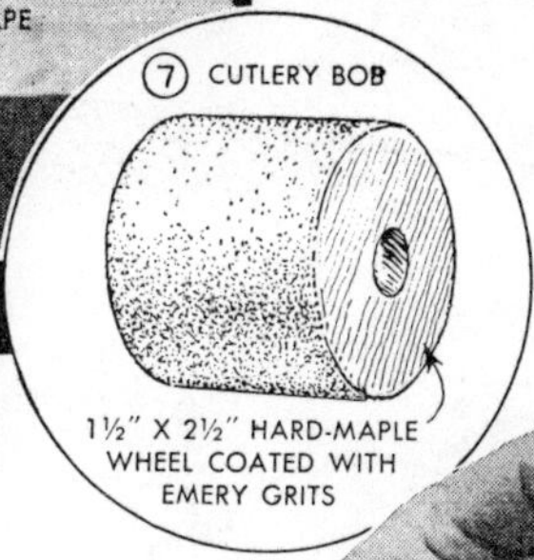

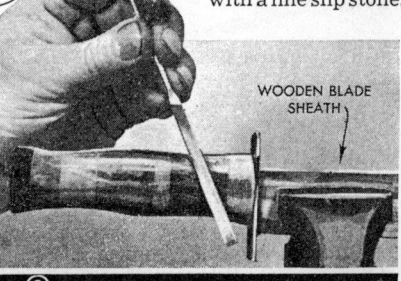

⑧ FINISH BY DRESSING WITH FINE FILE
— WOODEN BLADE SHEATH

chatter marks that are very difficult to remove. Usually five or six firm strokes are sufficient to produce a smooth surface.

From now on extreme caution must be exercised as the blade has become dangerously sharp. Slippery with grease, it can inflict severe injury if it gets out of control. Working the smooth surface of the blade down to a silvery sheen is done on a cutlery bob shown in Fig. 7. It is turned from hard maple, then treated with glue and rolled in very fine emery powder or silicon-carbide flour. No. 2F flour will produce a nice luster but for an extra-fine finish, make a second bob and coat it with No. 500 silicon-carbide flour. In doing this, use tallow and firm pressure, keeping the blade in constant motion. The knife edge will now be very sharp but will be "feathered," and will not stand up under any practical use. A slip stone rubbed lightly along the shoulders will alter the bevel of the edge to a point where the mechanical strength is sufficient to support the thrust, the bevel being determined by the work for which the knife is to be used. Bevel A in Fig. 2 is best for cutting raw meat where bone may be encountered; B for cooked meat, fowl, etc., and C for boneless meat, vegetables, etc. When dull, resharpen on the 500 bob, well-greased, and finish with a fine slip stone.

blade to eliminate grinding marks is done on an abrasive-coated cloth wheel, often referred to as a "grease" wheel because it is smeared with tallow. Such a wheel consists of a number of muslin buffs glued together after which the edge is treated with glue and rolled in emery powder. No. 120 powder will be suitable for the dressing operation. When dry, the wheel is smeared with tallow and is loaded with powdered pumice stone. Then the blade is pressed firmly against the wheel, dragging it the entire length on the wheel as shown in Fig. 5, the wheel rotating away from the edge of the knife. Avoid light pressure as it results in

It's Easy To Tie
KNOTS

ALTHOUGH a piece of rope or cord has no separate parts such as top, bottom, or sides, in knot tying one has to think of a length of rope or cord as having three sections. These are the two ends and the standing part, Fig. 3. No matter how complicated the knot it consists basically of three turns, the bight, and the overhand and underhand loops. Certain knots are formed on the ends of separate ropes, others are tied on the standing part alone and some are tied with the end and the standing part. Knots also are formed with the separate end strands of a rope. Skilled users always "work" a new rope before putting it in service. "Working" a rope is simply a process of pulling, stretching, and gently twisting it throughout the length to take out the stiffness. A cotton-braided rope of the clothesline

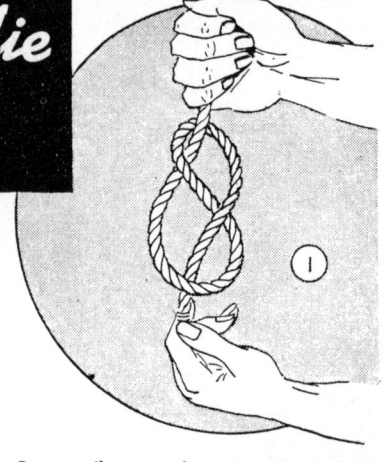

Do not coil or store damp or wet rope. Dry in the sun and then coil and store in a dry place. Rope that is not to be used for some time should never be allowed to tangle and kink. Always coil it when dry so that it will pay out smoothly and evenly

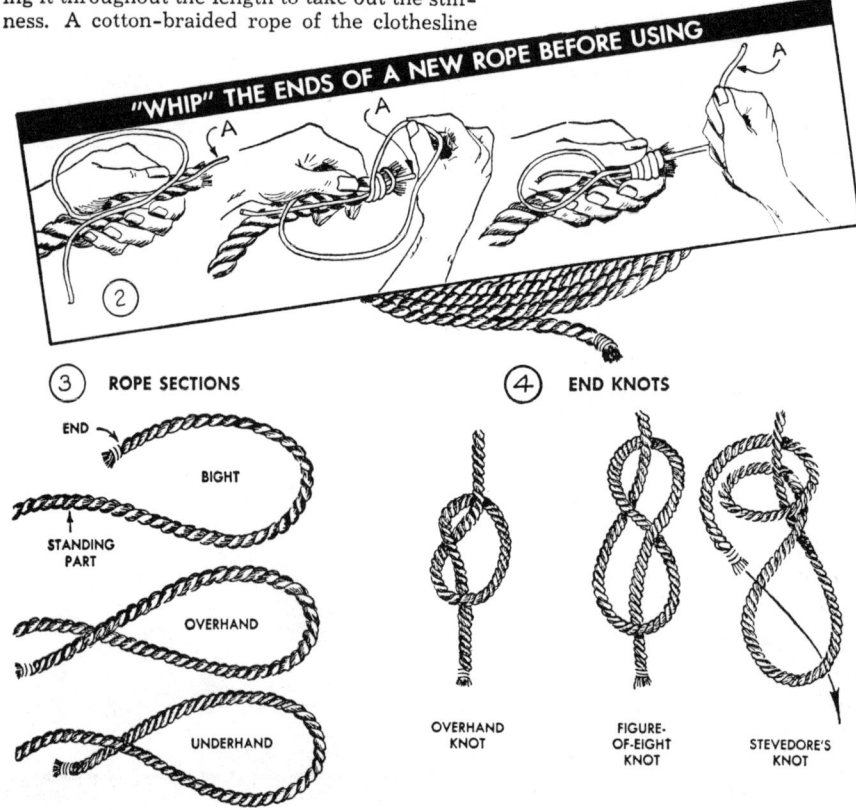

"WHIP" THE ENDS OF A NEW ROPE BEFORE USING

③ ROPE SECTIONS

END
BIGHT
STANDING PART
OVERHAND
UNDERHAND

④ END KNOTS

OVERHAND KNOT
FIGURE-OF-EIGHT KNOT
STEVEDORE'S KNOT

FLAT COIL, LONG COIL AND BACK SPLICE

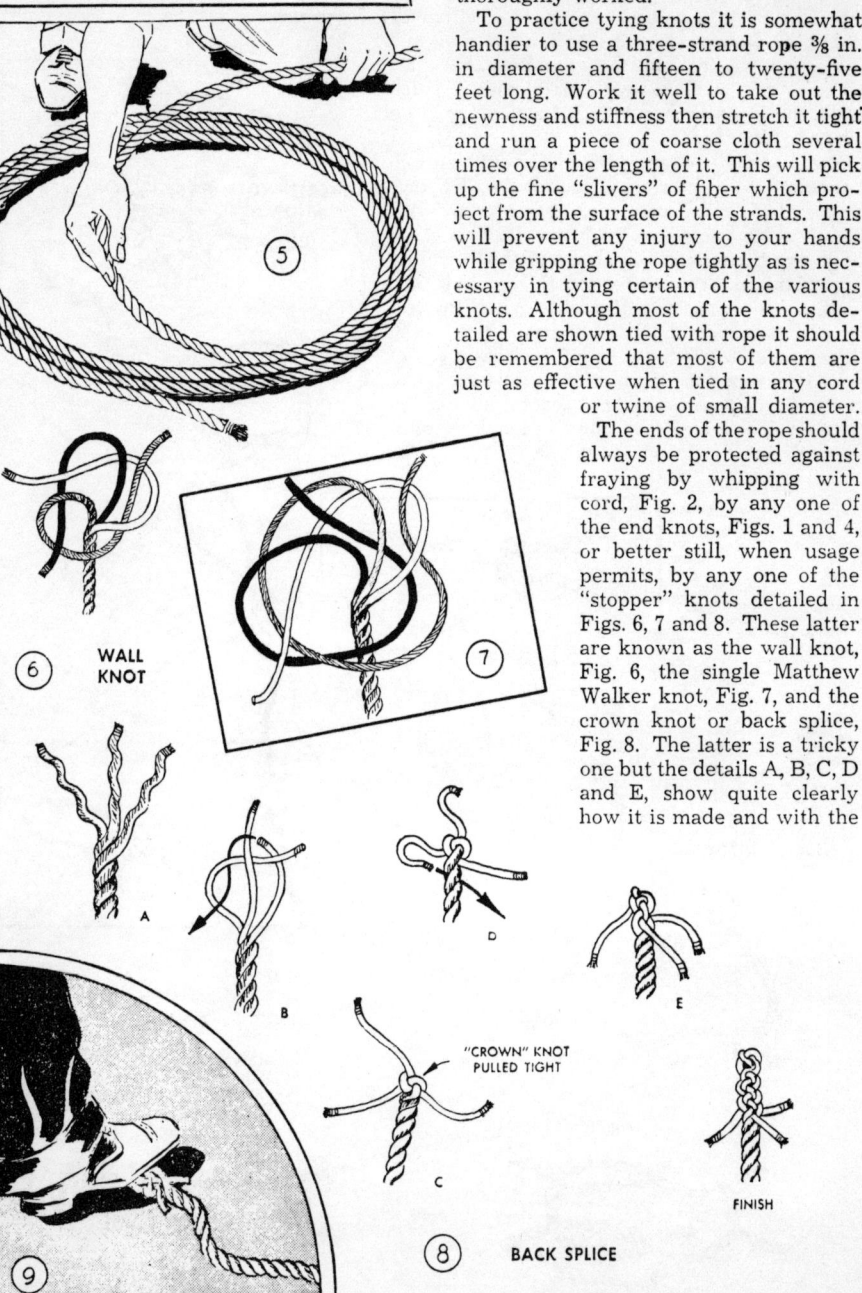

variety will quickly become soft and pliable in ordinary use, too soft for most rope work, but a hard-laid manila-fiber rope is not suitable for use until it has been thoroughly worked.

To practice tying knots it is somewhat handier to use a three-strand rope ⅜ in. in diameter and fifteen to twenty-five feet long. Work it well to take out the newness and stiffness then stretch it tight and run a piece of coarse cloth several times over the length of it. This will pick up the fine "slivers" of fiber which project from the surface of the strands. This will prevent any injury to your hands while gripping the rope tightly as is necessary in tying certain of the various knots. Although most of the knots detailed are shown tied with rope it should be remembered that most of them are just as effective when tied in any cord or twine of small diameter.

The ends of the rope should always be protected against fraying by whipping with cord, Fig. 2, by any one of the end knots, Figs. 1 and 4, or better still, when usage permits, by any one of the "stopper" knots detailed in Figs. 6, 7 and 8. These latter are known as the wall knot, Fig. 6, the single Matthew Walker knot, Fig. 7, and the crown knot or back splice, Fig. 8. The latter is a tricky one but the details A, B, C, D and E, show quite clearly how it is made and with the rope in your hands and the strands unlaid as in the first step A, it becomes easier still. When you end the splice lay the knot on the floor and roll it with your foot as in Fig. 9. If you're careful, it finishes off the end with a neat, professional job. The trick in getting a neat tie of either of the knots shown in Figs. 6 and 7, is to "snug up" the tucked strands separately and by stages until all three are in place and equally tight. If desired, the projecting ends of the strands may be whipped as in Fig. 2.

Right at the beginning it's important that one know how to coil a rope properly. Fig. 5 shows how to lay up an average length of rope in a flat coil, also known as the deck coil. You begin by laying the outer circle first and then winding inward in a clockwise direction giving a half turn to the rope as each loop is laid. When the full length has been laid, tighten the coil by grasping the edges and twisting it counter-clockwise. On very long ropes the same procedure is used, except that the rope is coiled in several layers. An outer wall is coiled first and the inside is built up with successive windings one on top of the

VARIATIONS OF THE SQUARE KNOT

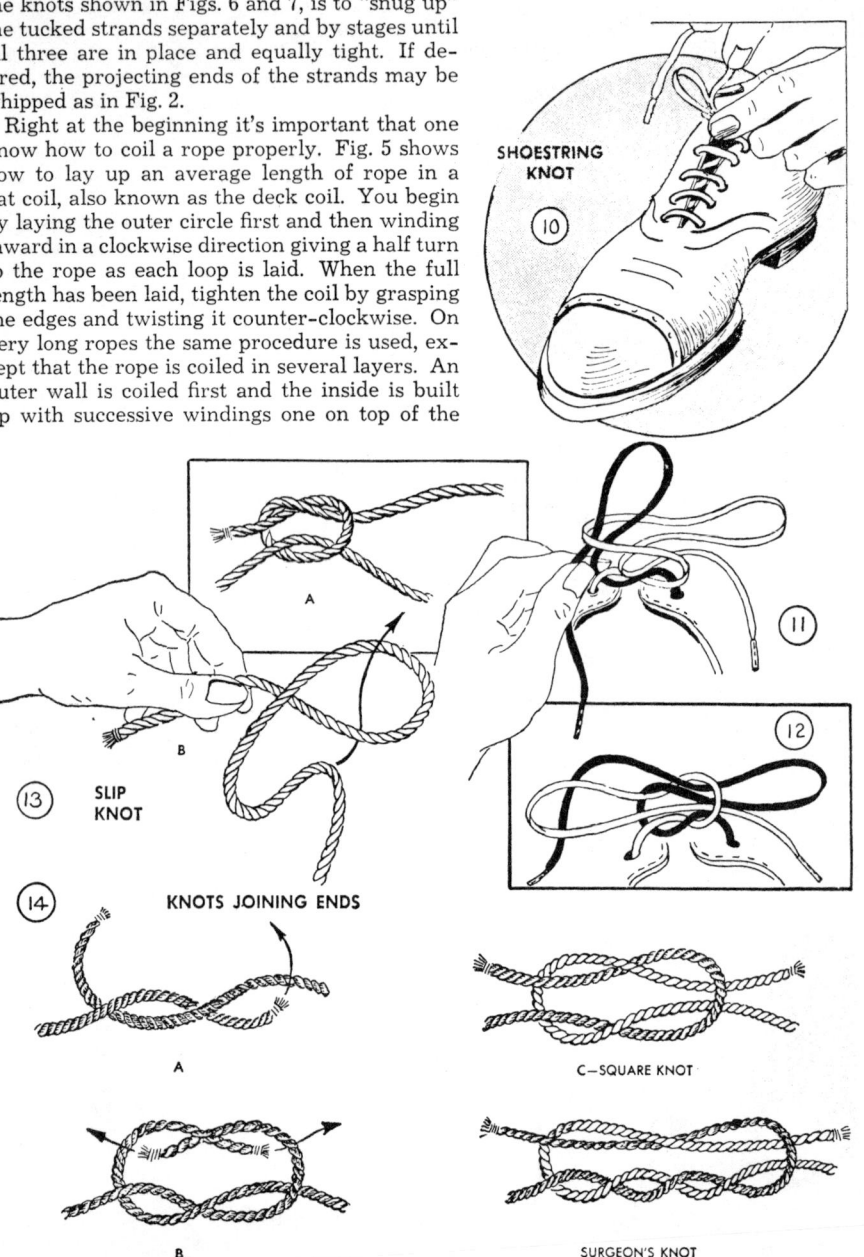

SHOESTRING KNOT
(10)

(11)

(12)

(13) SLIP KNOT

(14) KNOTS JOINING ENDS

A

B

C—SQUARE KNOT

SURGEON'S KNOT

"BENDS"

(15)

BINDER-TWINE BEND

WEAVER'S KNOT

SHEET BEND

"SLIPPERY" SHEET BEND

A

B
SINGLE CARRICK BEND

ENDS SEIZED TO STANDING PARTS

DOUBLE CARRICK BEND

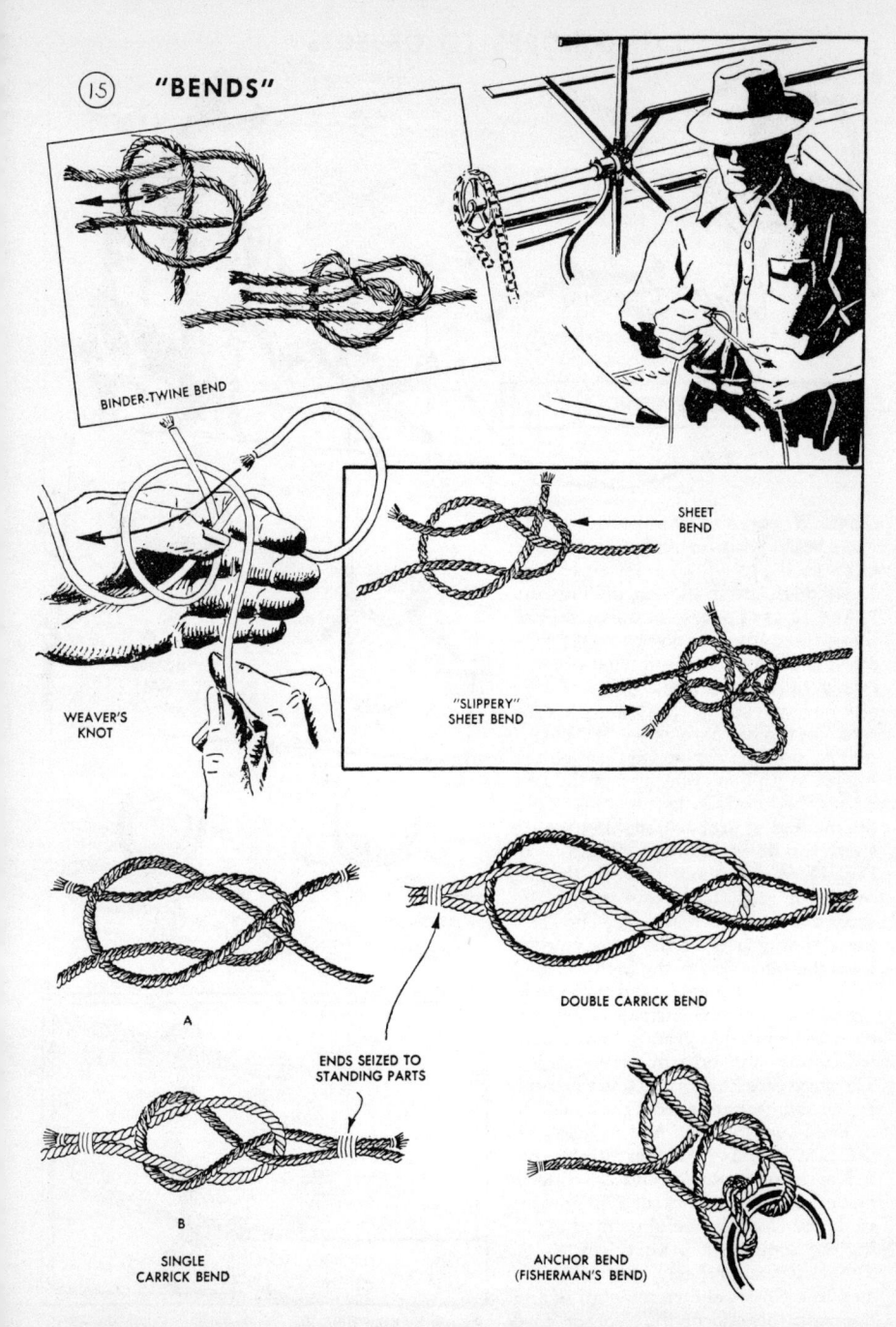

ANCHOR BEND (FISHERMAN'S BEND)

The single and double carrick bends are types of joining knots which are practical for use only on large-diameter ropes or hawsers joined for towing. The sheet bend is especially useful in joining ropes or heavy cord of different diameters. The slippery sheet bend is essentially the same thing except that one end is slippery, that is, the knot can be untied merely by a tug on the free rope end

TYING ROPES TO OBJECTS

(16) CLOVE HITCH

(17) HALF HITCH

DOUBLE HALF HITCH

(18) SLIPPERY HALF HITCH

(19) TIMBER HITCH

other. When you pay out the rope always begin with the end at the center of the coil.

Although the shoestring tie, Figs. 10, 11 and 12, is of course in universal use there are many who do not tie it properly. Too often it ends in what is commonly called a "granny" knot. Actually the shoestring tie is a square or reef knot with both ends "slippery," that is, the ends are looped through the bights. A pull on either end of the lace unties the knot. To better understand the method of properly making the tie study the three details A, B and C, in Fig. 14, which show how to tie the square or reef knot. Note that in the second crossing, B and C, the ends and the standing part of the rope emerge from the same side of the bight formed in the other. If the end and the standing part are on opposite sides then you have the so-called granny knot which will either slip or jam under strain. The surgeon's knot, Fig. 14, is the same as the square knot except that one additional turn of each end is made on the first crossing. Under strain this will hold until the second crossing is made. Fig. 13 shows one important variation from the usual method of tying the common slip knot, Fig. 13, B. The latter is properly tied with the standing part as shown in detail B, and not with the end of the rope or cord as is so often done.

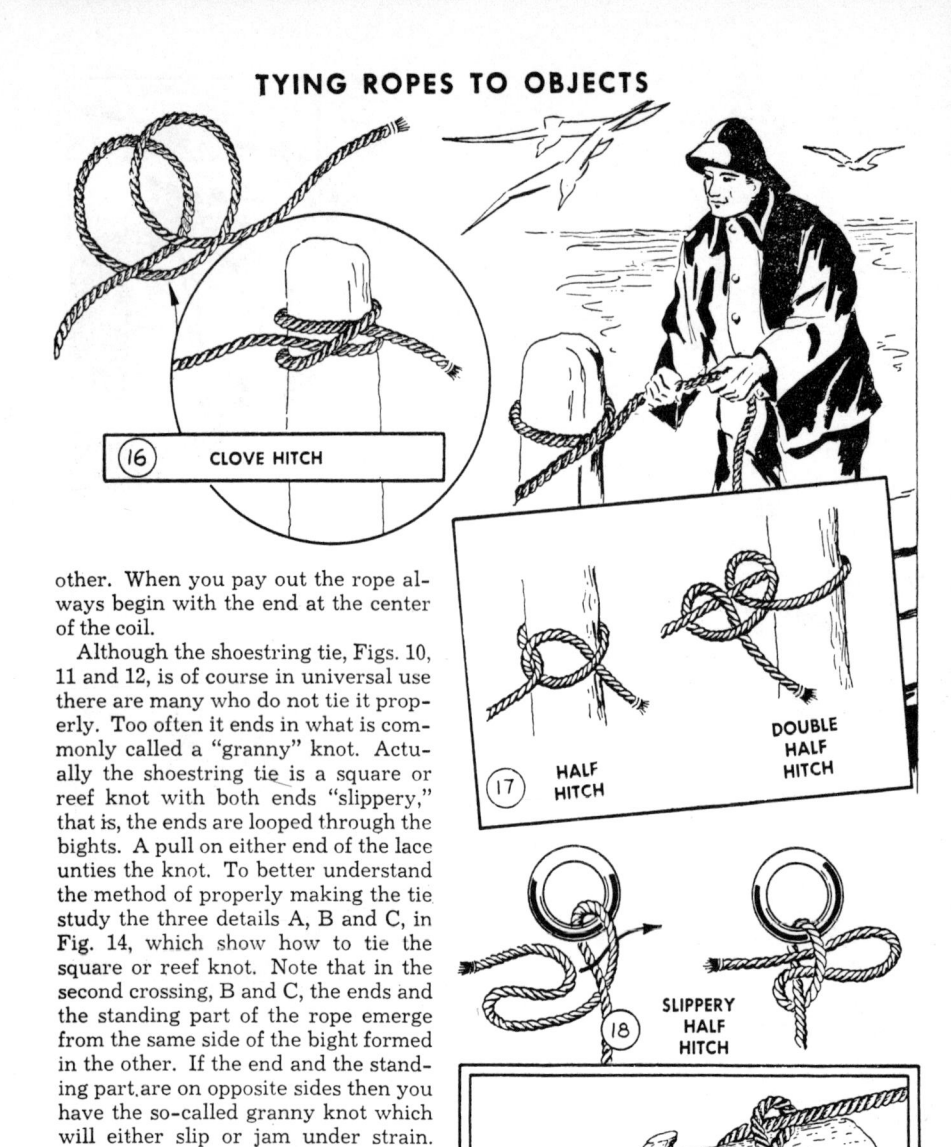

Fig. 15 details a number of "bends" which are used chiefly for joining ends

As will be seen from the above details, practically all the simple hitches are essentially the same, consisting fundamentally of turns about the object and loops in the free rope end. For the sake of clarity the timber hitch is shown above incomplete. It is ordinarily finished as in Fig. 22

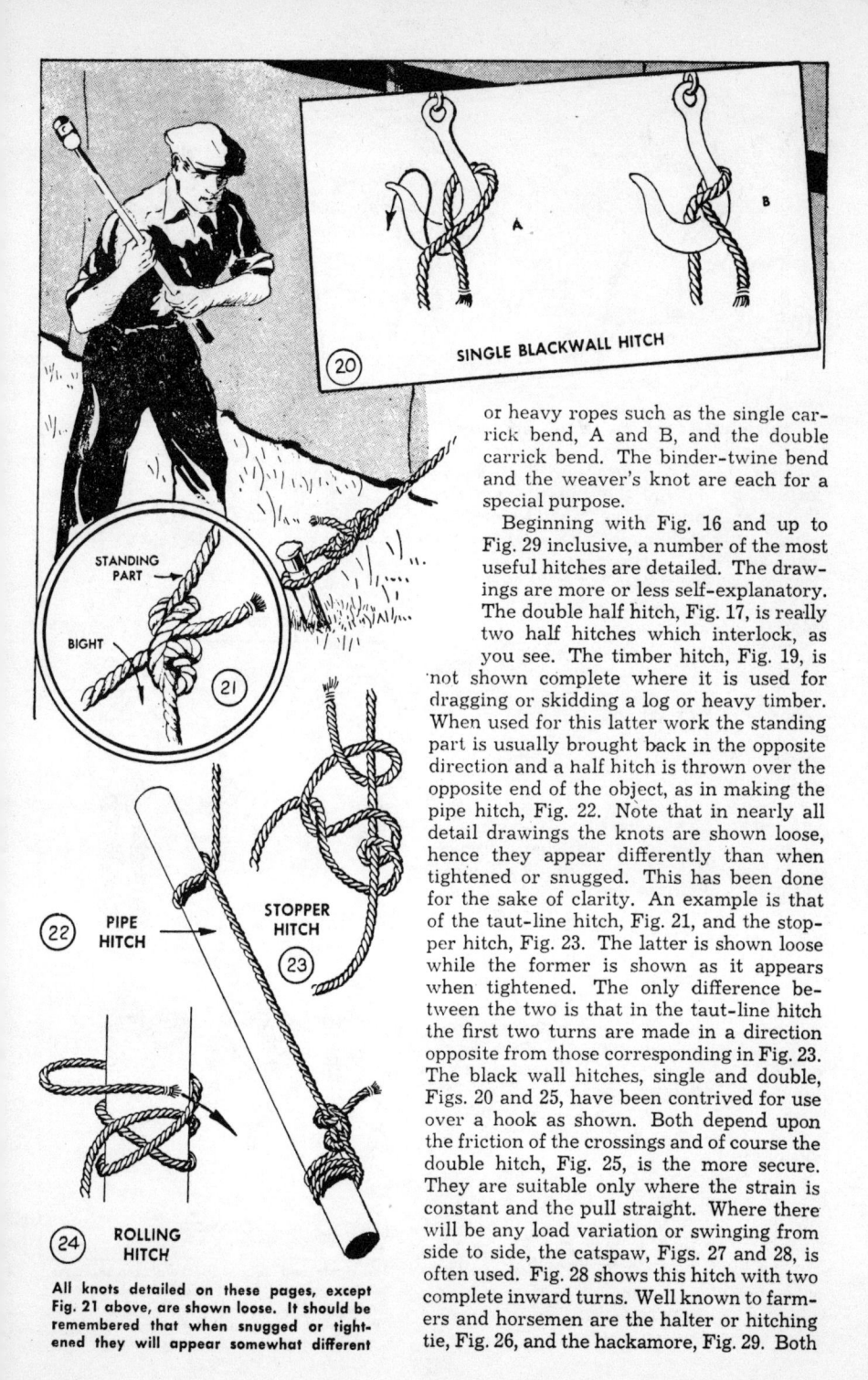

SINGLE BLACKWALL HITCH

STANDING PART

BIGHT

PIPE HITCH

STOPPER HITCH

ROLLING HITCH

All knots detailed on these pages, except Fig. 21 above, are shown loose. It should be remembered that when snugged or tightened they will appear somewhat different

or heavy ropes such as the single carrick bend, A and B, and the double carrick bend. The binder-twine bend and the weaver's knot are each for a special purpose.

Beginning with Fig. 16 and up to Fig. 29 inclusive, a number of the most useful hitches are detailed. The drawings are more or less self-explanatory. The double half hitch, Fig. 17, is really two half hitches which interlock, as you see. The timber hitch, Fig. 19, is not shown complete where it is used for dragging or skidding a log or heavy timber. When used for this latter work the standing part is usually brought back in the opposite direction and a half hitch is thrown over the opposite end of the object, as in making the pipe hitch, Fig. 22. Note that in nearly all detail drawings the knots are shown loose, hence they appear differently than when tightened or snugged. This has been done for the sake of clarity. An example is that of the taut-line hitch, Fig. 21, and the stopper hitch, Fig. 23. The latter is shown loose while the former is shown as it appears when tightened. The only difference between the two is that in the taut-line hitch the first two turns are made in a direction opposite from those corresponding in Fig. 23. The black wall hitches, single and double, Figs. 20 and 25, have been contrived for use over a hook as shown. Both depend upon the friction of the crossings and of course the double hitch, Fig. 25, is the more secure. They are suitable only where the strain is constant and the pull straight. Where there will be any load variation or swinging from side to side, the catspaw, Figs. 27 and 28, is often used. Fig. 28 shows this hitch with two complete inward turns. Well known to farmers and horsemen are the halter or hitching tie, Fig. 26, and the hackamore, Fig. 29. Both

DOUBLE BLACKWALL HITCH

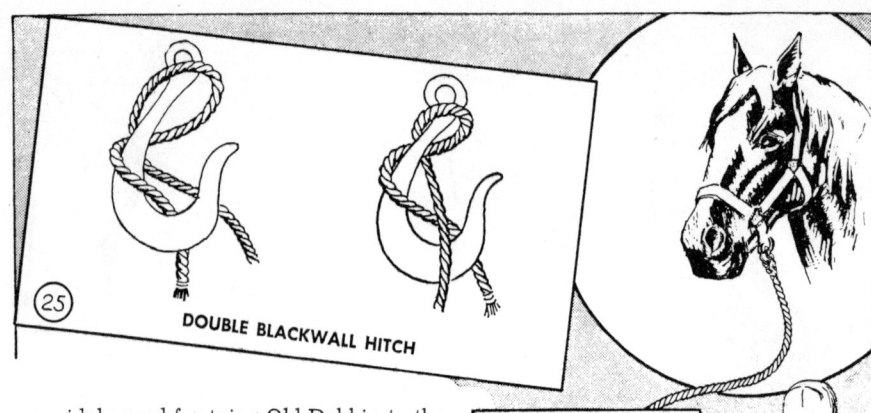

are widely used for tying Old Dobbin to the hitching post or the manger. Both have the feature of being easy to untie. The first is "locked" by passing the end through the bight as shown in the right-hand view. To untie, simply pull out the end and give it a light jerk. Note the similarity of the hackamore to the figure-of-eight knot in Fig. 3. Both knots are good ones to know as they are useful for other purposes.

When you need to shorten a rope without cutting it or you find a weak spot in a long rope which needs strengthening, then the sheepshank, Figs. 30 and 31, is the answer to the problem. Take up the slack as in the top detail, Fig. 30, then throw single or double half hitches over the loops as shown.

Everyone should know the trick of wrapping and tying a parcel post or express package securely. Fig. 32 details what is known as the packer's knot. It is simply a figure-of-eight knot with the end emerging parallel with the standing part. On square packages, place the loop over the package, center it, and pull tight to make the first crossing at right angles to the ends. Take the standing part down over one end and back to the first crossing. Take it over and under the first crossing, then around the opposite end and back to the starting point. To fasten the cord pull it tight and throw a series of half hitches as in Fig. 35. In tying a long rectangular package proceed as in Figs. 33 and 34, and finish as in Fig. 35.

If you ever chance to be faced with a sudden emergency where quick action with a rope may mean saving a life, you should know how to tie the various forms of the bowline knot, Figs. 36 to 40 inclusive. The single bowline or bowline loop is a most valuable and important knot as it forms a loop of any required size and the knot will

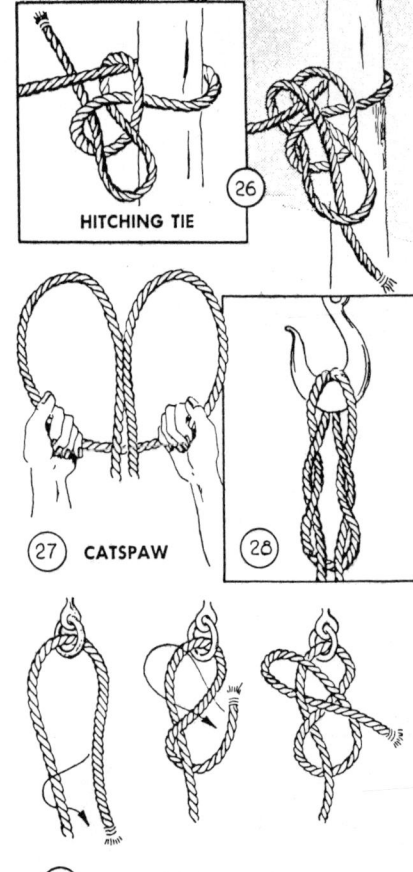

HITCHING TIE

CATSPAW

HACKAMORE

The single and double blackwall hitches hold securely when subjected to a continuous strain. However, they are safe for human life only when taken in the middle of a rope with both ends fast and supporting the load

SHEEPSHANK AND PACKAGE TIE

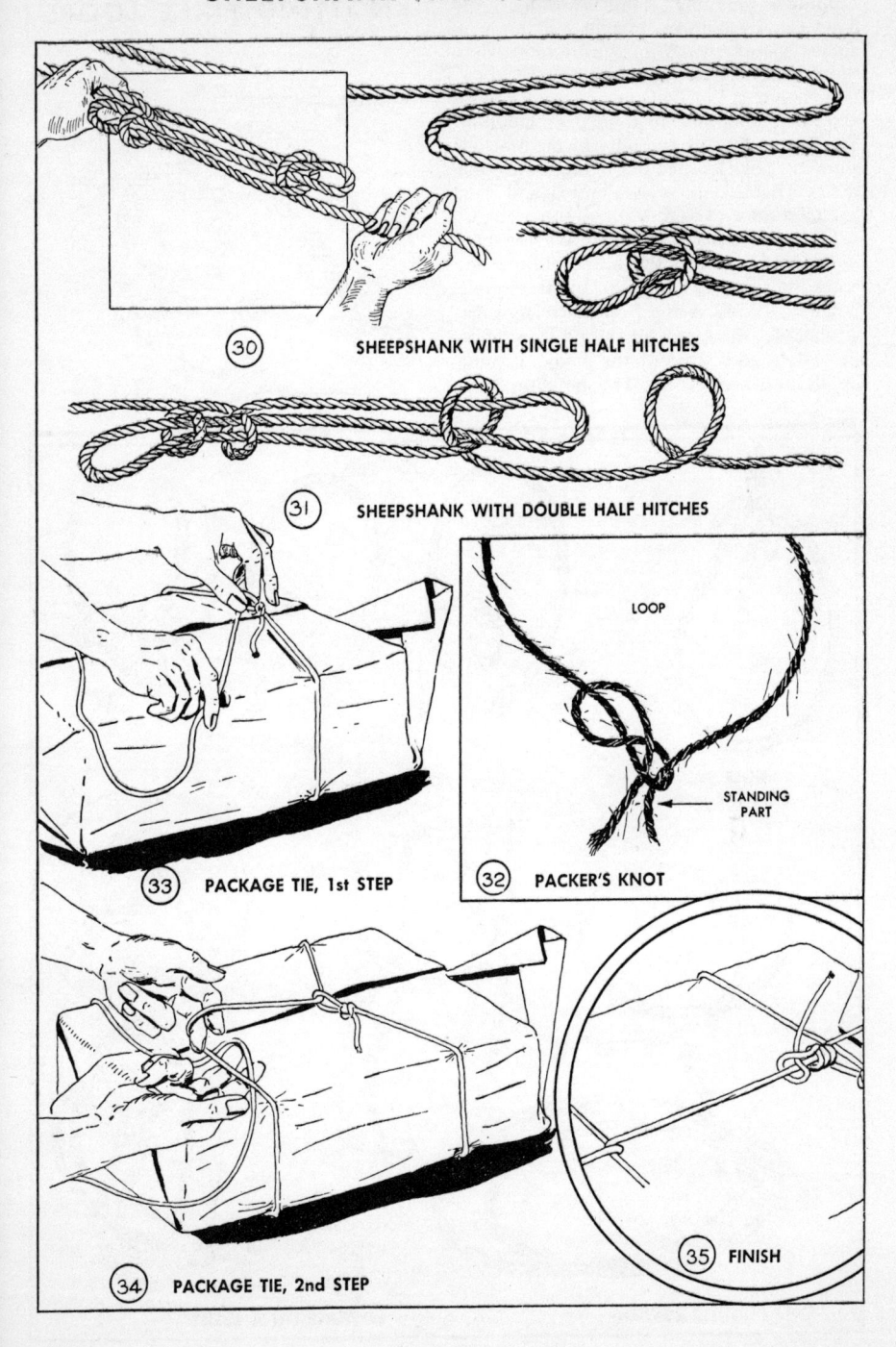

30 SHEEPSHANK WITH SINGLE HALF HITCHES

31 SHEEPSHANK WITH DOUBLE HALF HITCHES

32 PACKER'S KNOT

LOOP

STANDING PART

33 PACKAGE TIE, 1st STEP

34 PACKAGE TIE, 2nd STEP

35 FINISH

stand any strain the rope will bear without slipping or jamming. If one is familiar with it he can tie it in an instant and untie it almost as quickly. Although there are differing methods of tying the single bowline, that shown in Fig. 36 is one of the simplest and most common. In the Texas bowline, Fig. 37, the knot is partially formed on the standing part by passing a bight through the overhand loop, as in A and B, and then bringing the end back through the bight, C. A figure-of-eight or Matthew Walker knot in the end prevents the latter pulling out when the knot is snugged. In the double bowline, Fig. 38, A and B, the two loops are adjustable. One may sit in one loop while the other goes around the body, leaving hands and arms free. The bowline on a

KNOTTING FIXED LOOPS

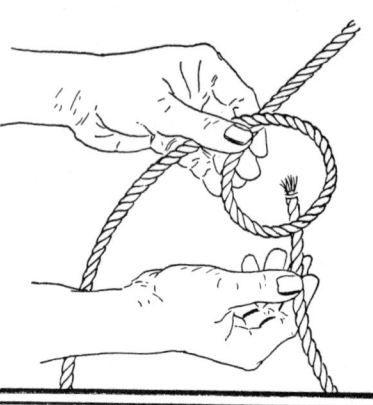

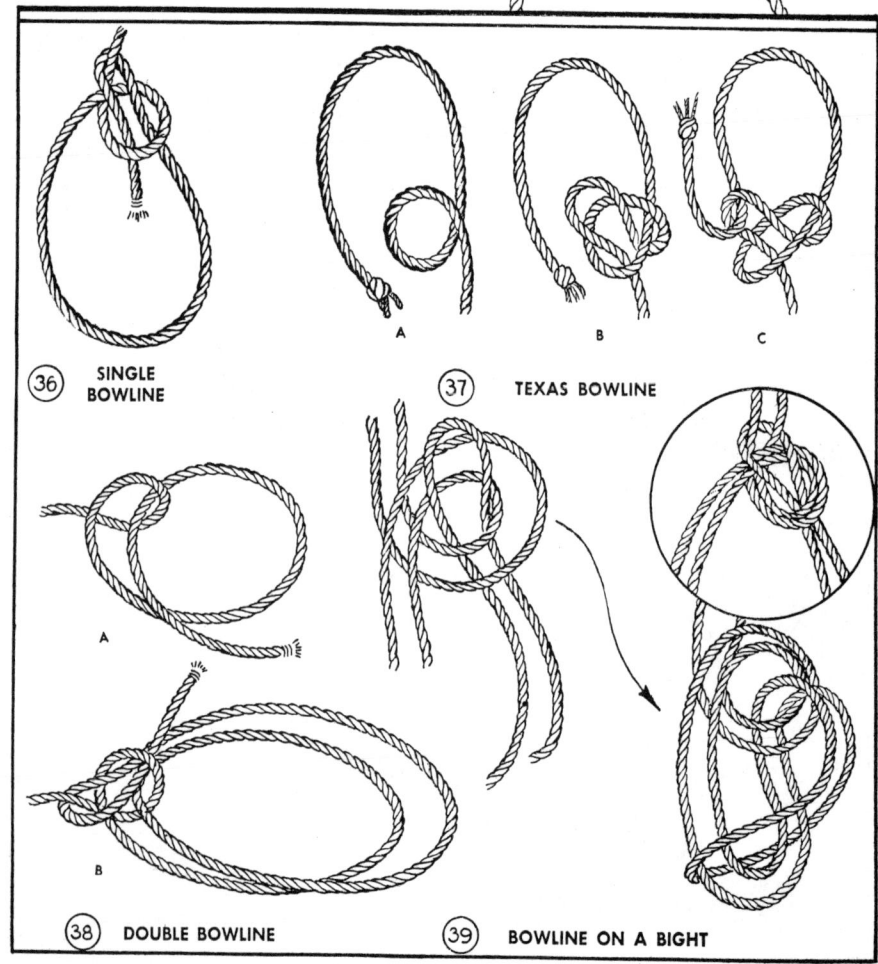

(36) SINGLE BOWLINE

(37) TEXAS BOWLINE

(38) DOUBLE BOWLINE

(39) BOWLINE ON A BIGHT

40 RUNNING BOWLINE

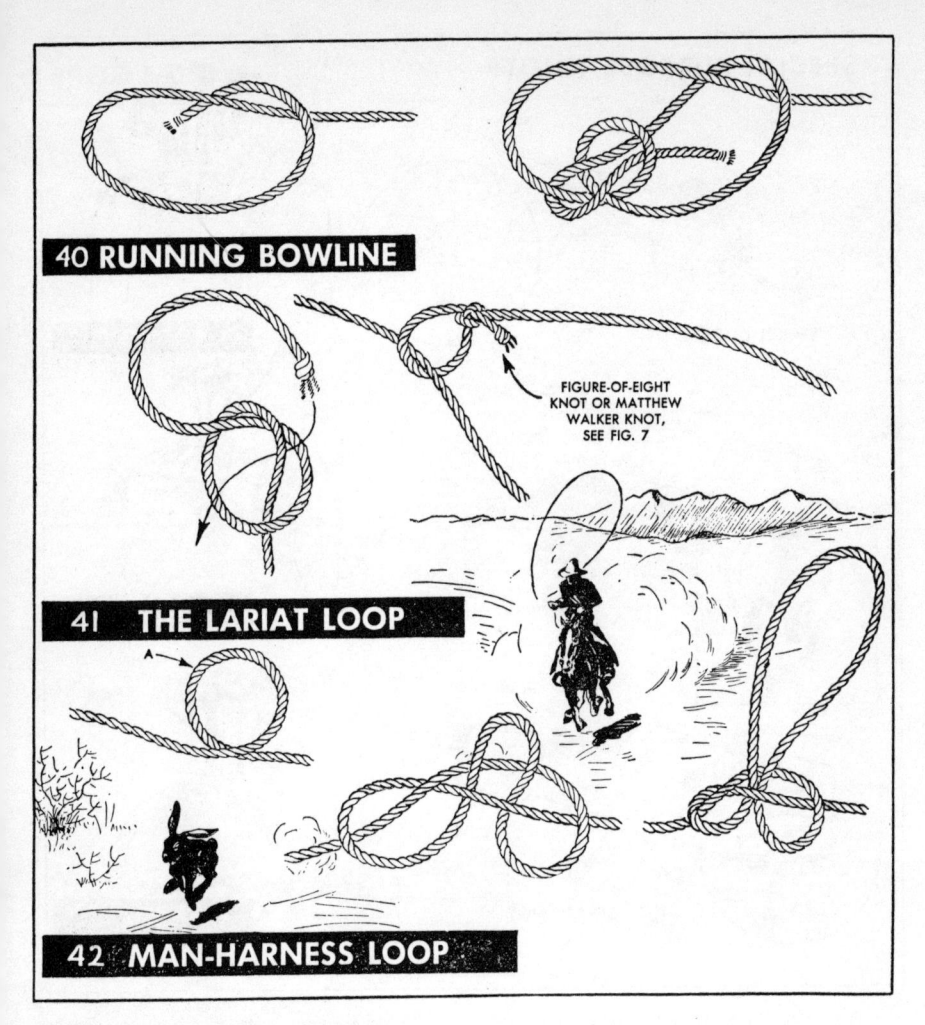

FIGURE-OF-EIGHT KNOT OR MATTHEW WALKER KNOT, SEE FIG. 7

41 THE LARIAT LOOP

42 MAN-HARNESS LOOP

bight, Fig. 39, is useful where two fixed loops are needed. The running bowline, Fig. 40, may be started as shown in the first detail or by making an overhand loop as shown at the right. Otherwise you have to tie an ordinary bowline and then turn the loop back over the standing part.

When purchased readymade lariats now are usually provided with a pear-shaped brass ring or honda spliced or seized into one end. Fig. 41 then, details what might be termed the old-fashioned lariat loop. It is also an excellent tie for other purposes as it forms a fixed loop of any practical size and is quickly and easily untied by merely loosening the overhand knot. If used for a lariat or any other purpose where the strain is great, the end must be finished with a stopper knot as shown.

Adding manpower to a rope can be done very effectively by tying a series of harness knots, Fig. 42, in the standing part. To tie this knot you form an underhand loop as shown in the first step, Fig. 42. Actually the loop must be much larger than that indicated, which is only for the purpose of illustration. Then grasp the rope at about the point A on the loop, and bend it down and to the left. Bring it up under the standing part and over that part of the loop which remains, as in the second step. Then pull out the loop and snug the knot before putting it under strain.

Figs. 43 to 46 inclusive detail a number

SPECIAL-PURPOSE KNOTS

(43) BECKET HITCH

(44) LARKSHEAD

(45) STRAP KNOT

(46) SACK TIE (MILLER'S KNOT)

1st TURN

2nd TURN

FINISH

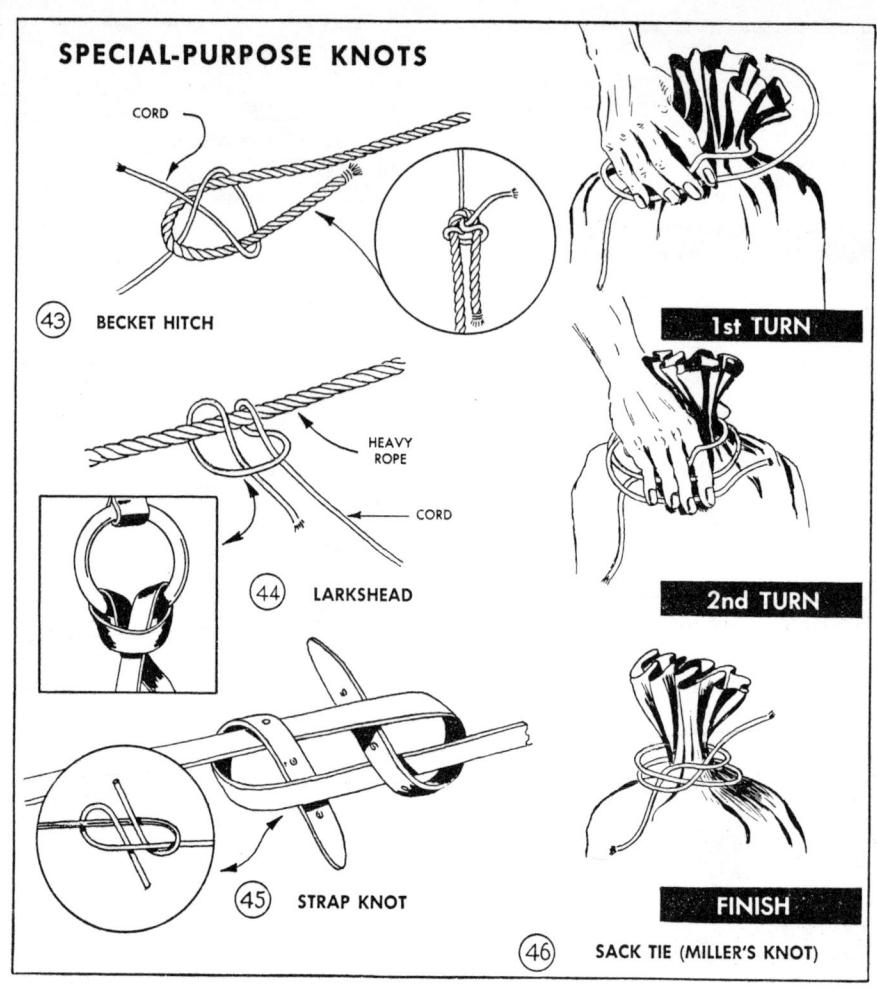

of useful special-purpose knots. Where a long rope must be taken up a ladder to the top of a building it's much easier to pull the rope up after you get to the top than to carry it up. Attach a cord to one end of the rope with a becket hitch as in Fig. 43. This simple hitch has many other uses as you can see. It's handy where it is necessary to join the ends of ropes of different diameters, or where attaching a cord to a rope will serve some special purpose. Of the several applications and forms of the larkshead, Fig. 44, only two are shown. It's handy where necessary to attach a small rope to a large one along the standing part for a pull at right angles. It also is used when tying a rope to a ring or post. The Western saddle-girth hitch is really a larkshead tied with the cinch strap, as you see in the left-hand detail, Fig. 44. Another strap knot, good to know in an emergency, is shown in Fig. 45. It consists of two interlocking half hitches and is particularly effective in joining the ends of flat straps. A wire splice can be made similarly as shown in the circular detail, Fig. 45. Most all farmers are familiar with the sack tie or miller's knot, Fig. 46, but campers, hikers, and others who handle sacks filled with fine material should know how to tie it. As you can see, it is similar to the clove hitch, Fig. 16, and is tied by first laying the starting end of the cord over the index finger as in the top detail. Two turns are then made, each passing under all four fingers as in the second view. Then draw the

SPLICING

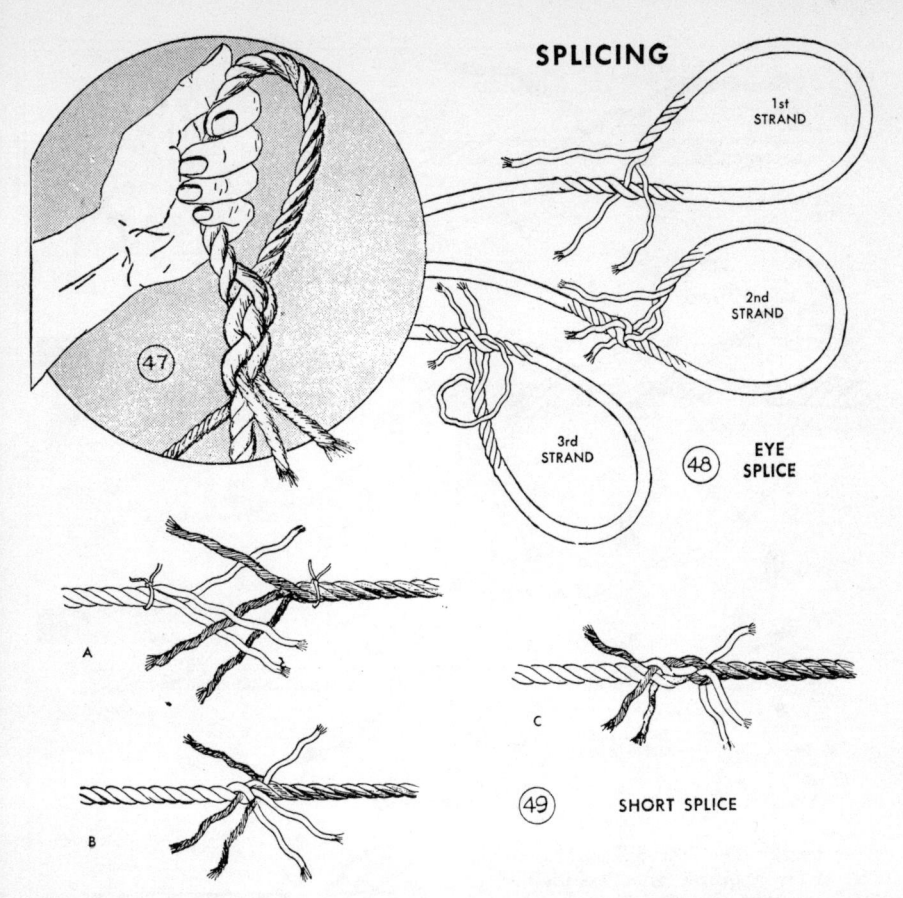

47

1st STRAND

2nd STRAND

3rd STRAND

48 **EYE SPLICE**

A

B

C

49 **SHORT SPLICE**

winding end, either straight or in a bight, up under the first turn just ahead of the index finger. Remove the latter, grasp both ends of the cord, or the end and the bight, and jerk the knot tight.

Now one thing to keep in mind: None of the knots described can ever be fully effective unless snugged up tight before putting under strain. Never trust a knot until you have made sure it is drawn tight. Remember, too, that the strength of the rope or cord in the knot is never as great as that of the standing part. All knots are shown tied on three-strand rope.

When rope is depended on to hold a given strain or load through knots it should be remembered that the strands and fibers within the knot tend to take a "set" where the rope is knotted for long periods of time. When untying such a knot be especially careful not to twist or kink the rope unduly, then carefully straighten the bends by gently pulling the rope from points on each side of the affected section. Finally lay the rope out on the floor and roll it under your foot to relocate the strands and yarns. The rope should never be dampened for the purpose of straightening it.

Of all the rope splices the eye splice has the greatest appeal because of its neat, professional look. Fig. 48 shows one simple way to splice an eye in the end of a rope. The drawings are self-explanatory except perhaps for one point. Before tucking the third strand the loop or eye is turned over. After you have made two or more tucks, over and under the rope strands, the splice will tend to become bulky, as in Fig. 47. Snug up the unlaid strands separately, pulling equally on each until you get the knot tight and smooth. Then separate each strand into its individual yarns and cut away half the latter. Finish the tucking with the half strands. This trick tapers the

1st STEP

2nd STEP

FINISH

ENDS TUCKED OVER - UNDER

OVERHAND KNOTS

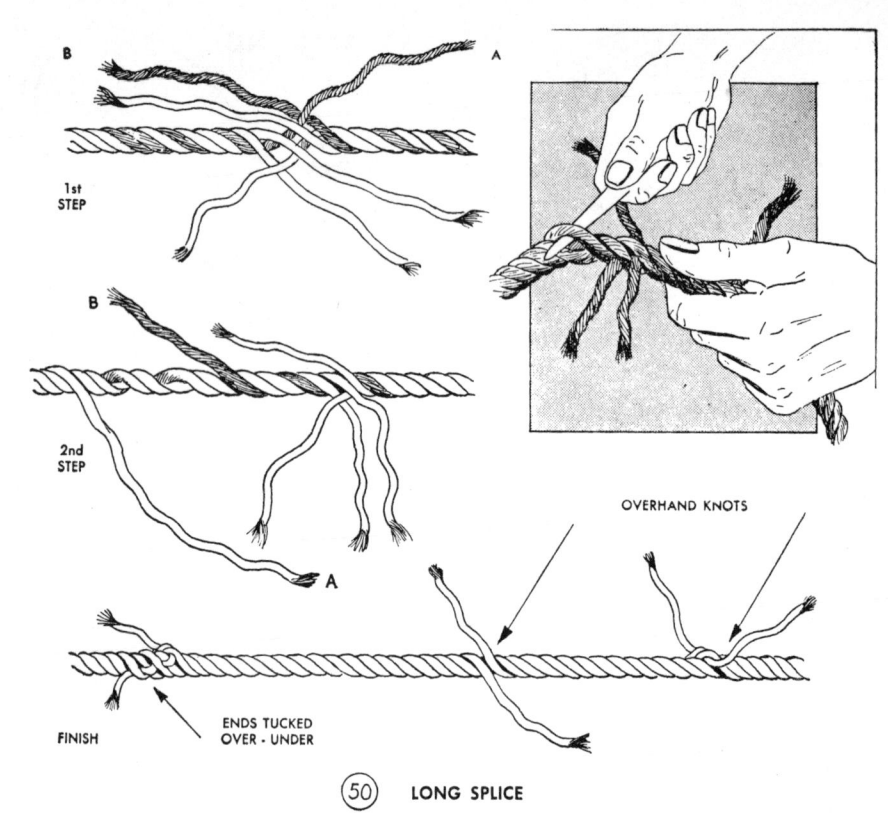

㊿ **LONG SPLICE**

splice neatly. The short splice, A, B and C, Fig. 49, is a quick, effective method of splicing a long rope for practical purposes. Unlay 10 to 20 in. of the end strands and tie with a cord as at A to prevent further unlaying. Then simply place the unlaid ends together with the strands in the relation shown and tuck the strands of the left-hand rope over and under the strands of the right. Continue the procedure with the right-hand rope strands. Smooth by rolling on the floor with your foot. In the long splice, Fig. 50, strands are unlaid for a distance equal to 8 to 12 times the circumference of the rope. Place ends together, then unlay strand A and fill the space with strand B. Do the same with corresponding strands in the opposite direction. Finish with overhand knots and tucks as shown. Fig. 51, A, B and C, show the method of waterproofing a rope. The canvas strip B, is coated with white lead after which the cord, serving, is wound on with a special fixture or serving mallet, as shown. Figs. 51 and 53 detail the making of round and

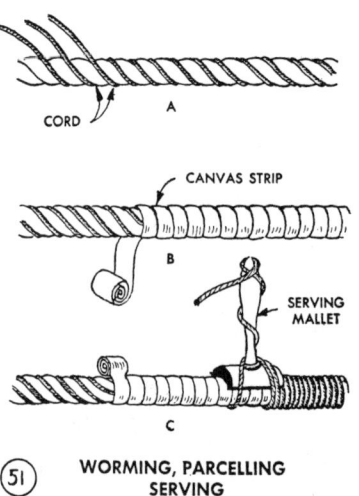

CORD

CANVAS STRIP

SERVING MALLET

㊶ **WORMING, PARCELLING SERVING**

racking seizing, the latter method being used when rigging ropes together where strain on one is greater than on the other. In the round seizing the first winding is

SEIZINGS

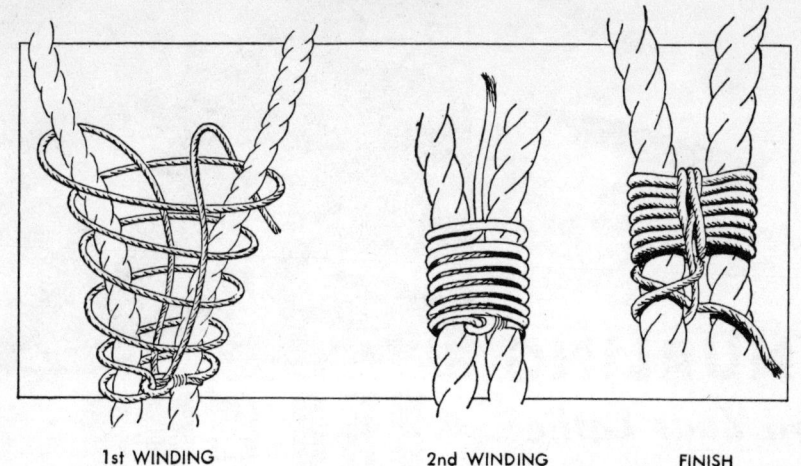

1st WINDING   2nd WINDING   FINISH

(52) **ROUND SEIZING**

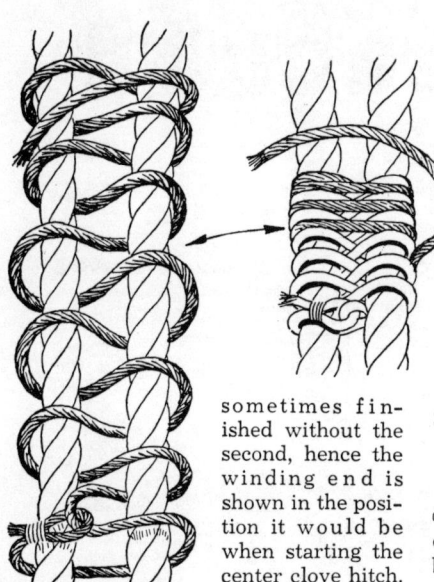

RACKING SEIZING (53)

Making a neat seizing requires much the same skill and attention to details as making either a short or long splice. In Figs. 52 and 53 the loops are shown loose. The trick is to keep snugging the loops individually and by stages in much the same way as you lace up a pair of high leather boots. No complicated series of loops of this nature can first be placed and then drawn up as a whole by simply pulling on the free end of the cord. The first few loops will always be loose and eventually the whole series will slacken under strain. In making any seizing it is usually best to use a cord not less than one-eighth of the diameter of the rope, although this is not a hard-and-fast rule. Although cotton cord may be used, cords of jute or sisal fibers are usually best for this purpose.

sometimes finished without the second, hence the winding end is shown in the position it would be when starting the center clove hitch. In either case, finish with a clove hitch around both the winding and the rope, as shown.

Singeing Hand Rope on Elevator Removes Projecting Fibers

Workmen who operate elevators of the type that have hand ropes will find it a good idea to keep them smooth and free of small projecting fibers. These can be removed easily and quickly with the flame of an ordinary blowtorch, which is passed rapidly along the surface of the rope, taking care not to burn it.

KNURLING
On Your Lathe

Photos courtesy Armstrong Bros. Tool Co.

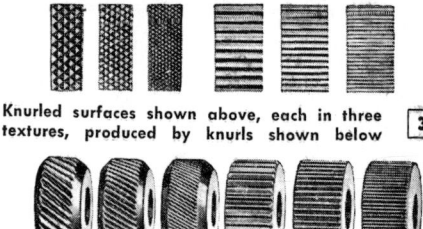

Knurled surfaces shown above, each in three textures, produced by knurls shown below

KNURLS + = PATTERN
WORK
A

DIFFERENCE IN RADII OF BEAD AND BASE OF TEETH

BEAD ON WORK

ANGLE OF TEETH | PITCH
25¾° FINE 20
29½° MEDIUM 12
36° COARSE 8
B

CROSS SECTION OF CONCAVE KNURL

FINE 80 | MEDIUM 50 | COARSE 34

NUMBER OF TEETH ON ¾" DIA. KNURLS

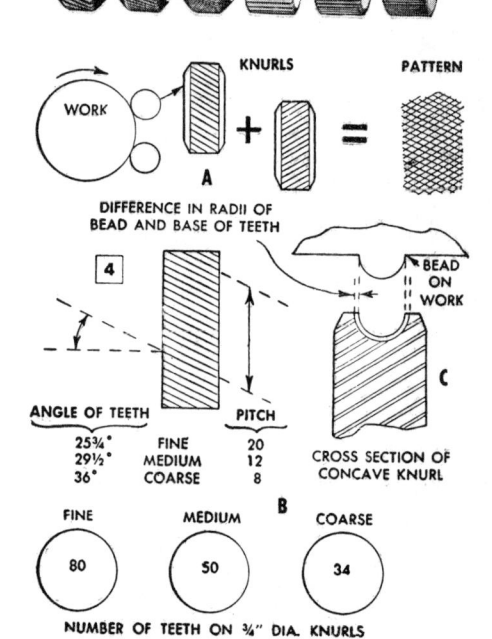

KNURLING is done in most cases to provide a more positive grip on turned parts, although sometimes it is used solely for ornamentation. On occasion, a straight knurl is used to prepare a round turning for a press fit, in which case the knurls are hardened so that they cut their own grooves when pressed into a mating piece. Knurling done on a lathe as in Fig. 2 produces embossed surfaces on the periphery of turned parts, such as tool handles, thumbscrews, nuts and sleeves, some of which are pictured in Fig. 1. Most knurled surfaces consist of tiny diamond-shaped projections in a continuous pattern. On narrow edges, such as screw heads, knurling usually consists of closely spaced parallel ridges. Both of these patterns are produced in fine, medium and coarse textures as shown in Fig. 3. The range in textures gives a selection that is best suited to the particular size of the work, as well as the amount of gripping power or friction required.

Knurling tools: A knurling tool consists of one or more hardened disks or knurls mounted in a holder, which is designed either to be held in the hand or clamped in a lathe tool post. Knurls vary in diameter from ⅝ to 1 in., and in width from 3⁄16 to ½ in. Teeth on the edge or face of knurls leave their impression on unhardened work when brought against the surface with sufficient pressure while the work is rotated. Some of the metal is forced up between the teeth and then forms projections or ridges above

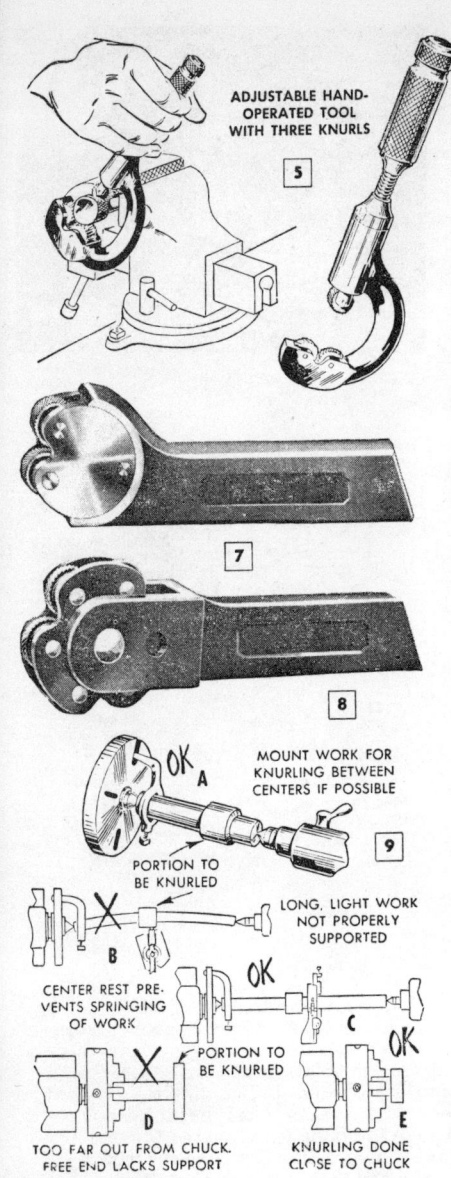

ADJUSTABLE HAND-OPERATED TOOL WITH THREE KNURLS

5

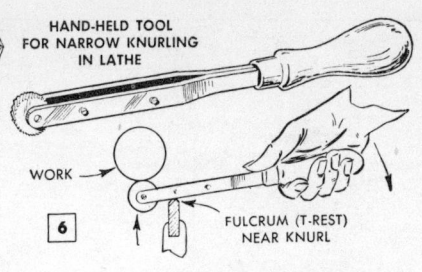

HAND-HELD TOOL FOR NARROW KNURLING IN LATHE

6 WORK — FULCRUM (T-REST) NEAR KNURL

7

8

9 MOUNT WORK FOR KNURLING BETWEEN CENTERS IF POSSIBLE

A — PORTION TO BE KNURLED

B — CENTER REST PREVENTS SPRINGING OF WORK / LONG, LIGHT WORK NOT PROPERLY SUPPORTED

C — PORTION TO BE KNURLED

D — TOO FAR OUT FROM CHUCK. FREE END LACKS SUPPORT

E — KNURLING DONE CLOSE TO CHUCK

knurled. If both were of the same radius, metal would be pressed on the adjacent surface or shoulder, spoiling the appearance of the work.

Hand-held knurling tools, Figs. 5 and 6, are used to knurl narrow edges of work rotated in a lathe, which do not require a traverse movement of the tool as is necessary on a surface wider than the knurl. The tool shown in Fig. 5 is adjustable to various diameters and can be used to knurl work held stationary in a vise while the tool is rotated by hand. It also can be used on flexible work. This tool has two stationary knurls and a single one on a threaded shank which is advanced on the work. The single-knurl hand tool shown in Fig. 6 must be supported on a fulcrum, such as a T-rest placed close to the knurls. In use, it is forced upward against the underside of the work with considerable leverage when the handle is moved downward. The knurls on this tool are interchangeable, one having a diamond design.

Fig. 7 shows a common type of knurling tool designed to be held in a lathe tool post. It has two knurls mounted on a semicircular holder that is self-centering. This equalizes the pressure of the knurls on the work which is necessary to assure uniform patterns. Similar distribution of pressure is obtained with the tool shown in Fig. 8, which holds three pairs of knurls. Such an arrangement is convenient when flat and concave knurls or different textures are required on one piece.

Setting up lathe for knurling: If possible, mount the work to be knurled between centers, Fig. 9, A, as this gives maximum support against the heavy pressure required for knurling. Small-diameter work may require the support of a center rest to prevent spring, details B and C. When work is mounted on a faceplate or in a chuck, the knurling is done as close to the headstock as possible, details D and E.

Set the compound rest at a 30-deg. angle to the right, as in Fig. 10, A, so that its screw handle will not interfere with turning the cross-feed screw. Slide the knurling tool into the tool post as far back as possible to obtain maximum rigidity, and set the tool post so the faces of the knurls are parallel

the surface. Using a pair of spiral-tooth knurls with the teeth slanting in opposite directions produces a diamond pattern as in Fig. 4, A. Detail B gives the angle and pitch of spiral teeth for fine, medium and coarse textures, also the number of teeth on ¾-in.-dia. knurls. When using a concave knurl, the radius at the base of the teeth, as seen in the cross-sectional view, detail C, must be slightly larger than the radius of the bead or rounded portion to be

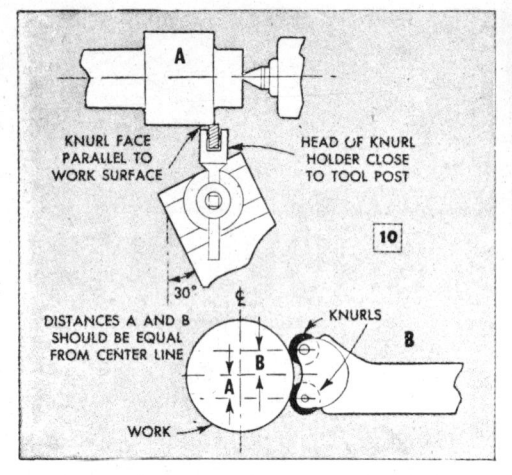

KNURL FACE PARALLEL TO WORK SURFACE
HEAD OF KNURL HOLDER CLOSE TO TOOL POST
DISTANCES A AND B SHOULD BE EQUAL FROM CENTER LINE
KNURLS
WORK

[10]

[11]

START OF KNURLING: ⅓ TO ½ OF KNURL FACE EXTENDS OVER WORK SURFACE

[12]

2nd PASS: KNURL DIRECTION REVERSED AT THIS POINT

IMPRESSION DEPTH VARIABLE

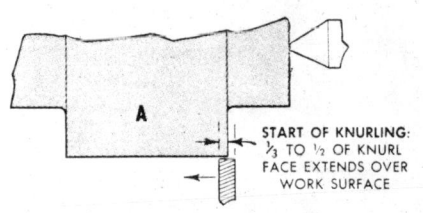

to the work surface. Adjust the tool for height to locate the knurls an equal distance above and below the center of the work, detail B. Then tighten the tool post. Knurling operations are done at slow back-gear speeds—not exceeding 50 percent of the regular cutting speeds recommended for the material being knurled.

Knurling: Adjust the lathe carriage so that the knurl face is positioned at the tailstock end of the work where knurling begins. In case it starts at the edge, as in Fig. 12, A, about ⅓ to ½ the width of the knurl should project over the edge of the work. Then with the work rotating and the knurls almost touching it, apply cutting oil liberally, Fig. 11, and feed the cross slide inward so that the knurls enter the work about 1/64 in. This will produce a well-defined pattern. Steel requires application of cutting oil to avoid overheating, but cast iron and soft metals like brass and aluminum are knurled dry. Stop the lathe after a few revolutions but without backing out the tool. If the design is not correct and double impressions split the diamonds, back out the tool and start over again at another point. One way to prevent double impressions is to feed the knurl with considerable pressure at the start and then relieve it slightly. Test knurls will be rolled out automatically in the knurling process. For surfaces wider than the knurl face, use the longitudinal carriage. Fine patterns generally are finished in one pass but coarse ones often require two and sometimes three passes. When a repeat pass is necessary, knurling should be continuous without removing the knurl from the impression, although the lathe is stopped at the end of a pass. The direction of the power feed is reversed and the cross-feed slightly just after the leading edge of the knurl has passed the headstock edge of the work as in Fig. 12, detail B. But the knurl must not be permitted to run off the work. For a third pass, the process of reversal is repeated at the tailstock end. Running the tool over the work too often may ruin the pattern. In addition, any loose particles of metal that happen to get between the knurls and the work will mar it. Keep the tailstock center well-lubricated. ★ ★ ★

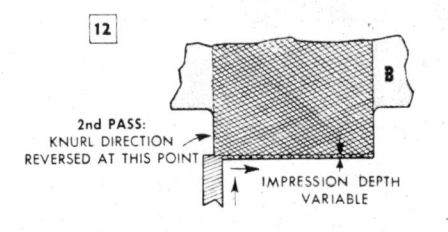

① DIAMOND (COARSE) ② DIAMOND (MEDIUM) ③ DIAMOND (ONE PASS) ④ DIAGONAL

KNURLING

THE SECRET of clean knurling, such as shown in Figs. 1 to 4, is pressure, which must be applied before the work or rolls can make one complete revolution. If you give this matter a little thought, you will see that the knurling tool is not perfect—if you run any toothed wheel around a piece of round stock, it is not likely that after making a complete revolution the teeth will again engage in the same place as before. However, if the initial engagement is under heavy pressure, the teeth will make a deep track; then on the second revolution the teeth will hit one side or the other of this track, maintaining the pattern.

The preliminary setup is to have the tool on center as in Figs. 5 and 6. It

Set the tool square with the work and check the tailstock support. Then apply oil to work surface and to knurling rolls

With lathe idle, advance crossfeed .050 in. Rock the work by hand and apply more pressure on tool without springing work

Run lathe at about 50 r.p.m., and make test knurling pattern. If satisfactory, engage long feed at about .030 in. per rev.

⑮ HERRINGBONE   ⑯ STRAIGHT LINE   ⑰ STRAIGHT (CUT WITH THREAD)

is a good idea to make a special tool-post ring, as in Fig. 11, to insure proper positioning on every job. The knurling rollers should be square with the work, Fig. 7. When you make the setup, put a light or a white piece of paper under the work, as shown in Fig. 12, in order to see plainly the parallelism between work and rollers. Now, advance the cross-feed by hand, Fig. 13, setting up a heavy pressure —about all your two hands can apply. The lathe is not running. Oil the knurling rollers and the work, using any kind of machine oil. Start the lathe. Do not make a full revolution, but stop about halfway and give an extra twist on the feed screw if possible. Start the lathe again and let it run for three or four revolutions, Fig. 14. Then stop the lathe and check the pattern. If you get a clean pattern on this first ring, the job is as good as done. You can then engage longitudinal feed to run the knurling tool any required distance along the work. At the end of the cut, reverse the feed, take another bite of about .005 in., and run back to the starting point as in Fig. 18. These two passes should bring up a full knurl, but another pass or two can be made if desired. You can reverse and stop the lathe as often as you like—the only rule you have to follow is that you should keep the knurls en-

⑱ TWO PASSES ACROSS THE WORK GIVES A FULL KNURL

POOR START → ← GOOD START

⑲

In a poor start the pattern is split. Try again somewhere else on the work

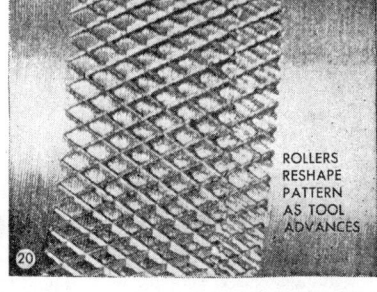

ROLLERS RESHAPE PATTERN AS TOOL ADVANCES

⑳

Knurling rollers will reshape the pattern after a good track is established

STRAIGHT KNURL

THREADING TOOL

㉑ STRAIGHT KNURL CUT WITH THREAD GIVES CHECKER PATTERN

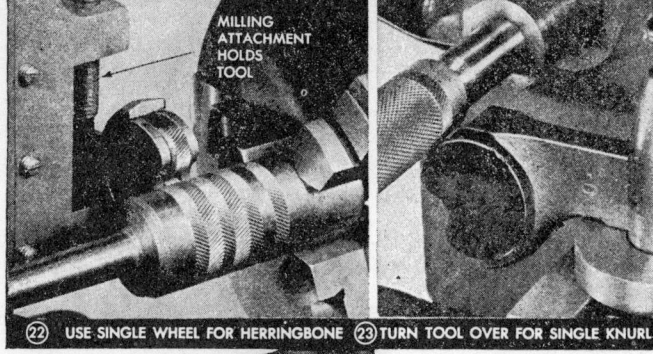

(22) USE SINGLE WHEEL FOR HERRINGBONE (23) TURN TOOL OVER FOR SINGLE KNURL

(24) KNURLING METAL SLEEVE OVER WOOD

(25) STRAIGHT KNURL (HARDENED)
KNURL INCREASES DIA. OF WORK ABOUT .010"

ONE WHEEL CONTACTS WORK

(26) Face knurling: One wheel contacts work for straight pattern, both wheels for diamond pattern

(27) STRAIGHT DIAMOND

gaged until the job is done.

Do all knurling at slow speed, about 50 r.p.m. The feed should be about .030 to .040 in. per revolution (24 to 32 threads). If the work is extremely hard, the knurling rollers can be set at a slight angle, Fig. 8, to assist penetration. Since this leaves a light cut at the end of travel, it is useful only when the knurled area can be run over length and trimmed to size. If the knurl runs from the end of the work, engage only the half width of roller, Fig. 9, the idea again being easy penetration. Another good method of sinking the tool is to turn a slight shoulder about half of the knurl width, Fig. 10, and make the initial run on this. This is good practice if you want just a light cross hatching of lines as shown in Fig. 3. Once the knurls are tracking, they will hold true in a light cut, but they will not start in a light cut. If the work is light of body and long, always use a center rest or follower rest as a support—pressure alone is useless if the work springs. The preferable mounting is between centers with feed on the first run toward the headstock. If you use a chuck be sure to stop the work against the chuck itself or against some kind of stop inside the spindle—otherwise it will creep and spoil the job.

A poor start on a diagonal or straight knurl often can be corrected with more pressure. However, this does not work with the diamond pattern. When you split the diamond on the initial pass, extra pressure does not help. In this case, check the setup carefully and try again in another place, Fig. 19. Once a good track is established, the knurling rolls will roll out almost any depth of pattern in a poor start, Fig. 20.

Sometimes you want a novelty pattern. Try the straight knurl cut with a thread, as shown in Figs. 16, 17 and 21. Another novelty pattern is the herringbone, Figs. 15 and 22. Make this with a diamond knurling tool, but use only one roller at a time. This setup is easily made by using the milling attachment as a holding device, permitting the necessary raising and lowering to engage one roller only. Another method of engaging one roller is to turn the tool over, as shown in Fig. 23.

While knurling is most used for traction and ornamentation, it makes a good holding device for fastening thin metal sleeves over wood, as shown in Fig. 24. Use it also for drive fits, Fig. 25. The knurl increases diameter of work about .010 in.—often you can save an undersize job by this method.

Both straight-line and diamond patterns can be run on the face of the work. The straight knurl should be worked with single roller only, Fig. 26, while the diamond permits engagement of both rollers, Fig. 27. For these setups, lock the carriage and use compound for feed. Pressure required is much heavier than for spindle work, and knurling of this nature is best confined to narrow rims or light impressions.

Ladder Secured to Slanting Roof Without Use of Nailed Brace

One homeowner avoided the use of nails in securing the base of a ladder placed on a slanting roof by anchoring it as shown inside another ladder laid flat on the roof and held in place with ropes. These were run from the supporting ladder to stout boards braced against the inside of windows near by. Another advantage of the arrangement is that the upright ladder can be moved along the supporting ladder as work progresses, thus eliminating the necessity of repeatedly securing the base each time the ladder is moved.

Abrasive Applied to Ladder Steps Provides Sure Footing

Here is one way to safeguard against the possibility of slipping and falling from a stepladder the next time you redecorate or wash windows. First, apply a coat of varnish to the treads and then, while it is still tacky, sprinkle it with either sawdust or sand. The particles will adhere to the varnish and create a surface sufficiently rough to provide a safe, nonslip footing.

Ladder Used as Painting Stand For Screens and Storm Windows

Your stepladder can be converted into a painting stand for screens and storm windows with the fittings shown in the illustration. Suction cups on the upper fittings hold storm windows for painting. The block is reversed for screens, which are hung on the finishing nails. In the lower fitting, a pointed finishing nail keeps the lower edges of the painted windows and screens from contacting any surface. Both upper and lower brackets are turned vertically to be out of the way when the ladder is stored.

Swinging Shelf at Top of Ladder Provides Extra Tool Space

The space on top of a stepladder, handy for holding tools and supplies while working, may be doubled in size with the addition of this shelf, which swings out of the way when not in use. Cut from a board to the size of the ladder top, the shelf is pivoted with a bolt and nut, as well as two large washers. The washers are placed between the board and the ladder top to allow it to swing freely.

Disappearing ATTIC LADDER

Critically counterbalanced and riding on ball-bearing skate wheels, this lightweight attic ladder is easy to raise or lower. By using a pail filled with sand as a counterweight, it's a simple matter to find the exact balancing point of the ladder. From the detail below you can see how sides are added to the trap door to which sets of skate wheels are fastened to engage hardwood cleats screwed to the sides of the ladder. The pulleys should be located high enough overhead to keep the pail of sand from touching the attic floor when the ladder is drawn up.

Photos courtesy United States Plywood Corp.

BEDROOM FURNITURE

RUMPUS-ROOM BAR

WINDOW SILL

WHAT YOU CAN DO WITH
Decorative Laminates

By Wayne C. Leckey

MOST EVERYONE knows what usually happens when a frosty drink is placed on a highly polished table or a burning cigarette accidentally falls from an ash tray—the table top is ruined and a costly refinishing job is in order. Homeowners who have experienced this will be eager to learn of a comparatively new plastic material that literally defies marring. Called decorative laminate, this amazing plastic material will withstand most acids, will resist heat and cold and endure chemical action. It is virtually unstainable. It won't chip, scratch, craze when wet, or corrode when soft drinks, food or liquor is spilled on it. It requires no refinishing or polishing — just a quick wipe with a damp

Applications throughout the house

cloth will bring back its original bright luster, year after year.

Familiar to many who first remember seeing it on restaurant tables and counter tops, decorative laminate, as its name implies, is a durable high-pressure plastic laminate with a decorative surface. It is made up of many layers of resin-soaked kraft paper and veneer which are fused together under intense heat and pressure. The laminations actually merge to form a new substance which is coated with a super-tough skin of clear plastic that locks in the color and pattern for life. Marketed under several different trade names, decorative laminate can be had in roll or sheet form and in a score of patterns and colors, including beautiful reproductions of rich mahogany, oak, prima vera and walnut wood grains. Available in various lengths and widths, decorative laminate also may be purchased at lumberyards, ready bonded to ¾-in. plywood. Most linoleum stores carry laminates and the cement that's needed to apply them.

Decorative laminates are at home in every room of the house. On kitchen counters, cabinets and other work surfaces where cleanliness is a must, this miracle material provides a most practical covering. Liquids, grease and food particles leave no residue on its impervious glass-smooth surface. On furniture in the living room and bedroom, laminate fills the double function of fashion and lasting utility. A more practical top for a cocktail table could not be found. In the bathroom it does triple duty in repelling water, medicines and cosmetics. On a refreshment bar in the recreation room it comes through without a scratch or stain. Dated panel doors can be converted to beautiful flush doors by

COFFEE TABLES

There's a practical use for decorative laminates in every room of your home. Whether used on counter tops, cabinets, dinette and coffee tables, recreation-room bars or bathroom vanitories, this amazing material outshines and outlasts any other surfacing product. Practically indestructible, decorative laminates retain their glass-smooth luster indefinitely with a mere wipe of a damp cloth

Photo courtesy Midcontinent Adhesive Co.

covering them with any of the true wood-grain reproductions to be had. Window sills will stay permanently protected from the hot sun when surfaced with laminate. A section of a wall can be paneled in wood grain to give contrast to a papered or painted wall.

While decorative laminates have an exceptionally durable surface, it is not so hard that it cannot be damaged under certain conditions, and while it can stand constant abuse, one should avoid giving it unnecessary punishment. It is neither recommended nor necessary to clean the surface with gritty scouring powders—only a damp cloth is needed to clean it. Also, avoid placing a hot pressing iron directly on the laminate, or using it as a cutting surface. Always place a pad under a toaster, waffle iron, percolator or other hot appliance.

Heretofore, the application of decorative laminate has been limited to the furniture manufacturer having facilities for gluing and clamping the material securely to the surface. Now, thanks to the development of special "no clamp" adhesives, the application of decorative laminate can be done right at home with a few common hand tools. Of the several adhesives available, Roltite and Tapon were successfully tried by the author in bonding several different decorative laminates.

While laminates can be had in both roll and sheet form, the roll type is $1/32$ in. thick and the sheet type is $1/16$ in. thick. The thinner material is cut by merely scoring the top surface with an awl and then bending it upward to snap it off, whereas the heavier $1/16$-in. material must be cut with a fine-tooth (metal-cutting) keyhole saw, using downward strokes to avoid chipping. It is always best to cut the material about 1 in.

While not a sheet laminate but of the same durable quality, molded counter is available in 6-ft. lengths

Above, laminate makes excellent durable covering for traveling case, and a most practical surface material for a vanitory in a powder room as pictured below

oversize to be on the safe side if chipping should occur and also to have enough waste to later trim carefully to line. When it comes to actual application, the work should be done in a room temperature of at least 70 deg. F., and it is important, too, that the laminate be of room temperature. The cement must be applied warm, heating it beforehand in a pan of hot water until it is warm to the touch. All wood surfaces to be covered must be flat, clean and dry. If the surface has been painted or varnished, better results are had if the finish is removed. If the edges as well as the top surface are to be covered, the laminate is always applied to the edges first, using a cement of thicker consistency which is made specifically for edge work. Assuming that only the top and not the edge of a work counter is to be covered, first brush a coat of cement on the wood with a clean paintbrush and allow to dry at least 30 min. Then apply a second coat. While this is drying, brush a coat on the back of the laminate and allow all coats to dry not less than 30 min. There's no need to hurry, as the cement can be left to dry up to two hours and still be workable. To determine when the cement is dry enough to bond the laminate, test it with a piece of wrapping paper. If, after pressing it on the cement, it has a tendency to pull the cement away from the surface, the cement is not dry enough. If it is necessary to allow the cement to dry for a longer period than specified, it can be reactivated by applying another coat on top of the first one. Try to get an even coating over the entire area; too thin an application will result in dull spots.

These decorative laminate samples show four popular patterns which are available in variety of pastel shades and wood grains to suit any interior treatment

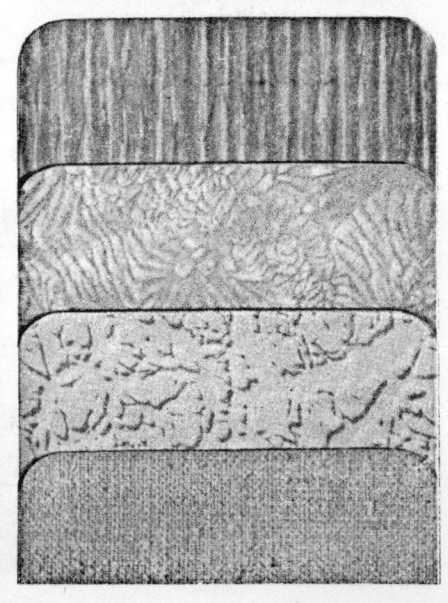

LAMINATES

Courtesy Marsh Wall Products, Inc.

Bonding is immediate and permanent when cemented surfaces come in contact, so place wrapping paper lightly over the cemented surface and position the laminate on top of the paper. When in proper position, the laminate is raised slightly and the paper is withdrawn 2 or 3 in. This allows a portion of the laminate to make contact with the coated surface, after which the paper is pulled out all the way. All that remains to be done is to roll the laminate firmly to assure over-all contact, and the bonding is completed. Trim the edges back to line with the square edge of a single-cut file, working carefully and slowly to avoid chipping. If the edge of the laminate is not to be covered, the laminate is filed at a slight bevel (with the flat side of the file), just enough to break the sharp corner. Any excess cement on the surface may be rubbed off with the fingers.

Remember that if the edge of the work is

Here are pictured the two methods recommended for cutting roll and sheet-type laminate. Roll type is scored with pointed tool like an awl or ice pick and bent upward to snap off. Sheet laminate is cut with a fine-tooth saw

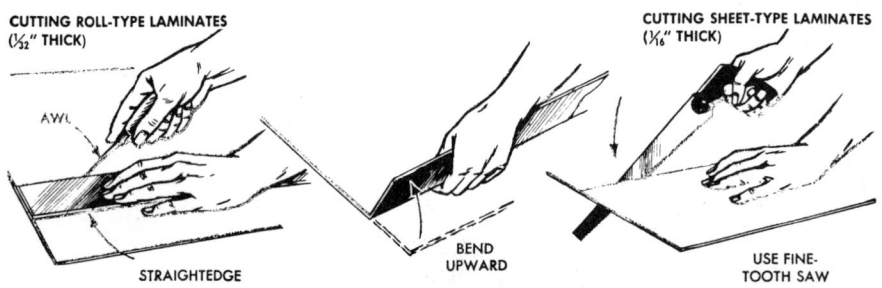

CUTTING ROLL-TYPE LAMINATES (1/32" THICK)

AWL

STRAIGHTEDGE

BEND UPWARD

CUTTING SHEET-TYPE LAMINATES (1/16" THICK)

USE FINE-TOOTH SAW

POSITIONING THE LAMINATE

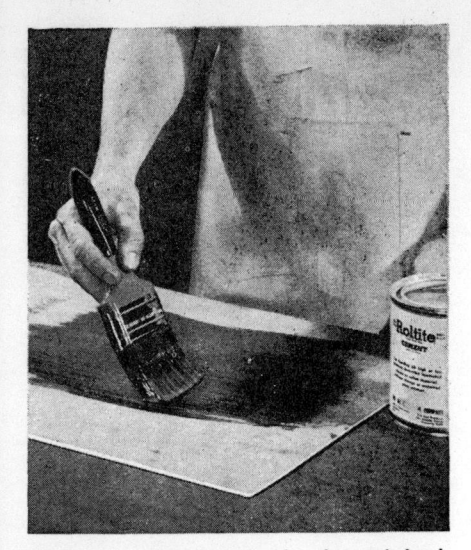

The cement is applied warm with a clean paintbrush and left to dry at least 30 min. Two coats are applied to the wooden surface and one to the laminate

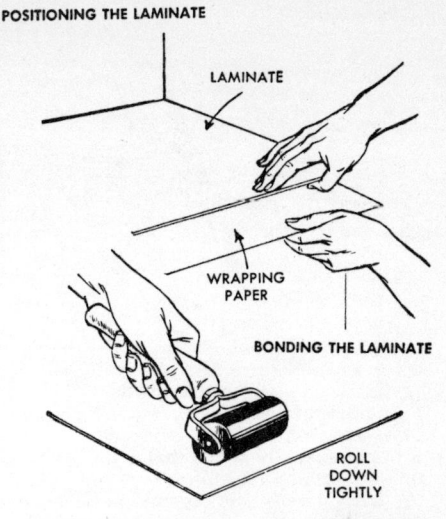

Wrapping-paper separator permits laminate to be positioned on cemented surface prior to final bonding. Rolling with firm pressure assures positive contact

FINISHING

CUTTING LAMINATE FLUSH — FINE FILE
DRESSING EDGE — DOWNWARD STROKE

APPLYING EDGE FACING

PROJECTS ¼"

MAKING SINK OPENINGS

HOLE FOR SAW BLADE
PRICK MARKS — FILE TO LINE WITH FINE RASP

REPAIRING POOR BOND

SOLVENT IN OIL CAN
HEAT LAMP

to be covered, the edge is done first and then the top surface. Cut the laminate so that it will be flush with the bottom of the edge and wide enough to extend about ¼ in. above the top surface. Apply the cement as before, using a special and thicker cement (Tapon), and let dry. Then apply by aligning the laminate in position and tapping it in place with a hammer and a block of wood. Finally, the waste is filed off as before, flush with the top surface. Laminate covering the surface should always be laid to extend over the edge of the facing strip. If it is necessary to bend the facing strip around a corner, heat it at the bend with a heat lamp (never an open flame) to prevent surface cracking. Do not attempt to bend it around a sharp corner.

Where a seam must be used, abutting edges of the laminate are first placed face to face and both cut at one time. This is done by clamping the material between two hardwood boards having perfectly straight edges, allowing the laminate to extend about ¼ in. Saw as close to the hardwood as you can without cutting into it and then dress down both edges of the laminate with a file before removing the clamps. Where the abutting ends of the laminate have a tendency to curl, place a piece of wrapping paper over the seam and press the laminate with an electric pressing iron set at 200 deg. F., or "silk" on indicator dial. Move the iron back and forth with pressure until the laminate becomes hot and has relaxed into positive contact at the seam.

If for any reason you obtain a poor bond

CUTTING A SEAM

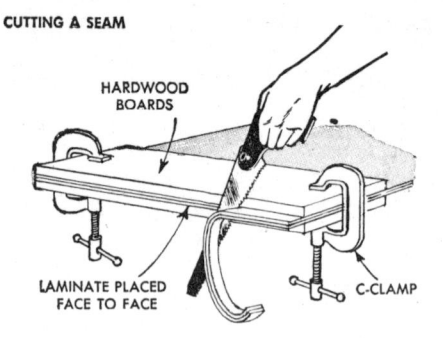

HARDWOOD BOARDS
LAMINATE PLACED FACE TO FACE
C-CLAMP

or if a blister should develop under the laminate, the defective area generally can be repaired by heating it until hot (130 deg. F.). This relaxes the laminate and reactivates the cement film, after which mere pressure usually will rebond the spot. If not, a C-clamp and block of wood will do the trick where it is possible to clamp the work. In the case of a blister, heat and press the surface as described and then quickly chill the surface with ice. If it is necessary to remove the laminate completely, cement solvent applied with an oil can to the cement will soften it to the point where the laminate can be peeled carefully from the surface. In the case of some cements, it is recommended that repairs must be made within a few days of the original application, as the cement tends to vulcanize with age and prevents any repairs being made.

Where walls are perfectly straight and flat, the laminate can be applied directly. Otherwise, it is best to provide a smooth base of hardboard. If desired, hardboard panels can be purchased prebonded with laminate and ready for direct application. The left-hand photo below shows a wall covered with prebonded laminate.

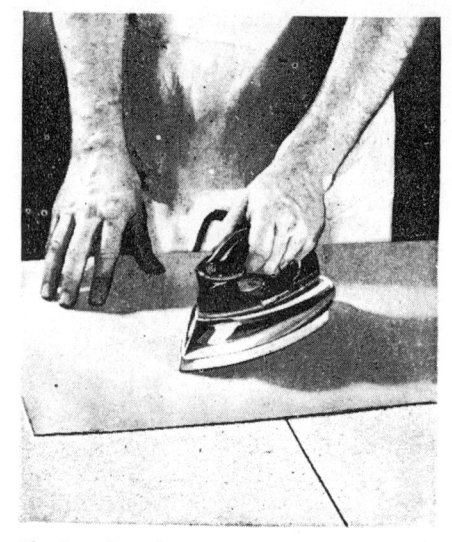

Abutting edges of a seam which have a tendency to curl, are heated with an iron to make pliable and pressed in contact with the cement-coated surface

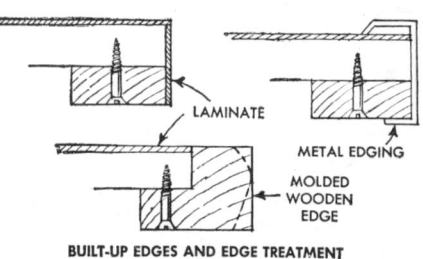

LAMINATE
METAL EDGING
MOLDED WOODEN EDGE

BUILT-UP EDGES AND EDGE TREATMENT

If you do not wish to face the edge of the work with laminate, the edge can be attractively concealed with either a metal or wooden molding. Note that laminate applied to the surface extends over facing strip

Photo courtesy Woodall Industries, Inc.

Photo courtesy The Formica Co.

Converting Jugs Into Lamp Bases

A jug, bottle or other narrow-necked container need not be drilled in order to convert it into a base for a table lamp. Instead, use a toggle bolt and a wooden or metal disk, and assemble them as detailed. To do this, remove the base from the light socket and drill a hole in one side to admit the lamp cord. Then pass the bolt through the original outlet and through the disk, and screw the end of the bolt into the toggle. Push the toggle through the opening of the jug and tighten bolt until the unit is held snugly.

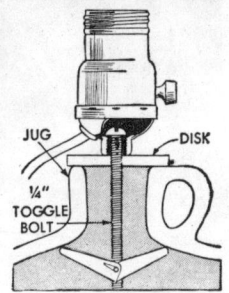

Rewire Ornamental Lamps Easily

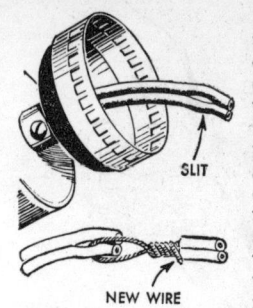

An ornamental table or floor lamp can be rewired quickly by using the old wire to pull the new wire through the base and socket. Pull up enough of the old wire to trim it off and cut a slit between the two wires about ½ in. from the end. Bare about 2 in. of the new wire and taper the insulation down to the wire. Twist the bared wire tightly, pass it through the slit in the old wire and fasten it securely in a loop. By pulling on the old wire at the lamp base, the new wire will follow through without catching.

Chromium Tubing Enhances Pin-Up Lamp

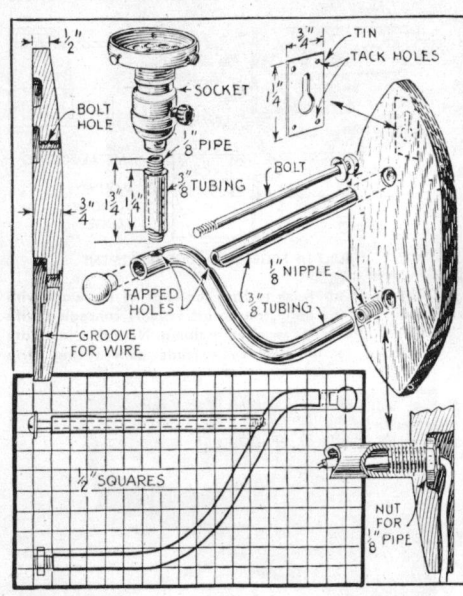

Here is a lamp you can hang any place on a wall. All you need do is drive screws into the wall at the desired locations, allowing the heads to project slightly to slip into the "keyhole" hanger on the rear side of the lamp base. Exposed metal parts are chromium-plated, including the lamp socket and reflector holder. The base is made of wood, finished natural or painted to suit. The holes in the base are counterbored to take the tubing.

LAMP SHADES

MAKING LAMP SHADES is one of the most enjoyable of homecrafts. It provides an opportunity to express your creative abilities in many different phases —color, fabrics, textures, proportion and balance. And the finished lamp shade will bring you much satisfaction and pleasure.

Selecting the frame: The first thing to consider in lamp-shade construction is the frame to be used. There are four common shapes in lamp shades— bell shape, Fig. 1; straight slanting sides, Fig. 2; drum shape, Fig. 4, and square shade with or without curved Chinese bottom, Fig. 5. Then there are two general kinds of top fittings, the clip-on type and the screw-on type. The band at the top and bottom of the frame can be either flat or round, although the flat band is easier to work with. Make sure your frame is strong enough to withstand the pressure to be exerted in stretching on the fabric linings.

In selecting the shape of the frame, you should consider the lamp base. The shape of the frame along with the color, design and texture of the fabric used and the trimmings applied must be in balance with the lamp base. Then, of course, the finished

Skirted shade above is made of taffeta, and trimmed with a double-box pleating of satin around the neck

This shade needed two overlinings, one of crepe faille and the other of nylon net piped with velvet ribbon

lamp shade and its base should be thought of in relation to your room furnishings.

Wrapping the frame: Tape is used to wrap the frame. A seam tape of the quality used for skirt hems or any cotton tape about 3/8 in. wide which is strong but not too heavy is recommended. Do not use cotton-twill tape as it is too thick for this purpose. First of all, wrap the struts, Fig. 6, securing the end of the tape with needle and thread. In wrapping the struts, overlap the tape about one half its own width. Be sure to wrap firmly as you will not want the tape to slip when you sew the fabric to the struts later on. The length of the tape required for wrapping each strut will be about three times the length of the strut if 3/8-in. tape is used. For the top and bottom bands you will need about 4½ times the circumference of each band.

Making pattern for linings: Although tissue paper may be used to make patterns for shade linings, muslin is more accurate and will save material in the long run. Clean unbleached flour-sack muslin is easy to work with and strong enough to stand a good deal of stretching. If you decide to use muslin, take one corner of it and pin at the top of the frame. Then find the two struts exactly opposite each other and pin down the muslin, as in Fig. 7, for the sides of the pattern. Next pin the cloth to the bottom of the frame. When you have a good fit, take a pencil and mark the outline of the pattern, leaving 1 in. beyond the top and bottom lines and 3/4 in. on the sides. You are now ready to use the pattern on the fabrics selected for the underlining and the overlining. The same pattern is used for both.

Selection of fabrics: Almost any fabric of firm weave can be used for lamp shades. Fabrics suggested for outside lining are Indianhead, crepe faille, shantung, dotted swiss, organdy, eyelet, chambray, pique, satin, chintz, glazed chintz and taffeta. Because of its inelasticity, taffeta is one of the most difficult fabrics to work with. Some of the sheer fabrics, such as dotted swiss and organdy, require a second top lining. Textured fabrics like novelty weaves, shantungs, slub broadcloths and slub meshes give an interesting effect on shades when the lamp light is turned on. For underlining, crepe faille, crepe, or firm, fine marquisette are recommended.

Cutting and fitting linings: The pattern should always be laid on the bias of the material unless a special effect is desired as in covering a square frame. In this case, the straight of the material may be used. To use the bias of the material, fold the fabric once to form a triangle. Put the pattern in place and cut on the double thickness of the fabric. Then, with right sides together and allowing ½ in. for the seam, baste the sides and fit on the frame. If the fit is snug, remove the fabric from the frame and sew a double line on the sewing machine. Fit again; if still snug, trim the seam to 1/8 in.

The underlining should be cut first using the muslin pattern, and fitted on the inside of the frame. Overlap the cloth over the top and bottom bands, allowing at least ¼ in. of material to hang over the top and bring the rest of the fabric under the bottom

BELL SHAPE
Fig. 1

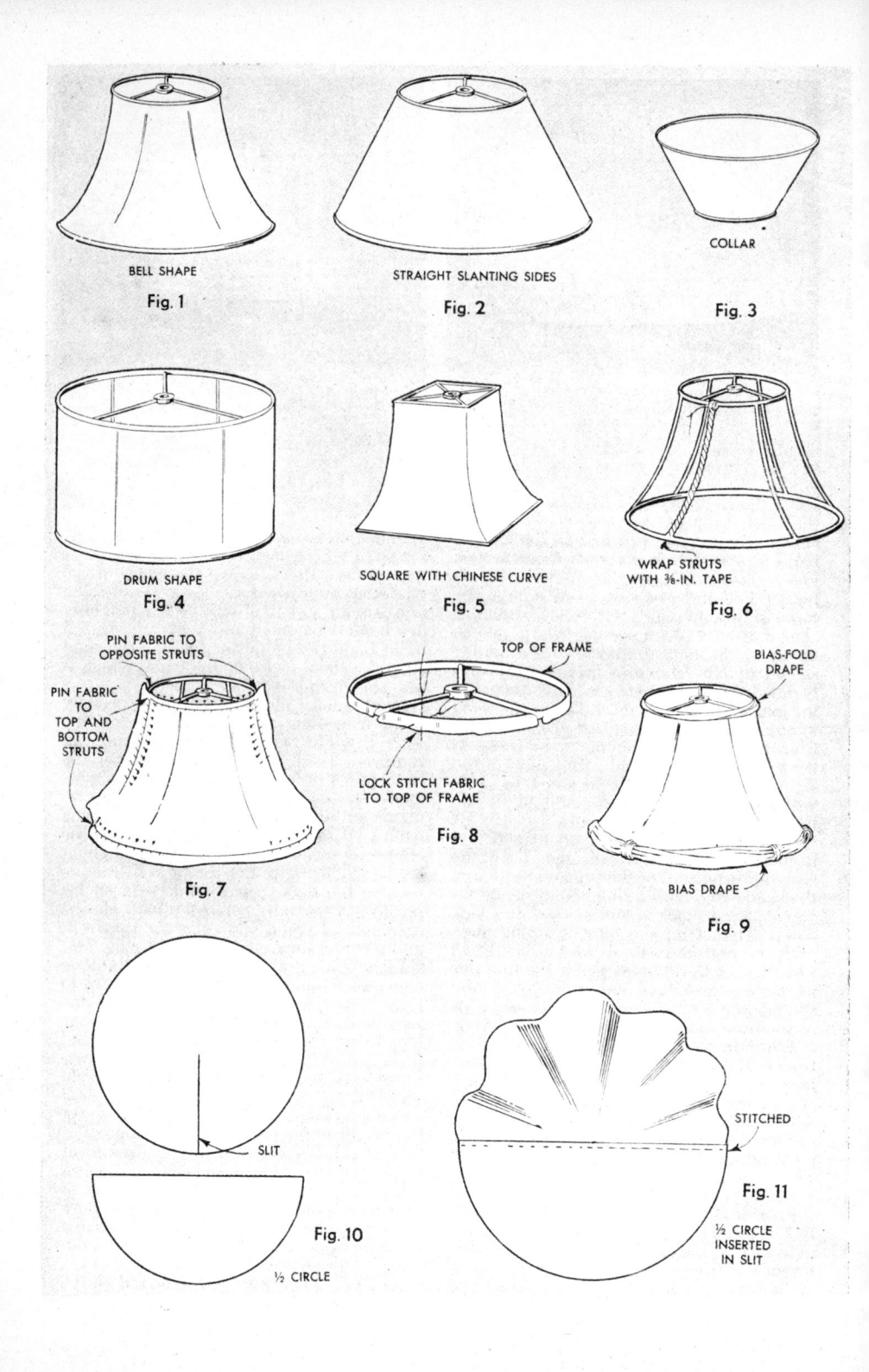

STRAIGHT SLANTING SIDES
Fig. 2

COLLAR
Fig. 3

DRUM SHAPE
Fig. 4

SQUARE WITH CHINESE CURVE
Fig. 5

WRAP STRUTS WITH 3/8-IN. TAPE
Fig. 6

PIN FABRIC TO OPPOSITE STRUTS
PIN FABRIC TO TOP AND BOTTOM STRUTS
Fig. 7

TOP OF FRAME
LOCK STITCH FABRIC TO TOP OF FRAME
Fig. 8

BIAS-FOLD DRAPE
BIAS DRAPE
Fig. 9

SLIT
Fig. 10
½ CIRCLE

STITCHED
Fig. 11
½ CIRCLE INSERTED IN SLIT

Here are two examples of bias-drape trimming. The shade shown at the left is trimmed this way only at the top, with a bias-fold strip at the bottom. Velvet bias drape was used to trim the other lamp at both top and bottom

band. Pin the cloth on the top at each seam, slitting the material about ¼ in. where the wires go across the top of the frame. When working on the top band, rest the bottom of the shade on a flat surface. Then turn the shade upside down to work on the bottom band. Pin the cloth down so there are no loose places. When the underlining is pinned to your satisfaction, sew it on to the top and bottom bands with a lock stitch as shown in Fig. 8. A strong sharp needle about 1¼ in. long threaded with a double strand of No. 24 white thread is recommended. Now cut bias strips of underlining material 1½ in. wide and 3 in. long. Fold in edges of the material and place it under the wires that cross the top of the frame at the edge of the top band. Bring ends of the strip in and over the band and tack. This will conceal the slashes made to fit the lining around the cross wires.

Now cut the top lining and fit and sew it to the outside of the frame. Turn the seam allowance over the top and bottom of the shade, and with a simple one-strand diagonal stitch, tack material over the lock stitch. Then trim the outside lining even with the bottom edge of the top band and top edge of the bottom band. For the finishing touches, tack onto the struts and the top and bottom bands folded bias strips of the material used for the outside lining.

Trimmings: The type of trimming to be used will depend on the kind of lamp shade being made. Tailored shades would require such trimmings as draping, cords, braids, double-box pleating or moss fringe. More decorative shades usually take ruffles, pleating, lace, velvet, gold or silver braid. The variety of trimmings is unlimited, but here are a few suggested trimmings.

First of all, there is the **bias drape**, shown in Fig. 9. To make this, prepare a bias strip at least 7 in. wide and as long as the circumference of the band which is being draped. With the outside of the strip against the outside of the shade and the width of the strip extending beyond the band away from the body of the shade, sew one edge of the bias strip to the band. Then fold the strip over to reveal the right side and gather the strip at each strut, where it is to be fastened with a loop of braid or whatever trim is used there. Leave the edges of the bias strip unfinished as they will not show and provide a more attractive drape. If only one band is trimmed with a bias drape, the other band can be finished with a bias-fold strip the same width as the band, which is also shown in Fig. 9.

Double-box pleating is an especially fancy trim. Bias strips 1 in. wider than the desired width of the finished pleating are required. To determine the length of the bias strip needed to go around the shade, measure the circumference and figure upon getting 7 in. of double-box pleating from 1 yd. of bias strip. For example, if the circumference is 20 in., you will need about 3 yds. of bias strip. For the most attractive results, the bias strip should be faced for pleating. Especially colorful effects can be obtained by using one color for the front and a contrasting color for the facing. Machine-stitch the strips together on the long edge, allowing a ¼ to ½-in. seam. Turn to right side. Beginning at one end of the strip with the length of the strip in your lap, bring the material toward you and form a pleat ⅝ in. wide. Then bring the material forward again and form a second pleat on top of the first pleat. Stitch these down almost to the end of the pleat, leaving the needle in the fabric and releasing presser foot. Then form a similar double pleat away from you to meet the edge of the first double pleat. Stitch almost to the end of the pleat. Keep repeating this procedure until the whole strip is pleated. Tack the pleating to the bottom or top, or both if desired, of the shade.

Another fancy trim is the **shirred ruffle**,

The double-box pleating is used effectively at both the bottom and the neck of the shade pictured above

The bias-drape and shirred-ruffle trimmings are combined for the finishing touch on shade shown above

which is made in the same way as the strip for box pleating except that the strip is shirred. The amount of bias strip needed for the shirred ruffle is about three times the circumference of the area to be covered. In shirring the bias strip, use a heavy thread in the sewing machine bobbin so that there is no danger of the thread breaking when the gathers are pulled up. An interesting effect can be obtained by using velvet-covered cording in the seams of the bias strip. Cording the ruffle is done by sewing the cording on the right side of the bias strip facing the strip and using the cording foot. Then when you face the strip, use cording foot as you seam the bias strips together.

The skirted shade: This is one of the most difficult shades to make. It requires a diagonal frame, Fig. 2, and a collar, Fig. 3. Wrap and cover the frame and collar with inside and outside linings. The collar is treated exactly like the frame when you are wrapping and lining.

To make the skirt, take a piece of tissue paper and lay it over the top of the frame, matching the center of the tissue to the top center of the frame. Draw the tissue down and pin at quarter intervals, then pin all around on the bottom band easing in all fullness evenly. Trim bottom edges even with the bottom band of the frame. Unpin and use this circle of tissue as your pattern. Cut one circle of buckram and one of the material you are using for the skirt. Fold the circle pattern in half and cut a half circle of the buckram and a half circle of material. Slit the buckram from the edge to the center of the circle on the straight of the material. Now insert the half circle in this slit and overlap the edges ¼ in. and machine-stitch. See Figs. 10 and 11. Do the same with the material for the skirt; however, make the seam by putting the two right sides of material together. Then press.

Next, match the seams of the skirt to the seams of the buckram. Baste edges of circles together and baste a circle 2 in. from center of circles. Now cut a 2-in.-wide bias strip of the contrasting material the circumference of the skirt bottom. Seam the ends. Turn under ⅛ in. of one edge of strip and stitch. With raw edges together and right side of bias face down on material side of circle, machine-stitch ⅜ in. from edge the entire edge of the circle. Bring bias over edge to the buckram side. Then place circle buckram side down on top of frame. Pull opposite edges of circle to bottom of frame and pin. Pin quarters down, then eighths, then sixteenths. It is a good plan to measure the edge of the skirt in inches and divide by 16. Pin at each sixteenth and catch skirt to the bottom band of the frame. Proceed in same way for collar skirt. Then circle neck of shade with double-box pleating.

The square shade: The square frame is unique in that each side must be put on singly. It may be possible to cut and machine-stitch the lining pattern from a muslin pattern if all the sides have identical dimensions. Any great deviation will require a separate pattern for each side. Sew top lining to each strut, one side at a time. Lock-stitch the material to the top and bottom bands and to both side struts. Proceed with other sides, keeping material straight.

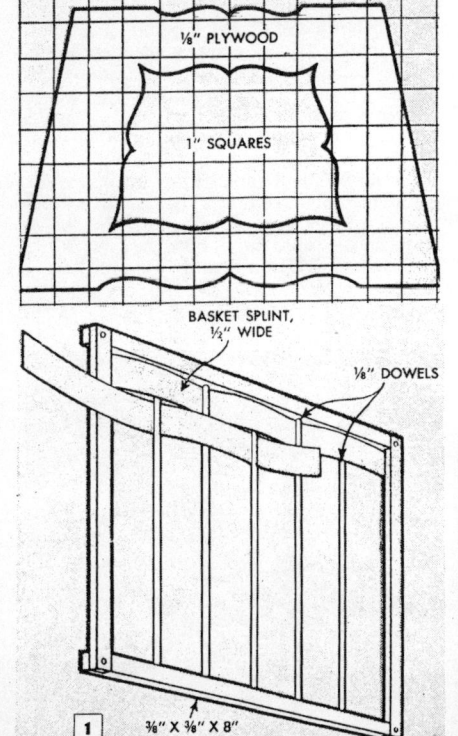

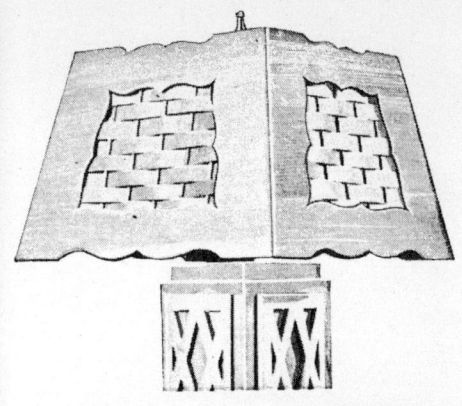

Basket-Weave Shade

LAMP SHADES

ANY OF THE three lamp shades pictured above will add a distinctive touch to your living room or bedroom. Although they are entirely of wood, the shades provide a soft, side-lighting effect resulting from the novel use of translucent basket splint and louvers made of thin slats. As detailed, these shades are too small for the massive lamp bases frequently used in modern furnishings. However, they easily can be enlarged and their proportions varied somewhat if desired. The plywood portions of the shades may be stained, finished natural with clear varnish or shellac, or painted, depending upon your likes and the color scheme of the room.

The basket-weave shade, detailed in Fig. 1, consists of four scrolled sides of 1/8-in. plywood. Each side is backed with an insert

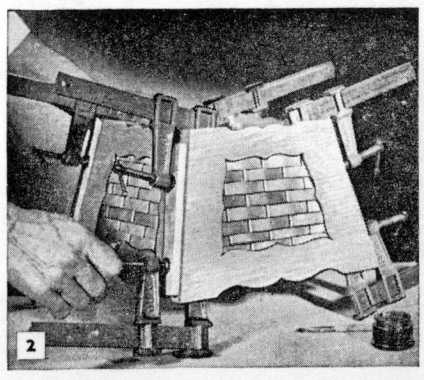

Louver Shade

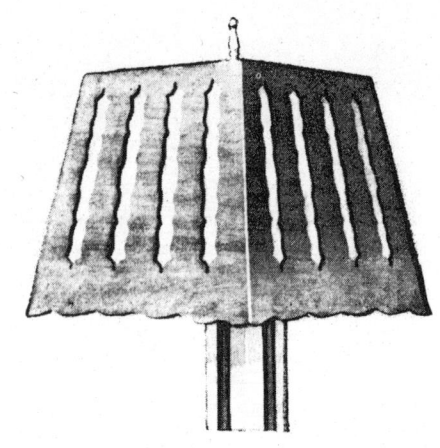

Scroll Shade

made by weaving basket-splint strips through dowels which are mounted vertically in a wooden frame. The sides are cut to size and jigsawed according to the squared pattern. After the basket splint has been cut into 1/2-in. strips, it is soaked in water to make it pliable. The strips are woven through the dowels while still wet and the ends of the strips are tacked to the sides of the frames. Note that the top and bottom pieces of the frames are notched so the side members will fit flush. The completed inserts are glued to the inner faces of the shade sides directly over the scrolled cutouts. The sides of the shade are mitered and glued together as in Fig. 2, glue blocks being installed along the joints, as in Fig. 3.

The sides of the louver shade consist entirely of thin wooden slats which are edge-glued in saw kerfs cut at a 60-deg. angle on adjacent faces of the corner members. To make the shade, first cut the kerfs in the corner pieces as detailed in Fig. 5. This can be done by kerfing one face of a 5/8-in. board and then ripping four 5/8-in. strips from the board. To kerf the adjacent faces of the strips, join the four side by side with a couple of cleats and cut the kerfs as before. Finally, cut off the ends so each member contains 16 saw kerfs on adjacent faces.

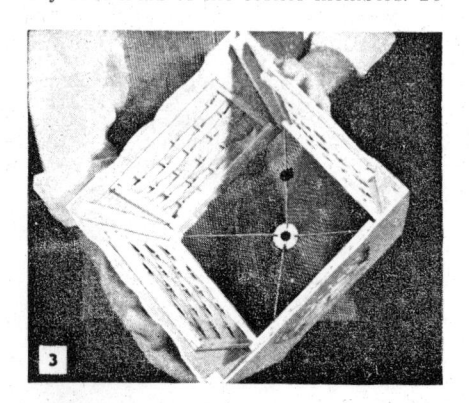

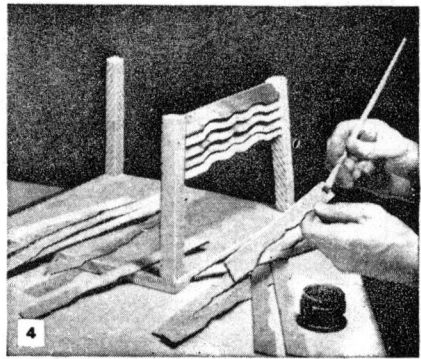

The slats are ripped just thick enough to fit snugly in the saw kerfs. As the scroll, which is shown in the squared pattern in Fig. 5, is the same on all the slats, several of them can be jigsawed at the same time. This can be done by stacking the strips for jigsawing, or by cutting the scroll on one face of a block before ripping off the strips. Note in the upper detail how the 16 slats for each side are laid out and cut, one longer than the other, to give the desired taper to the shade. Gluing the slats in the saw kerfs is done easily by mounting the four corner members on a wooden base as in Fig. 4. These can be held at the correct angle by setting pegs in the base to engage holes drilled in the lower ends of the corners. Or, the members can be supported with nails driven through the underside of the base. To attain the proper angle for the corner members, temporarily fit the top and bottom slats of the four sides in the saw kerfs.

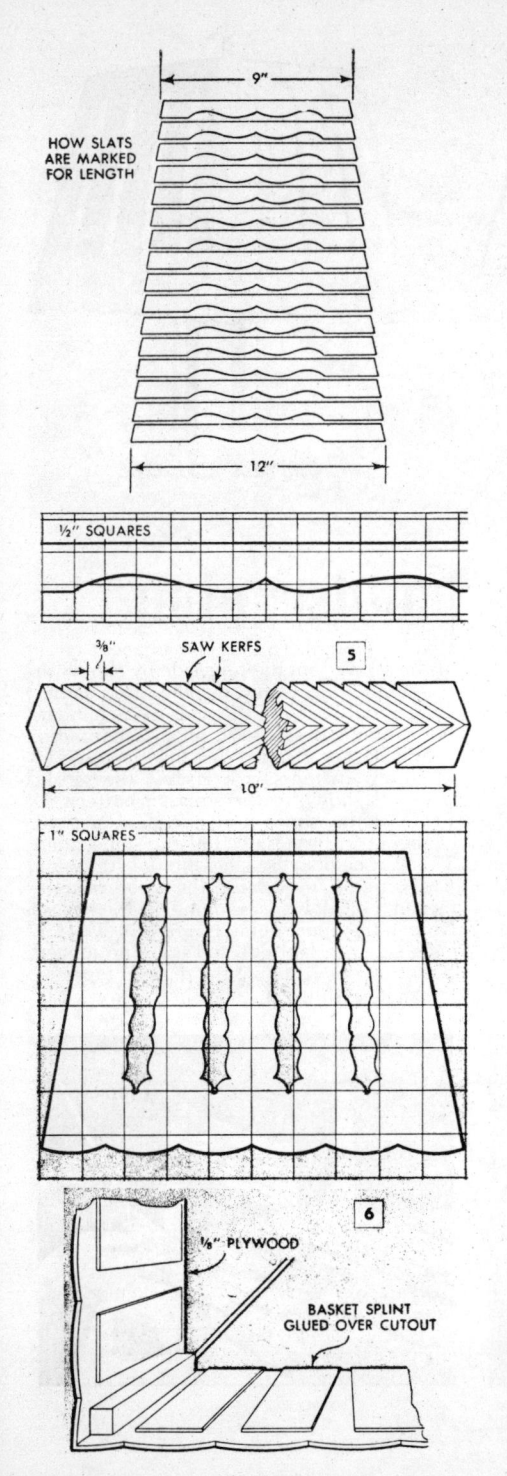

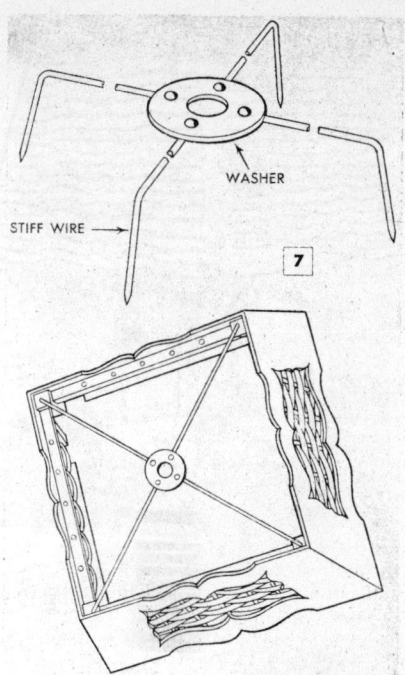

The scroll shade, detailed in Fig. 6, is made in the same way as the basket-weave shade. Strips of basket splint are glued over the scrolled cutouts, and the sides of the shade are assembled as before. Excellent splint for this purpose can be obtained from the thin wooden containers used for packing vegetables and berries, and the translucence of the wood can be increased appreciably by applying a thin coat of oil over the inside surfaces of the strips.

The finished shade is supported on the lamp harp by four lengths of stiff wire held together with a metal washer as in Fig. 7. The inner ends of the wire are inserted through four holes drilled in the washer and are peened over to lock them in place. The outer ends of the wire are pointed and bent downward at right angles so they can be driven into the corner members. Coat-hanger wire is satisfactory.

Charles G. Curtis Co.

Home Landscaping . . .

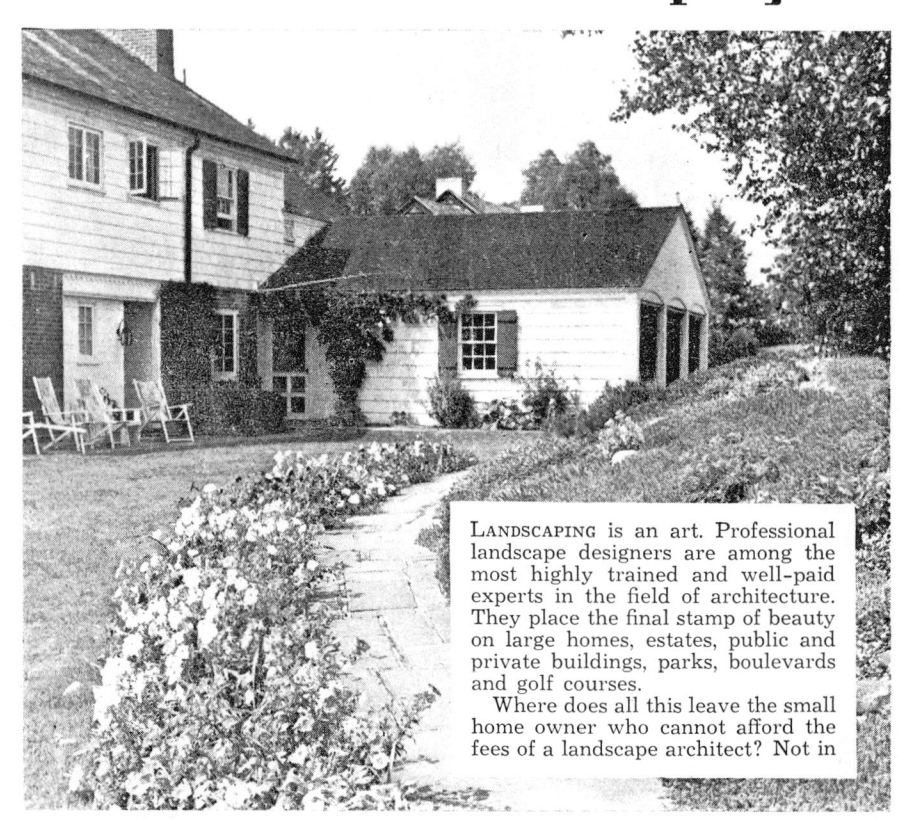

LANDSCAPING is an art. Professional landscape designers are among the most highly trained and well-paid experts in the field of architecture. They place the final stamp of beauty on large homes, estates, public and private buildings, parks, boulevards and golf courses.

Where does all this leave the small home owner who cannot afford the fees of a landscape architect? Not in too bad a position, really. Landscape artists are sympathetic with the desire of the average homeowner for beauty on his small plot of land. They have offered generous advice and counsel, and set down general rules for the guidance of gardeners who wish to apply some of the principles of landscaping.

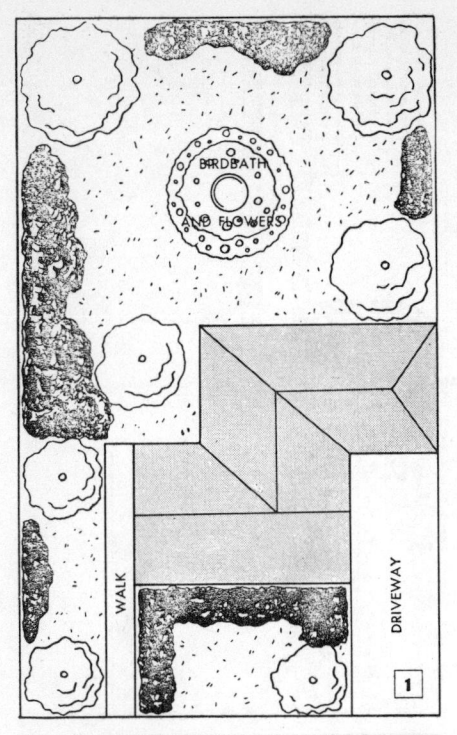

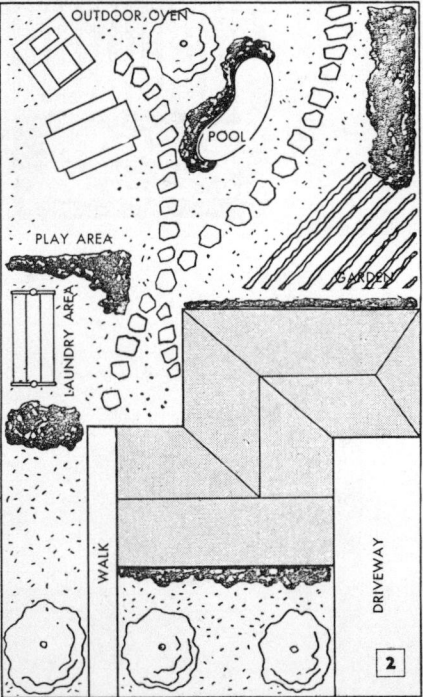

Everyone at one time or another has noted the raw and ugly appearance of a newly built house. However charmingly designed, the beauty of the house is marred by the bare scarred earth around the house and the blank skyline about and behind it. Let a little grass grow, however, and let but a few trees and shrubs be placed about it, and the house immediately acquires the beauty its designer intended to be revealed. Only a little haphazard growth will achieve much. Think, then, how much more will be done for such a house if the landscaping is well planned and well carried out!

Landscaping has a twofold purpose. The first—beauty—is obvious. The second—utility—will become apparent upon examination. Every home needs different treatment because the people who live in it are different from their neighbors in more or less degree. One family has small children, and will want at least some part of the home grounds devoted to space for playground equipment. One family is intensely interested in gardening; another is only mildly so. There will be a vast difference in their landscaping needs. The family requiring laundry space will have a problem not faced by the family owning a mechanical clothes drier or using a commercial laundry. Outdoor laundry-drying space must be planned carefully so that it is conveniently at hand and yet not too obtrusive. It must be attractively screened by some kind of planting, so as not to become an eyesore.

It is good to think of the home grounds as an outdoor living room. In it you will want everything that interests you in gracious outdoor living. In a limited space there will have to be compromises. Mother may have to sacrifice part of her flower-garden plans to space for a shuffleboard court or an outdoor oven. Father's allotment for a vegetable garden may have to yield to the desire of other members of the family for a rock garden or a garden pool.

To reach a compromise in this conflict of interests, careful planning is necessary. This planning should be done in several stages. The first step is the drawing of a rough sketch.

The rough: Begin your landscape plan with a blunt black pencil and a piece of paper on which you have drawn the boundaries of your lot in proportion. Make this space as small as practical—say about 2 by

4 in. if your lot is 50 by 100 ft. This won't give you any room for details, which is a desirable limitation at this stage. What you want to achieve is the "large form," or the general appearance your landscape will present. This, in the end, will be a more important factor for pleasing or unpleasing results than any of the details.

Block in the space taken by the house, garage, walks, driveways and other permanent fixtures. The area that remains is to be divided into two parts—planted space and free space. The planted space is, of course, for trees, shrubbery, flowers and vegetables. Free space will be allotted to lawn, play areas, laundry space and for similar purposes.

Using free movements (remember that here you can spoil only a small piece of paper), try to work out interesting shapes in and around the free space. Don't bother about professional landscape symbols—just use your own way of showing trees, shrubbery and other details.

Once started, you'll make a lot of these sketches. They will be very revealing. They will show weaknesses and impracticalities in the ideas you may have had before you put pencil to paper. And, as your pencil moves freely within the sketching space, other ideas will come. Save all the sketches, as you probably will want to combine two or more in your final rough plan. Here's an important tip—don't use the boundaries of your lot as *landscaping* boundaries. The shape of your lot is a space *within* which you work and not *about* which you plan. Try to keep away from straight lines in planning shrubbery and other plantings. Nature never plants in a straight line. And there will be enough planes provided by house, walks, driveway, etc., to guard against the possibility of curves becoming monotonous.

The plan: Now you're ready to take your plan out of the rough stage and into details. And we hope, for the sake of final effect, that you resolve (and stick to it) not to depart from the general curves of your rough as you plot the details.

Figs. 1 to 4 show four landscape plans drawn about similarly sized and placed house plans. Each plan places different emphasis on certain areas, although there are points of similarity which are made necessary by the layout of house and grounds. The number of variations is practically limitless, and there is no reason for your home, even though it may be of the same design and material as every other in your block, to present the same appearance as any neighbor's house.

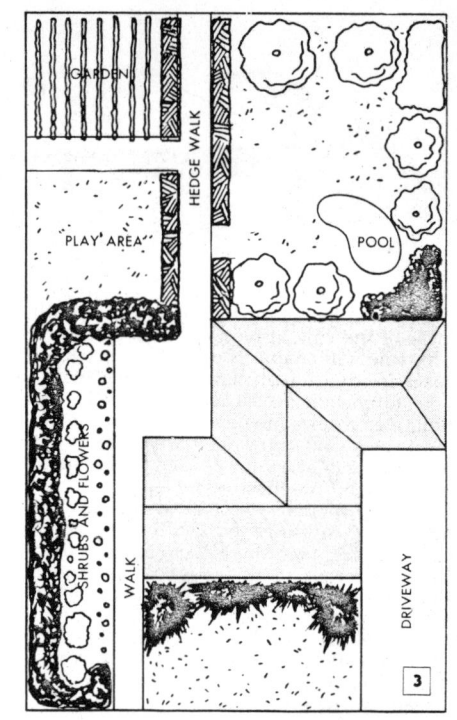

The formal plan will be larger than the rough sketch—a scale of ¼ in. equals 1 ft. is suggested. Make an L-shaped scale ruler of a strip of heavy cardboard, marking both arms of the rule in feet at ¼-in. intervals. This rule will enable you easily to mark off space in square feet that you intend to devote to certain areas. Again block in the buildings, walks and drives, and outline the lawn space and other open areas with light pencil lines. Make your house plan detailed, with rooms, porches, doors and windows indicated, so that you can plan plantings near these carefully. Now treat each special area of planting space individually. You can forget about the whole plan now, since your rough sketch has assured the over-all effect.

In working out details within an area, remember that balance is wanted, but also that good balance is not always obtained by mathematical precision.

Planting trees in front of a picture window, for instance, does not mean that two trees of equal height should be placed at exactly located spots on either side, with a uniformly spaced line of shrubbery between. Two or three trees at one side of the window, with a shrubbery mass curving before the window and sweeping to meet a porch column, will provide more pleasing balance.

Shrubbery is more attractive in varying masses than in straight rows of uniform height. Two or three varieties of shrubs are better than a single kind if their shapes and textures harmonize. Choices for planting will be given later, but the way in which plantings are massed should be given some attention on your scale plan.

The model: You can, of course, proceed directly from plan to planting. But professional landscape artists recommend one more planning step—the construction of a scale-model house and grounds. A scale model gives a much more realistic forecast of the final effect than a two-dimensional plan, and may reveal defects in the original plan that will save you money.

To lay out a plan accurately to scale you will need a draftsman's T-square, a 45-deg. triangle, a pair of dividers, a small drawing board, thumbtacks and a triangular scale showing common scale reductions.

Begin by thumbtacking a piece of illustration board to the drawing board. Then, with the scale, lay out the size of the lot, using a scale of ¼ in. equals 1 ft. For some types of construction a scale reduction of ⅛ in. equals 1 ft. is practical. Mark the property boundaries on the board. Then, using the same scale reduction, lay out the floor plan of the house on another sheet of illustration board. Scale the thickness of the walls and mark the inside over-all di-

Lay out the exact scale size of your lot on heavy illustration board and mark boundary lines in pencil. Use the common reduction of ⅛ or ¼ in. to 1 ft.

Cut out the lot, using a metal-edged ruler and a sharp, thin-bladed knife. Hold the blade in a vertical plane so the edges of the board will be square on all sides

On another board lay out the house plan. Mark the partitions. Plan should be an exact scale reduction of the inside dimensions measured from wall to wall

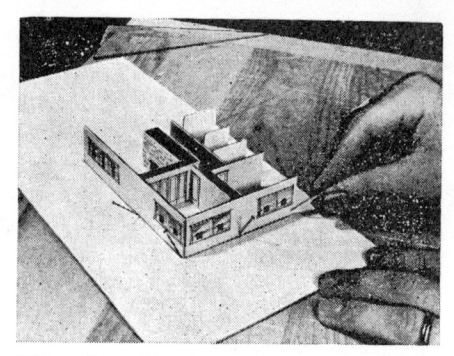

With walls and interior partitions cut to scale, the model may be assembled. Hold walls with pins while joining the partitions with airplane cement

mensions of the floor plan. Then cut on the inside lines. Cement this cutout in place on the plan of the lot in exactly the same location as the full-size house is to be. This method of cementing the floor-plan piece to the board on which the lot plan is laid out makes it easier to position the walls. Use your T-square and triangle to get parts laid out exactly at right angles and be careful about the measurements.

Next, determine the scale height of the outside walls and partitions. Mark off and cut strips of the illustration board so that you can cut pieces of partition and wall stock as required. Scale-sized windows and doors are represented by drawing the outlines directly on the walls with black ink, or with pencil. Assemble the parts and fasten them together with airplane cement, reinforcing the outer walls with common pins. On most plans you can cement all the partitions in place before setting the outside walls. At this stage, with the exterior walls in place, cut a ceiling piece of the exact outside dimensions of the floor plan so that it overlaps the thickness of the walls on all sides. This gives you a foundation on which to build the roof. If the house has a boxed cornice, then the ceiling piece should extend beyond the outer walls a scale distance equal to the overhang of the cornice. Build the roof in sections, using small blocks of wood to elevate the sections to the correct pitch. Carve the chimney top from a small block of wood to scale size, score it, or line with ink to represent bricks, and cement it to the roof. Finally, cement the model house in proper position on the illustration board. Next, make and place the garage if this is separate from the house.

Cut green blotting paper to represent the open lawn areas of your landscape. Cement this in place. Hedges and shrubs can be worked out to scale with green modeling clay. Use crumpled colored crepe paper to represent flower beds. Trees can be represented by tiny artificial trees such as can be obtained in any dime store, or by sprigs of evergreen. Trellises, archways and fences may be carved from pine or balsa wood. Walks, drives and other concrete surfaces will be represented by the white

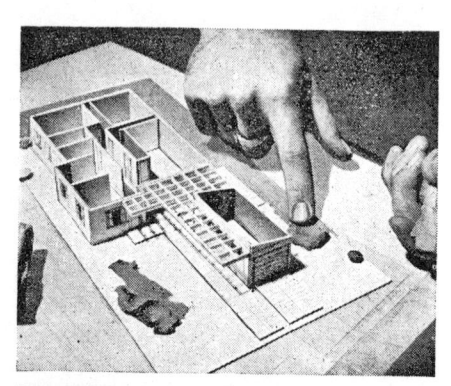

Colored modeling clay is used to form shrubbery and hedges. Strips cut from a sponge also can be used

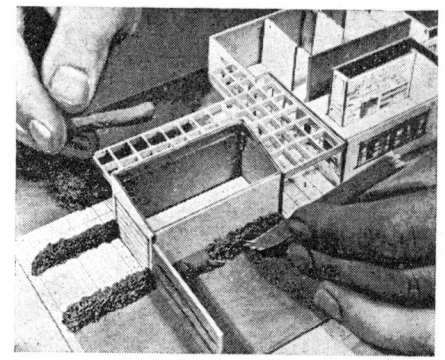

Finishing the modeling-clay shrubbery with the splintered end of a stick gives a realistic effect of foliage

Here trees and shrubbery have been combined to set off the house without spoiling architectural details

unadorned area of the illustration board. Paint the model house and other appropriate parts of the layout with painter's oil colors thinned to an easy brushing consistency. Use soft-bristle brushes to spread the paint in a uniform coating on the smaller surfaces.

The scale layout may seem to be a lot of trouble. But it can be a lot of fun, and may save you some unpleasant shocks in the future. Don't forget to make your pieces scale correctly in height as well as in length and width.

Trees should be the first concern of your landscape plan. Trees will frame your picture, and can ruin it if located incorrectly. A small ranch house, for instance, will be dwarfed and its pleasing rambling effect destroyed by plantings of Lombardy poplars that eventually will grow to 70 ft. or higher. Smaller trees, placed advantageously, will make the house appear larger than it is. The shapes of trees also will have an important over-all effect on the picture. With a tall, narrow house, a broadening effect should be sought for in the over-all landscape plan. Trees of rounded form, in plantings leading the eye away from the corners of the house, will achieve this.

Tree shapes generally fall into one of the following classifications: pyramid, inverted pyramid, oval, columnar and clump. Pyramid trees include the cedar, fir, hemlock, pine, spruce and larch. Inverted pyramid shapes are assumed by the elm, honey locust and Japanese pagoda tree. The oak, maple and tulip are oval-shaped trees. Trees of the columnar type include the

This is wrong. Trees are planted too closely for future growth. The center one will block the window

Doorstep plantings must be chosen with care. Plant trees that will spread, rather than attain height

Paul Hadley

For the formal house—formal plantings. The several kinds of evergreens here have been trimmed to shape

KEY
1. Globe arborvitae
2. Pyramid arborvitae
3. Dwarf juniper

KEY
1. Pyramid arborvitae
2. Pyramid juniper
3. Dwarf juniper
4. Globe arborvitae

KEY
1. Pyramid arborvitae
2. Taper queen juniper
3. Globe arborvitae

Lombardy poplar, cypress and eucalyptus.

Clump trees are those which naturally grow with multiple trunks, or in close-growing groves, such as the birch, ash and willow. In addition, there are trees which are especially suited to being trimmed into fancy artificial shapes, such as the laurel, linden, plane and horse chestnut.

The shape and the eventual size of the mature trees will determine the general effect and character of the landscape picture. Be careful to allow for future growth, so that the trees will frame the house and not hide it. Locate the larger trees near the rear corners, so that they will extend their branches above the house, softening the harsh lines of roof and corners and showing the home to advantage. Trees should always appear to be associated with some other part of the grounds—either the house, other trees or a shrubbery mass. Landscape artists rarely put a lone tree in the center of open lawn.

Trees fall into two main groups—deciduous and evergreen. Evergreens keep their leaves or needles during winter, so that a continuously green foliage is maintained. But they present a cold and formal appearance in spring and midsummer, when other plantings are brilliant with blossoms and sparkling with foliage. So it is doubtful that you will want to confine your tree plantings solely to evergreens.

Most of your plantings should be native trees. This is important because native trees are used to the soil and climate of the region, and because they are less expensive. By keeping a lookout along the highway or in the woods, you can often find good specimens suitable for transplanting to your grounds in early spring or fall. (Don't be guilty of trespassing on private land or of robbing state preserves.)

With these general matters in mind, you are ready to choose your trees.

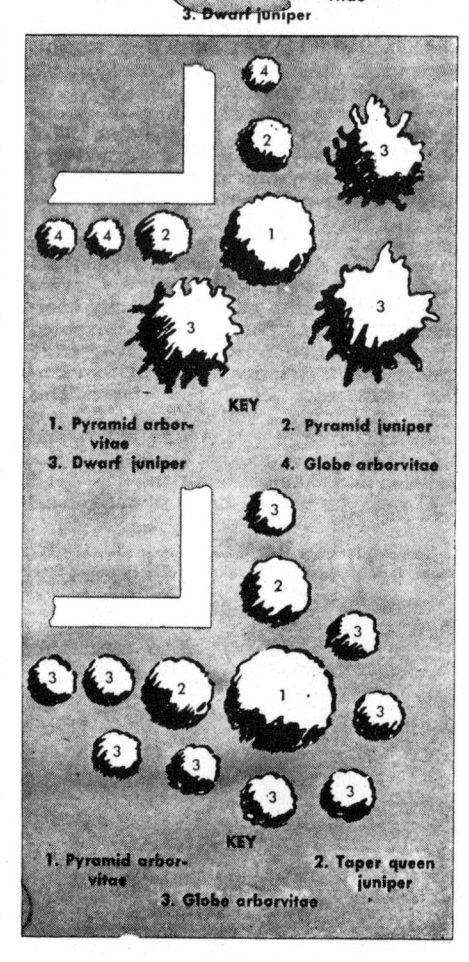

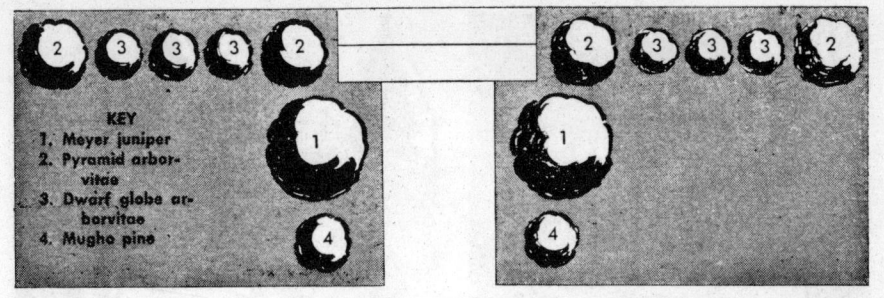

KEY
1. Meyer juniper
2. Pyramid arborvitae
3. Dwarf globe arborvitae
4. Mugho pine

When making a landscape plan, number the duplicate shrubs of a given variety. Where there are duplicates in a single group or in adjacent groups, it's quite important that the duplicates be of the same age and training

Shrubbery plantings not only add materially to the value of any home, large or small, but they give it that natural appearance so difficult to achieve in any other way. In making a selection for a given location, keep in mind the exposure, type of soil, appearance, and especially the spread and height of the shrubs when they have attained maximum growth. In some locations and for certain specific purposes, the relative rates of growth of a group of shrubs must be considered in the planning. Most of the extremely hardy shrubs grow very slowly, taking years to attain maximum height or spread. Others grow rapidly, taking only two or three seasons to reach tree size from a single nursery "whip." Certain varieties of slow growers have been designed by nature to withstand very severe cold without harm and without any additional protection. Some thrive in exposed positions in thin soils, while other more tender plants require sheltered locations and moist soils rich in plant foods. This wide variety gives you a choice of plants with size, color of foliage and bloom, and growth habits suitable for almost any plan, location and climatic conditions.

Corner plantings, using both tall, conical shrubs and the low-growing types, which spread a thick foliage only a few inches above the ground, are especially effective around small homes. Plan views of such plantings are shown in the keyed drawings. Usually, tall-growing shrubs are planted in the background and the lower varieties in the foreground in those arrangements where close grouping is necessary. Where shrubbery borders a walk, driveway or a terrace stairway, those bordering the stairway should be of the low-growing varieties, all specimens being of equal age at the time of planting. Where the stairway leads from a lower to a higher level of the garden or lawn and not directly into the house, the lower-level terminal planting usually is a low, spreading evergreen, the spreading junipers being especially suitable. At the top of such an arrangement the stairway row planting may properly terminate in a single taller specimen on each side. When making any closely grouped planting requiring the use of duplicate specimens, be sure that the duplicates are of the same age.

Border plantings, on the other hand, should not achieve the more geometric symmetry required near a building. Instead they should show the rambling out-

In corner plantings, left, low-growing shrubs with thick foliage ordinarily are planted in the foreground. Right, formal plantings require regular seasonal care. Trimming must be done repeatedly as growth progresses

Above, a one-plant garden of assorted chrysanthemums. Below, a bed devoted entirely to floribunda roses

Totty's
J. Horace McFarland

line often found in natural surroundings. The one exception, perhaps, is a hedge which also serves as a boundary between two properties. Here the planting may consist of individuals of a single variety of the same size and age, or two varieties of contrasting foliage if there is room.

It pays to buy balled stock for both individual plantings and also for the larger specimens in a group planting. Balled evergreens should be at least three years old. The five-year specimens of most varieties of evergreens cost more but these have had the benefit of two or three years' training in the nursery.

Smaller specimens dug from the nursery row are supplied with open-root systems packed in damp moss. These should be handled with the greatest care after unpacking to prevent the roots from drying out before the shrubs can be planted in a permanent location. If there are a number of shrubs such as a single lot of one variety for a long hedgerow, all those which cannot be planted immediately after unpacking should be "heeled" into moist soil. To do this dig a V-shaped trench in the garden and lay the shrubs side by side against one side of the trench with the roots at the bottom. Cover the roots and trunks some distance above the ground line with soil and pack it lightly. Soak the trench thoroughly. When you remove the shrubs from the trench, or the original packing, place the roots in a pail or tub filled with water.

Tall flowers can be blended effectively with a hedge to make a colorful background for informal gardens

Flowers are the jewels in the setting provided by your trees and shrubbery. One important fact should be grasped by every home landscape designer—flowers have a place in every part of the home grounds, but they must be wisely selected and properly located to give the most pleasing effect. At the front of the house they should be used for foundation and shrubbery plantings. Almost never should they be placed in a bed in the middle of the yard. At the front, flowers are especially useful during the early days of a landscape. Use them to fill the bare spaces around and between plantings of shrubbery, which often are thin and ragged when new. Most flower plantings in the front should be those of long-season blooming habits, with a few tulips, daffodils and crocuses for early spring, and a few late flowers such as chrysanthemums for late fall. This will provide color for your front from early spring to the onset of winter.

You will use flowers sparingly in your front yard, but in the back you may splurge with blossoms. Plan your beds formally or informally according to the general theme of your landscape plan.

Formal balance is achieved by a geometric layout of beds or pathways between beds. Plantings should be balanced on either side of an imaginary center line through the whole garden area. An informal garden can be an irregular border containing masses of varieties of flowers which are backed up by a shrubbery border or a boundary fence. A massed flower garden should be near the living room of the house or an outdoor terrace. This will add greatly to the beauty of the surroundings. If the land area is large enough, several flower gardens may be planted. That nearest the house can be formal while those at greater distances can be informal. The shape of a flower garden is most pleasing when it is a little longer than wide (about one and one-half times the width). But longer rectangles, or even a square, can be made interesting with a little planning.

Use low-growing flowers to border walks and the open boundaries of a lawn. Use of flowers here is much better than the common practice of spading out the grass and leaving raw, open trenches at the sides of walks. Border flowers can also be combined with shrubs, either by planting the flowers in front of the shrubs or devoting sections of the shrub border to flowers.

Tall-stemmed flowers, such as the castor bean, cosmos, sunflower and basketflower, can be used as screens for temporary fences, or to hide rubbish burners, garbage cans and other unsightly objects.

The simplest garden to design and maintain is the one-plant garden. This may be a rose garden, an iris garden, a zinnia garden, a petunia garden or any one variety. Equipment, garden practices and controls can be standardized in a one-plant garden.

The one-color garden is a popular idea among home gardeners. Is blue your favorite color? Imagine, then, the pleasure you'll get from a flower bed the edging of which begins with pansies, violets and forget-me-nots, which rises in the middle ground to bachelor buttons, cornflowers and columbines and is topped by tall sweetpeas, delphiniums and asters. Or visualize multi-shades of yellow made by California poppies, dwarf marigolds, strawflowers, double buttercups, dahlias, nasturtiums and black-eyed Susans.

You should plan a separate area for cut flowers for indoor decoration. The grounds which do not furnish beauty for the home

Take advantage of nature's oddities. Here a large rock has been left to set off shrubbery plantings

Climbing vines add charm to any garden plan. They are especially useful for breaking up wall expanse

A flagstone walk flanked by flowering shrubs leads to a rock garden. The treatment here is informal

interior are performing only half their function. The vegetable garden is an ideal place for growing cut flowers. The cultivation and plant food which a vegetable garden receives also makes flowers grow better. Another argument for a separate area for cut flowers is that the beauty of the decorative garden and border plantings may be destroyed by cutting blooms for the house.

Other landscape areas: With all the "greens" in your landscape picture taken care of, you are ready to plan in detail the other areas of your home grounds. A close budget may not permit the completion of all your wants within a year or two, but sooner or later you'll want to add lawn furniture, some garden ornaments, perhaps some playground equipment or possibly an outdoor kitchen. You'll have to decide whether you want flagstone walks or a grass path in and around your flower and vegetable gardens. Whether you want a fence or not will depend on whether you actually need one to confine pets or small children and on whether it will or will not add beauty to the grounds. On small lots, landscape gardeners sometimes are reluctant to recommend a fence unless it is a prime necessity. You may want to build a rock garden or a garden pool. There will be spots on your ground which can be enhanced by trellises and pergolas, although care must be taken that these garden furnishings are not overdone. If they are, the whole landscape scheme can be made to appear in extremely bad taste. Remember that outdoor cooking areas and game courts get hard use—for the sake of the rest of your lawn you'd better plan to pave these areas in concrete, asphalt, gravel or tanbark.

Latticework Relieves Plain Exterior-Wall Area

Bare, unbroken, exterior-wall areas between widely spaced windows in a house can be made interesting by the addition of latticework designs similar to those shown. Such a design also will serve to make the house appear longer, lower and "tied to the ground," an effect long considered very desirable in a house by most architects. Made from ¾ x 1-in. stock, the members used in the overlaid design are half-lapped and joined with nails and waterproof glue. The latticework is completed by finishing with several coats of exterior paint, either in the same color as the windows or in a contrasting color. If 1-in. blocks are inserted between the latticework and the wall of the house, the former can be used as a trellis, thus providing an additional decorative effect for the wall.

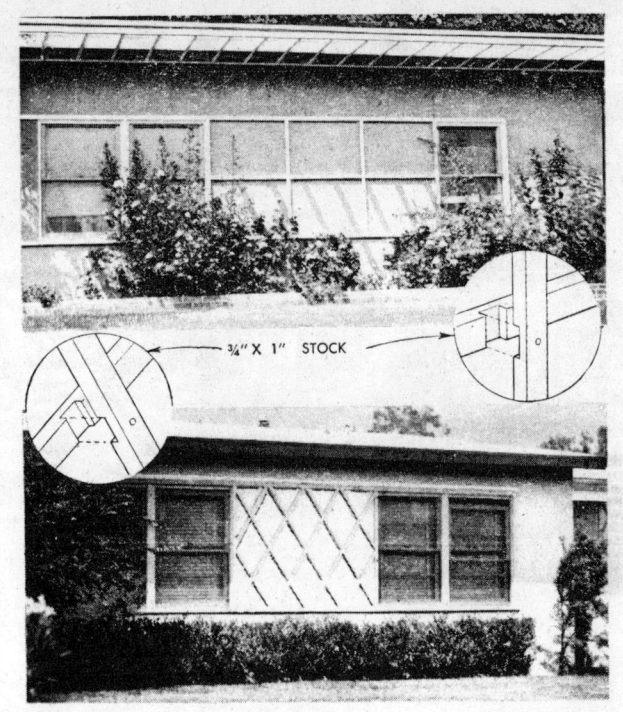

LAWN AIDES

Crabgrass Seedlings Killed by Thick Tall Lawn Grasses

One of the best ways of keeping crabgrass out of a lawn is to provide a heavy turf. Seedlings of crabgrass are extremely sensitive to shade, and if the turf is thick and allowed to remain at a height of 3 or 4 in. to provide dense shade during the germination period of crabgrass, which is early May in the northern states, the seedlings are likely to be killed. As lawns produce most growth during spring and fall months, condition them early in the spring. Lime applied at the rate of 50 lbs. per 1,000 sq. ft. in January will be beneficial. Fertilizer should not be applied later than early in March as later applications are beneficial to crabgrass. If there are spots that need reseeding, do it early so there will be a thick turf before the crabgrass germinates. Avoid watering during the summer months until the lawn starts to wilt; then water until the soil is soaked to a depth of 4 inches.

LAWN CARE

THE MAIN INGREDIENT of an attractive landscape that brings out the beauty of your home is an evenly smooth and green lawn. It can be thought of as the canvas upon which the picture of your garden is to be created. The flowers and hedges that border the lawn, the trees which are surrounded by the green turf and the ornaments which are placed on the grass—all these beautify the landscape only when an attractive lawn serves as a background. Since it is such an essential and permanent source of beauty, the construction and maintenance of a lawn deserve careful attention.

Establishing a New Lawn

Grading: If you are establishing a lawn for the first time on your property, your first and foremost task is to obtain a properly graded area upon which your lawn is to grow. Should the job of grading appear to be a difficult one, it is advisable to hire a landscape contractor to do the heavier work. These contractors are equipped with bulldozers, rototillers or similar cultivators with which they can accomplish the job in short order. You can usually hire these operators and their machines by the hour at reasonable cost.

One of your first considerations in grading is direction of slope. Generally speaking, northern and eastern exposures are the most desirable. Hot dry weather affects southern and western slopes more, especially if water drains off the surface quickly.

In grading, it is best to have natural and gradual contours. Severe grades (Figs. 1 and 2) should be avoided wherever possible. It is difficult to establish and maintain a lawn on steeply graded ground because of washing out in wet weather and extreme drying in periods of drought. Moreover, steep grades complicate mowing. Every lawn should, however, have some slope, no matter how slight it may be, to insure runoff of excess water. A pitch of at least 6 in. to every 100 ft. of lawn is generally recommended. Ideal grading provides a slight slope away from the house in all directions, just enough to permit good drainage. Any change in grade should be as gentle as possible, unless an abrupt terrace is to be a part of the scheme of the landscape.

If a severe drop in grade is absolutely necessary, a dry retaining wall at the foot of the slope helps, as shown in Fig. 4. Otherwise, terracing is recommended so that the grass slopes will not be greater than one foot of fall to three of horizontal distance. Where there are several terraces, a series of dry wall embankments with grassed terraces between provides a suitable arrangement. To make the mowing of terraces easier, slopes at the top and bottom should be leveled off gradually. The top of the slope or terrace should be convex, and the bottom concave (see Fig. 3). With this arrangement the lawn mower will cut without scalping the crest or skipping the base.

The actual grading should be done in two parts. The first part consists in working with the subsoil grade. However, before the subsoil is touched, scrape off the topsoil and pile it to one side. Then grade the subsoil, giving it the same contour as the final grade. It will, of course, be lower than the final grade to allow an even depth in which the topsoil will be restored or added. The time to provide a good drainage system is when you are working on the subsurface grading to prevent subsequent damage to the topsoil layer. This is also a good time to install any underground utilities such as a sanitary sewer, water lines to the house, major supply lines for lawn irrigation purposes, etc.

When you come to the surface grading, distribute the topsoil as evenly as possible, making sure that there are no thin spots to cause poor areas in the lawn. Special care should be given when grading slopes as there are often places that are cut and scraped severely, removing all topsoil and not replacing it in sufficient quantities. Never allow the subsoil to peek through. It is particularly important to have good depth of topsoil in lawn areas having considerable slope.

If there are large trees on the area to be graded, you will want to protect them during grading operations. Consult an experienced tree surgeon if the new grade is to be decidedly lower or higher than the natural grade under the trees. If the grade is to be lower, leave a sloping mound of soil around the base of the tree, extending out as far as the drip of the branches. The same amount of soil as was there originally should remain over the root zone of the tree. In the event that the grade is higher, provide some arrangement for a dry well or a fill of gravel and stone under the tree so the roots continue to get the necessary air and moisture.

Preparing the soil: The importance of good soil in producing healthy vegetation was discussed extensively in the previous section. This is as true for grass as it is for rare and delicate flowers. A good rule to follow in preparing the soil for grass is to prepare it as if it were to be a flower bed. Grass grows best in a light loam soil. A soil of this type absorbs and holds the necessary amount of moisture but is porous enough for the excess water to drain away and the air to penetrate it freely.

Before anything can be done to improve your soil, all rubbish—chunks of concrete, boards, plaster, etc.—should be removed.

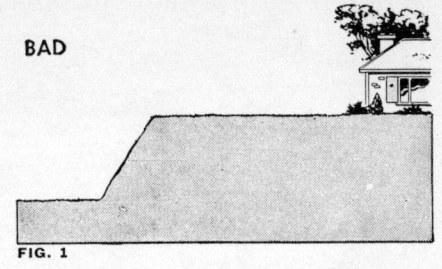

BAD

FIG. 1

A sharp grade, like the one above, should be avoided when establishing a lawn. There will be some washing away in hot weather and drying during drought

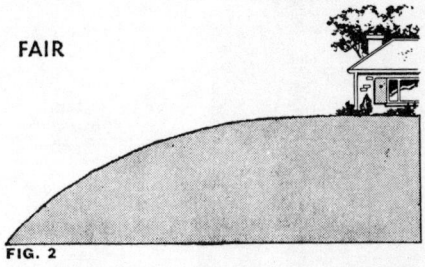

FAIR

FIG. 2

The grade of the lawn above is slightly better. However, mowing this lawn will still provide a problem, as the top will be scalped and the bottom ignored

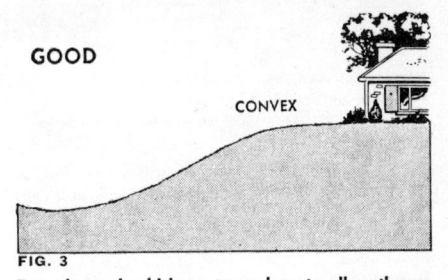

GOOD

CONVEX

FIG. 3

Every lawn should have some slope to allow the excess water to run off. The above lawn has an excellent slope with a convex top and concave bottom

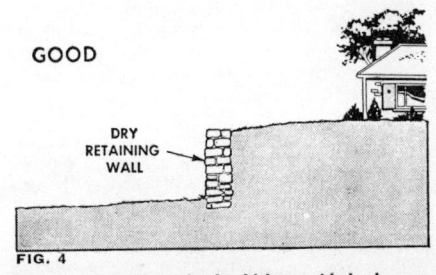

GOOD

DRY RETAINING WALL

FIG. 4

A severe drop in grade should be avoided wherever possible in making a lawn. But if an abrupt terrace cannot be avoided, a dry retaining wall at the foot of the slope should be constructed to aid drainage

FIG. 5

BEFORE SEEDING TOP SOIL SHOULD BE FINELY PULVERIZED BY RAKING

TOP DRESSING, 1 PART TOPSOIL, 1 PART PEAT TO DEPTH OF 2-3"

HEAVY MULCH WILL SMOTHER THE GRASS

FIG. 6

SPREAD THE TOP DRESSING

Using a bamboo rake, if possible, spread the top dressing over the ground and work it into the soil. Fall is the ideal time for this operation several days before seeding

Such trash will interfere with the movement of water in the soil and good grass will not grow.

For the best results in establishing a new lawn, prepare the ground some time before the grass is to be planted. Early summer preparation for fall planting and autumn cultivation for spring sowing are recommended. By preparing the seedbed in early summer for fall seeding, there will be time for the weed seeds to germinate. These weeds can then be killed by subsequent cultivations before seeding. Likewise, fall preparation for spring planting allows enough time for many of the annual weed seeds in the surface soil to germinate before fall is over and be killed by winter freezing. Settling of the ground will take place over the winter and the alternate freezing and thawing of the fallow surface soil will make it more mellow.

The ideal soil for a lawn can be established by beginning to build up the soil a year prior to seeding. This is done by sowing the area that is to become a lawn in soybeans, allowing the crop to grow for a season, then plowing the soybeans under in late summer.

If the soybean treatment cannot be given to the soil, it should receive a dressing of about 2 in. of well rotted manure. Then it is to be plowed or spaded to a depth of about 8 in. and thoroughly raked and rolled. If the soil is dry after it has received these treatments, it should receive an occasional watering, but otherwise should be allowed to lie idle until just before seeding time. Remove all weeds that have popped up, after which cover the entire surface of the soil with a topdressing (one part good topsoil to one part peat) to a depth of 2 or 3 in. Details of this operation are shown in Figs. 5 and 6. The mineral requirements of the soil should be met by adding 3 lbs. of nitrogen, 1½ lbs. of phosphorus and a still smaller amount of potash for every 1000 sq. ft. of soil area. This will amount to about 20 to 30 lbs. of complete fertilizer of an analysis of 10-5-3 or 12-6-4 per 1000 sq. ft., as shown in Fig. 7. Superphosphate is a good fertilizer which contains the three important elements. It should be mixed in the upper 4 in. of soil at the rate of 25 lbs. per 1000 sq. ft. This fertilizer will add enough phosphorus to the soil to last for a number of years. Other good sources of nitrogen which also carry small amounts of other elements are cottonseed meal, sheep manure, soybean meal and sewage sludge. Bonemeal is a suitable source of phosphorus and also contains a little nitrogen.

Now that the soil has been properly fed, it is ready for cultivation. The necessity for thoroughness here cannot be overemphasized. A finely pulverized seedbed is a must, and hand-raking will do the best job when the soil is in the best condition to be worked—that not-too-wet, not-too-dry soil should crumble easily. When all the stones and pieces of sod have been removed and a smooth seedbed is obtained, the soil

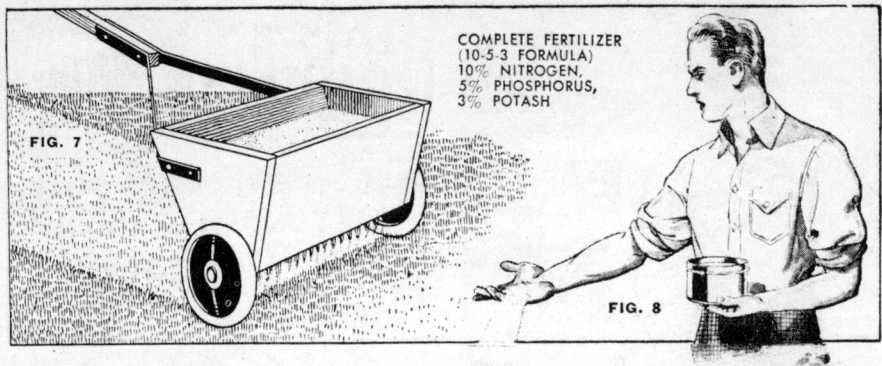

FIG. 7

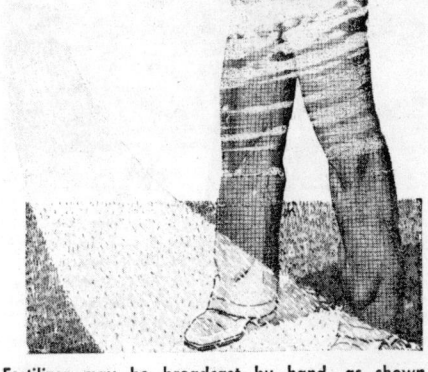

COMPLETE FERTILIZER
(10-5-3 FORMULA)
10% NITROGEN,
5% PHOSPHORUS,
3% POTASH

FIG. 8

should be well watered. Then, when the surface has become dry again, it is time to seed, providing that at least a week has elapsed since the plant food was applied.

Seeding: The time element in establishing a new lawn is also very important. Even the most suitable seed mixture will usually lose out to a crab-grass lawn if it is sown in late spring or early summer. Crab-grass seeds are practically always present on the soil surface, and they are more suited to the dry, hot conditions of summer weather than are the more desirable grasses. The better lawn grasses, on the other hand, have the upper hand during the cool, moist months of early spring and early fall. If these grasses can get going during these seasons, they will have a head start on the crab grass and other weeds and can easily crowd out most of the competing weeds when summer rolls around.

Fall or spring—which is the best time for sowing grass seed? In practically every instance, fall planting wins out over that of spring. Fall is nature's chosen time for grass planting. In the first place, grass roots have a chance to grow more extensively in the fall. Moreover, weeds stop growing as the weather becomes cooler. That's when the young grass seedlings have the ideal moisture and temperature conditions for germination and growth and can grow without danger of being crowded out by the weeds.

However, if you do not have the opportunity to do your sowing in the fall, early spring is a good second choice. The earlier in the spring you seed, the better your results will be. You can start spring planting even if there is still snow on the ground, but it will have to be that late and light snow which is the last stand of winter. It is best to wait until the frost is out of the ground, since seed sown on snow or frozen ground may be carried away if there should be a quick thaw.

When you are ready to seed, choose a day when the air is quiet. The grass seed

Fertilizer may be broadcast by hand, as shown above, or with a seed and fertilizer spreader. The latter method gives a more even distribution

may be sown either by hand or with a spreader. You can obtain a more even and quicker distribution with a spreader (see Fig. 10), but hand sowing is almost as satisfactory, especially if your lawn area is small. If a seed mixture is used, it should be thoroughly mixed and divided into two portions. One part should be broadcast over the soil while walking lengthwise over the area; the other portion is to be sown while walking at right angles to the direction of the first sowing. From 3 to 5 lbs. of seed per 1000 sq. ft. will be needed to sow a new lawn.

It is difficult to state definitely the best procedure for covering the seed. The big question is whether raking or brushing is better than allowing nature to bury the seed by the alternate wetting action of rain and drying process of wind and sun. To be on the safe side, the gardener would be wise to give an assist to nature by a light hand raking or by dragging a flexible object, such as a weighted sack or steel door mat, over the area. This will spread the grass seed more evenly and at the same time cover it lightly with loose soil. The seedbed should then be rolled to bring the

FIG. 9

The above seed and fertilizer spreader is also a yard cart. The hopper is removable and made of plywood reinforced with cleats at the joints, and fits snugly into the spreader hopper. The bottom is flat so that when mounted on the spreader it is high enough to clear the agitator. Rope handles on both sides of the hopper make it easy to carry to and from the cart

FIG. 10 National Garden Bureau

Although a seed spreader is a relatively expensive item, it is well worth the price because of the even distribution of grass seed you can obtain with it. Divide your grass seed mixture into two portions. Spread the first portion in horizontal strips, as shown above. Then operate the spreader over the same area at right angles to the first distribution

soil particles in close contact with the seed. A good practice to follow is to rake after the first part of the seed is sown and roll after the second sowing. A light roller is advised, just heavy enough to firm the soil but not pack it.

A top-dressing for the newly seeded area is optional. If you do use top-dressing, ⅛ in. of screened topsoil or compost will serve the purpose. Avoid the use of raw peat moss, as the grass roots may grow up into the peat instead of down into the soil. Manure should not be used as a covering unless it is well rotted or composted. Straw may be used after summer seedings or to protect slopes from washing. Not more than 3 in. of loose straw should be used and this is to be removed as the grass gets well started. Covering the grass seed becomes more important when it is planted on a slope, as the seed is likely to wash away before it can take root. Burlap, cheesecloth or heavy fiber paper can be pegged down over a seeded slope (Fig. 11), providing that the cover is removed as soon as the seed sprouts so that the grass will not be smothered. Erosionet, a loosely woven open-mesh material, is another good covering. It can be left on the ground to rot away or taken off as soon as the grass is well started.

Moisture is essential to germinate the seed as well as to keep the grass seedlings from perishing. The newly seeded area should be lightly sprinkled after it is rolled. Otherwise, seeding before a rain is a good procedure. Keep the newly planted lawn moist down to a depth of several inches at all times. In sprinkling the soil, use a fine spray so that the seeds will not be washed away or made to grow in patches. Even after the grass is up, it is important to prevent the soil from drying out. Many plantings are lost at this stage because the tender seedlings dry up easily and disappear after but a few hours' exposure to a hot sun. Two or three waterings with a fine spray may be required on bright days until the grass gets a good start. The new lawn has to be kept soaked so that the grass will be deeply rooted. Mere surface-watering will result in surface-rooting, and the first drought that comes along will kill much of the grass. The best way to water the lawn is to use a spray or canvas soaker. Fig. 12 shows an improvised sprinkler you can use. You can also prop the end of a hose on a forked stick, or on a device such as the one shown in Fig. 13, and let it spray. To test the depth to which the water has soaked, slip a knife into the soil and see how far down it slides readily. The sprinkler will have to be kept in one spot for at least two hours to give most soils a good soaking. Another way to test the amount of water reaching the grass roots is to set cans at varying distances from the sprinkler and later see how much water collects in each can.

With plenty of moisture and moderate temperatures, the new grass seedlings will appear in about 7 to 10 days if there is redtop (previous pg.) in the mixture. When the grass has reached a height of about 2 or 3 in., it is time to mow the new lawn. Never let the new grass grow longer than 3 in., because this can cause the lower part of the plants to turn yellow and the grass will be weakened. In clipping the new lawn, do not cut the grass blades shorter than 1½ in. If the grass is cut shorter than this, the development of the root system will be discouraged, and this increases the possibility of the plants being injured by extreme summer heat. By setting the mower blades at a 2-in. height, you will obtain the best results. Since new grass is thin and fine bladed, use the mower as gently as possible, and make sure that the cutting blades are well sharpened. Avoid damaging the lawn with the mower wheels. It is particularly important to stay off a new lawn when the ground is soft and wet as the grass seedlings are torn out easily.

After the first mowing, bare spots may show up in the lawn. These areas should be resown with seed immediately, regardless of the season, to prevent troublesome weeds and wild grasses from getting started. To assure that you reseed with the same blend of grass, it is wise to save some of it when seeding your lawn for the first time. A good way to store grass seed is suggested in Fig. 15.

If you choose to plant your lawn in the fall, you may wonder how you should protect it during the winter frosts. The answer is simple—you need do nothing. Young turf is not benefited and may even be harmed by protection from cold, so no winter covering is recommended. The young grass can withstand cold weather better than you think.

Sodding: This is an expensive way to establish a new lawn, but this practice is becoming more common and satisfactory around many new homes which have limited lawn areas. Sodding is desirable (1) on steeply sloped areas, (2) on borders of lawn areas, (3) at the edges of drives and walks, (4) on borders of planting areas to prevent soil from washing into them and to compose neat edgings, (5) where an immediate lawn effect is wanted and (6) upon small areas where subsequent care of a newly seeded lawn would be neglected or become too troublesome.

Good sod should be made up of densely rooted grass of the desired kind, and free

FIG. 11
Newly seeded grass on slopes and terraces should be covered so that the rain will not wash it away. Burlap bags can be pressed into the earth, as shown at the right. It will not be necessary to remove the cloth. It will rot

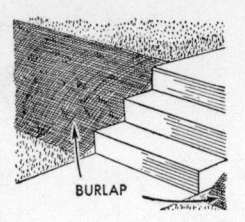

FIG. 12
Here is a good substitute for a garden-hose nozzle. It is made from a beverage can of the type that has a narrow neck. Solder a pipe coupling to the neck of the can and screw it on the hose. Punch several tiny holes into the can, the number depending on the spraying stream desired when the water is turned on. In watering a newly seeded lawn, only very minute holes should be used

FIG. 13
Your lawn hose and nozzle can be used to water a newly seeded lawn without the aid of a sprinkler if the arrangement pictured at the right is used. Merely bend a piece of heavy wire to support the hose in a vertical position. The device holds the hose so that it may be moved easily

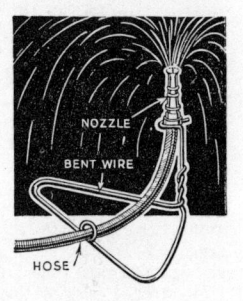

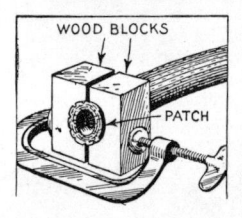

FIG. 14
A leak in a garden hose can be patched with the type of cold patch used on inner tubes. Use a clamp to put pressure on the patch while the cement is setting. If the hose is ribbed, it should be smoothed

FIG. 15
If grass seed is stored for any length of time, it should be protected from the moisture in the air to prevent spoiling. This is done by placing the sack of seed in a barrel and filling the barrel with ground feed. Pour a layer of feed over the bottom of the barrel before the seed sack is inserted

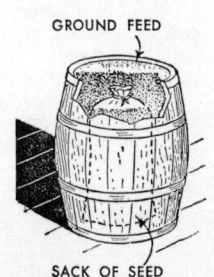

FIG. 16
After the ground has been well cultivated, lay the strips of sod snugly together but without crowding

from weeds and undesirable grasses. The strips of sod should be cut into pieces of convenient length and width and of a uniform thickness to insure an even surface when laid upon a well-prepared base. A width of one foot is about right with the thickness of sod varying from 1 in. for bentgrass sod to 1½ to 2 in. for bluegrass. Bentgrass sod can be thinner because of the extremely shallow-rootedness of this type of grass.

The best time of the year to lay sod is in the spring about the time when grass starts to grow. The seedbed is prepared as for seeding, and the strips of sod laid as in Fig. 16. They should lie together snugly but not be crowded. Then spread over the entire area a sandy loam topsoil, enriched with fertilizer and containing the seed of the native sod. About 1 cu. yd. of topsoil and 1 lb. of seed per 1000 sq. ft. of lawn area is the correct proportion. This mixture will settle into the crevices between the sod pieces, fill in the low spots and stimulate the grass growth until the roots become established in the soil below. The next step is to roll the newly sodded lawn to assure contact between sod and soil. The sodded area should then be watered. When sufficiently dry, roll the sod again to develop a smooth surface and to press the turf into contact with the soil below.

In sodding banks where there is a danger of the sod slipping, set the pieces lengthwise across the bank and hold them in place with wooden pegs. Peg each sod strip 1 to 2 in. below each upper corner and also at a point midway between corners. Drive the pegs through the sod and into the bank, deep enough so that the tops of the pegs will be slightly below the cutting blades of a mower.

Growing grass under trees: Establishing and maintaining a lawn under trees present a special problem. Here there are limitations in sunlight, plant food and moisture. Certain species of grasses, such as chewings fescue and woodlawn meadow grasses, can thrive with a minimum of light.

Grass under trees suffers from lack of moisture and plant food because shade trees withdraw these elements from the surrounding soil very rapidly. Therefore, the soil under trees has to be heavily and frequently enriched and watered to maintain grass.

Winter lawns: In the southern part of the United States, it is possible to maintain green growth on the lawn the year around. The winter grasses are, of course, different in species from those that grow in the warmer seasons. A winter lawn is, therefore, treated as a separate and annual crop. The grasses used are shallow rooted and will not forage deep enough into the ground to injure or disturb the permanent grasses. A complete plant food is needed to supply the winter grasses with nutrients.

These steps should be followed to establish a winter lawn. First, cut the grass on your permanent lawn close and remove the clippings early in the fall. Then apply a complete plant food evenly on the surface, using about 4 lbs. per 100 sq. ft. of area. Sow winter rye grass seed evenly, rake it into the soil lightly and soak it down thoroughly. Daily watering is required until seed germination. When spring arrives and the permanent grasses start to grow, the winter grasses can be easily raked out.

When you are raking, watch for weeds that were hardy enough to last the winter. These weeds are harmful to a healthy soil and lawn.

Renovating an old lawn

IF THERE IS A LAWN already established on your property, you may be faced with the task of trying to improve it. It may be that your lawn is so poor that you are debating with yourself whether to dig it all up and start from scratch or to try to build up what you have. Your final decision will have to rest on what attempts you have made in the past to improve your lawn. If repeated attempts to bring about a healthy, attractive lawn have failed, there is probably something fundamentally wrong. It would be useless to continue the same practices to improve your lawn when they have repeatedly failed you in the past. Instead, you should make an effort to find out what is wrong and correct it.

Poor soil might be the cause of a poor lawn. Therefore, your first task would be to examine the soil. You can do this by merely digging up a sizable chunk, going down into ground at least 10 or 12 in. and spading up the sample. Examine this chunk of soil carefully. Is it yellow or almost white when dry? If so, you probably have a stiff clay soil on your hands. Grass roots would make little headway in such soil because of poor drainage and aeration. On the other hand, the soil may turn out to be stony and gravelly. This is the type of soil from which the water drains away so rapidly that neither moisture nor nutrients can be absorbed by the grass roots.

If your soil turns out to be one of the two extremes mentioned above, it might be best to remove 6 to 8 in. of the soil and import good topsoil. Or, you can make your own topsoil by preparing a mixture of soil, leaf mold, fertilizer, etc., with a sifter-mixer of the type shown in Fig. 17. However, this may be impractical where the lawn area is large. Your alternative then will be to work in soil of opposite texture and some organic matter with the soil.

Here is a combination soil sifter and mixer that you can build yourself to prepare topsoil for your lawn and garden. It will accommodate a fairly large quantity of soil, leaf mold, fertilizer, etc. in one mixing. The frame is mounted on four legs set into the ground and diagonally braced. The screen frame rides on four roller-skate wheels and the track is protected by a sheet-metal guard. After good soil has passed through the screen, the siftings are dumped into a container in the manner indicated in the drawing. Flat-iron stops prevent the skate wheels from running off the end of the sifter track
FIG. 17

FIG. 18
Reseeding to patch thin spots in the lawn is hard to do without disturbing the grass. A household meat tenderizer is ideal for this purpose, as shown above. When pressed into the soil, the tenderizer loosens the soil and leaves small holes for the seed, but does not uproot the grass presently growing in that spot

FIG. 19
When tamping grass seed into the soil, a tamper often becomes difficult to manage if caked with wet soil. Burlap wrapped over the head of the tamper, as shown above, will help this situation by keeping the wet soil from sticking to the tool. Once in a while the cloth may have to be removed and cleaned

Improper grading also may cause you to have a poor lawn. A flat lawn may sag in spots so that puddles collect and drown the grass. On the other hand, the slope may be so severe that soil washes away during rains and droughts also take their toll. The solution here would be to spade up your old lawn and regrade the area for a new one. Install a drainage system if necessary and provide a sufficiently deep topsoil of good texture.

Cultivation: In time, many good soils become too compact because of trampling and settling. In such situations, spade up or plow the ground, incorporate material to lighten the soil and feed it with plant nutrients. If only limited spaces, such as low spots or areas under trees, need to be treated, the soil can be loosened and cultivated by 6 to 8 in. deep perforations. Drive an ordinary spading fork down as far as it will go into the ground and work it back and forth to enlarge the openings. The fork should be inserted at close intervals, the openings being made every few inches. Brushing sharp sand or compost into the holes will assure better movement of air and moisture through the soil.

Having decided to turn over your old lawn, you may wonder how deep you should spade. This will depend on the depth of the topsoil. Only go as far down as your topsoil extends, taking care not to bring up any subsoil. As long as you are going to the trouble of renovating your lawn, do a good job of it. Fill in all low spots with good topsoil and remove any high spots by taking away some of the subsoil and then replacing the topsoil.

Maybe you won't have to renovate your lawn completely. The poor condition of the lawn could be due to improper management in feeding, watering, mowing, weeding and controlling of pests and diseases. If this is all that is the matter with your lawn, your problems can be solved by proper surface treatments. These operations are discussed a little later on.

Fall is also the best time of the year to begin work in building up a lawn which is already established. Do not start off on the wrong foot by trying to rake off the dead crab grass. Experiments have shown that raking grass often does more harm than good, especially on certain species, such as bluegrass. Gardeners are apt to rake the crab grass vigorously, trying to check or eliminate the growth of this undesirable grass. If all the crab-grass seed could be removed along with the dead leaves and stems, raking would be justified. But this is not possible. Instead of removing crab-grass seeds, raking merely threshes, and a large number of seeds are actually helped to be well planted by this process.

It is true that dead crab grass on the lawn produces unsightly brown patches, but this looks no worse than the bare spots in the lawn that would result if the dead grass were removed. Moreover, the presence of the dead material can be an advantage in reseeding. You can sow your grass seed right into the dead crab grass, which acts as a moisture-conserving mulch during the winter, thus aiding germination of the seed. By the time the new grass starts to grow rapidly the following spring, the dead crabgrass plants will have practically disintegrated and the soil profits from the resulting addition of organic matter.

Where there are bare spots to reseed, loosen the soil about 2 in. deep and rake the surface to a fine, even texture. As Fig. 18 shows, a household meat tenderizer can be used to prepare the lawn for reseeding. Sow the grass seed evenly and lightly over the thin and bare spots. Then rake the area very gently and tamp the loosened soil down a little with the back of a spade or a tamper. To keep the soil from sticking to the tamper, wrap a piece of burlap around it, as shown in Fig. 19. Whether you are sowing seed into patches of dead crab grass or onto bare spots, this area must be watered thoroughly. Not until germination of the seed is complete should the surface be allowed to become dry again. This will require at least two sprinklings daily.

The addition of plant food is of first importance in improving an established lawn. As has been pointed out before, the natural source of humus in the soil is constantly being exhausted and must be replaced if vegetation is to grow successfully.

A relatively new treatment for grasses is a hormone feeding, which strengthens root systems. Lawns may be watered with hormone solutions during the spring and fall months, using a siphon attachment on the garden hose as in Fig. 20. To assure uniform results by this method it is necessary to apply a given amount of the solution per 1000 sq. ft. This amount, varying with the brand, will be recommended by the manufacturer. Measure the lawn and locate the areas with markers. Then sprinkle uniformly, taking care not to soak or flood the areas. When working on a steep, sodded slope or terrace, keep the bulk of the water on the higher levels as it will seep down the slope. Use a light, spreading spray under low pressure and work on a still day. The two sample sods in Fig. 21, from treated and untreated plots, show the superior root system of the treated specimen. The time interval was the same for both specimens.

There are a number of other plant foods that can be recommended. A complete fertilizer of a 10-5-3 or a 12-6-4 analysis will serve the average lawn. It should be applied at the rate of about 20 to 25 lbs. per 1000 sq. ft. Dump the fertilizer on the lawn in small piles here and there. You can either broadcast the fertilizer with a shovel as in Fig. 22 or spread it with the back of a rake as in Fig. 23. Work the plant food down into the turf with a rake or a stiff broom. Dragging the area with a flexible

FIG. 20

To treat the lawn with a hormone solution, extend a siphon attachment from the bucket containing the solution to the garden hose, as illustrated above, and spray lightly. The lawn should be sprinkled uniformly, taking care not to soak or flood any areas. The operation is best done on a quiet day

The illustrations below show the results of an experiment in which one patch of bluegrass was given the hormone treatment while the other was left alone. The superior root system, seen in the treated grass at the right, will produce better grass over the years
FIG. 21

FIG. 22
An effective way to broadcast top-dressing over the lawn area is with a shovel, as shown above. The plant food is to be distributed as evenly as possible

FIG. 23
The back of an iron rake can also be used to spread the fertilizer, as above. The fertilizer should be dumped on the lawn in small piles here and there

metal door mat is another satisfactory method. Watering the fertilizer into the lawn is the next step.

Ammonium sulphate is a good nitrogen fertilizer to use where the soil is already supplied with phosphorus. It comes in concentrated form and so should be either dissolved in water and sprinkled over the lawn or mixed with a little soil or top-dressing and broadcast over the area. Use about 5 lbs. of ammonium sulphate to 1000 sq. ft. of lawn. If allowed to remain on the grass blades, this fertilizer will cause some injury, so it should be spread just before a rain or washed into the soil with a spray from the hose. If it is not convenient to water the application, the ammonium sulphate or other commercial fertilizer may be scattered when the grass is dry and brushed into the turf with a broom or bamboo rake. Ammonium sulphate is used chiefly because of its quick action. New and vigorous growth can be observed in a lawn 10 days after it has received an application of this fertilizer. This stimulation of growth continues for a month or six weeks and then becomes exhausted. The only disadvantage in the use of ammonium sulphate is the acid condition that develops in the soil if used continually. Also, certain experiments have shown that the use of ammonium sulphate alone has a somewhat harmful effect on bluegrass turf.

Applying fertilizer in the fall encourages the growth of desirable grasses when crab grass and other weeds offer no competition, and this helps to strengthen the turf. In order to produce prompt effect, the fall feed should be in chemical rather than organic form.

Early spring feeding of the lawn is considered the most effective. At this time of the year in many parts of the country the alternate freezing and thawing leaves the soil pitted and honeycombed. This leaves the ground in a perfect condition for the plant food to seep down to the roots where it is needed. If fed just before the final thaw, the lawn won't even require watering to dissolve the plant foods. Spring rains, or tardily melting snows, take the food deep into the ground, encouraging the grass roots to delve down after it. This early root activity before top growth starts assures the lawn of a root system deep enough to sustain it during the hot, dry periods of summer.

If you are not able to get that early spring start in feeding your lawn, it is still not too late to do so later on in the spring. It is advisable to apply plant food before the grass starts to grow, but it is still permissible to feed the lawn after leaf growth has started. Sprinkle the fertilizer on when the grass is dry; then soak the lawn good with a hose or sprinkler to wash the plant food off the grass and into the soil. This is done so that the plant food will not "burn" the grass leaves, a condition which is not actually serious except that it turns the grass brown temporarily. Whenever it is that you apply the fertilizer, see to it that the lawn is cut very short so that the plant food will contact the soil easily. Feeding the lawn in the summer is not advised, because this would encourage the growth of crab grass and other weeds rather than the desirable grasses.

Rolling: An important operation, but one which many gardeners neglect, is rolling the lawn. This process is often overlooked because a roller is an expensive item for the amount of use it gets. Nevertheless, a roller is almost essential in producing a good lawn. By rolling the established lawn in the spring, you can press back into the soil the grass roots which have been heaved up by the freezing and thawing. This makes for better growth of the grass and smooths the surface at the same time.

The correct time for lawn rolling is just after the frost has left the ground. Then the surface is soft and moist but not wet. Wait until all danger of alternate freezing and thawing is past. Under no circumstances should you roll the lawn when it is too wet, for this is far worse than no rolling at all. The earth should be about the same consistency as it is for spading—a soil that is firm when picked up but will crumble easily when tapped lightly. One rolling in the spring is sufficient for the year. The correct rolling technique is shown in Fig. 24. A soil which tends to be loose and sandy can stand more rolling than a heavy soil. Where the soil is a heavy clay which forms a crust as it dries, rolling may do more harm than good.

The roller should be only heavy enough to firm back any grass which has been heaved up. About 100 lbs. of weight for each foot of roller width is recommended for ordinary use. Either a concrete roller or one of hollow metal half-filled with water or sand will serve the purpose well.

Automobiles backing into grass edges can leave ugly-looking gouges. These can be mended by cutting out the section marred by the gouge in a square or oblong piece. Reverse this section and replace it, as in Fig. 25. This leaves a neat edge on the outside. The damaged section is to be filled in with sifted loam and reseeded with the proper kind of grass seed.

If the ruts in the lawn are not too deep, they can often be removed by the use of a lawn roller, a thick plank and a sledge hammer, as shown in Fig. 26. When the ruts are too deep for this treatment, cut along both edges of the rut with an edge cutter. Roll back the turf in the rut, cutting carefully under it with a spade to loosen it. Then loosen the packed soil underneath, add some topsoil and replace the turf.

FIG. 24
In rolling a lawn, push the roller firmly in a forward rather than a downward position. Under no circumstances should the lawn be rolled when it is wet

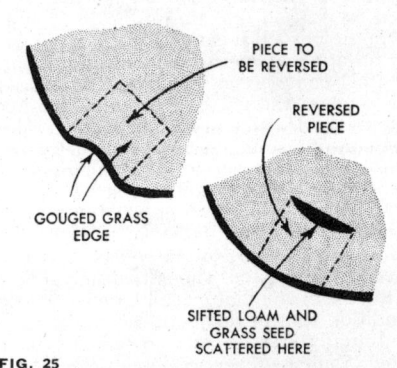

FIG. 25
To mend a gouge at a lawn edge, cut out a square piece of sod, as shown above, and reverse it. Fill in the damaged section with sifted loam and reseed with the same grass mixture as used for the lawn

FIG. 26
Ruts made in a lawn by the wheels of a car or truck can be removed without leaving any noticeable trace if the defaced area is rolled to make it compact, after which the ruts are pounded out with a thick plank struck by a sledge hammer as illustrated

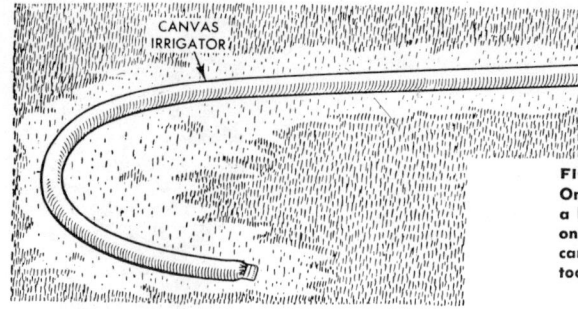

FIG. 27
One of the most satisfactory ways to water a lawn is with a canvas irrigator, like the one shown here. The water oozes out of the canvas hose and can be absorbed without too much being wasted by surface drainage

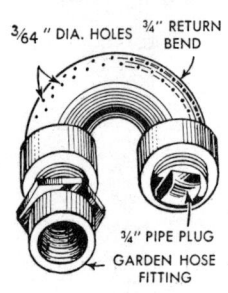

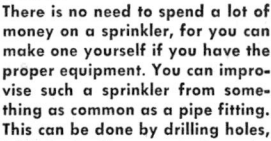

FIG. 28
There is no need to spend a lot of money on a sprinkler, for you can make one yourself if you have the proper equipment. You can improvise such a sprinkler from something as common as a pipe fitting. This can be done by drilling holes, 3/64 in. in diameter, to form the sprinkler head. One end of the pipe is sealed with a ¾-in. plug and the other end is fitted with an attachment for the garden hose

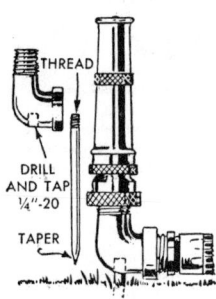

FIG. 29
A nozzle and hose can also serve as a sprinkler with an attachment like the one shown at the left. It is made from a ¾-in. street elbow and a ¼-in. metal rod which serves as a spike. After pointing and threading the rod, drill and tap the elbow as indicated. Then turn the rod into the elbow tightly to insure a leakproof fit. Screw the elbow on the garden hose and attach the nozzle as shown. To use, adjust the nozzle for a fine spray and force spike into ground

Here is a multiple lawn sprinkler which permits the watering of large areas at one time, and the wooden rollers allow it to be moved without damaging the lawn. The unit is assembled from pieces of pipe and standard fittings, and the rollers are turned from wooden blocks. Outlets are spaced at intervals along the length of pipe so the spray will cover the largest possible area
FIG. 30

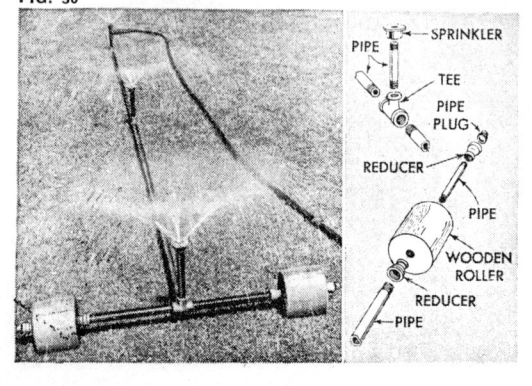

Lawn Maintenance

Now that you have a new lawn under way or have revived an already established one, you still have a steady job on your hands in taking care of the lawn. This means a proper program of watering, cutting and raking.

Watering: About an inch of water per week is required by most grasses to grow well. If the natural rainfall does not provide for this amount, which is usually the case during the summer in most climates, water has to be applied by artificial means. You can find out when this point has been reached by examining your soil. Using a knife or trowel, cut and dig up a small plug 2 or 3 in. deep. If the upper inch is dry, it is time to start watering. Replacing moisture to an inch or two of surface soil is a simple matter; but if the soil has been allowed to dry much deeper, the task is that much bigger. So don't wait until your grass starts to wilt before you decide to give it some water.

You can water the lawn by means of a sprinkler (Figs. 28, 29 and 30), a canvas hose (Fig. 27) or allowing the water to run *slowly* from the end of a hose. The last mentioned method is not recommended as it takes too long to water a large area properly.

The whole key to proper watering is in the word "soak" rather than "sprinkle." It would not be surprising if a survey revealed that more lawns are spoiled by light sprinkling rather than by drought. As has been pointed out previously, the roots of the desirable grasses delve deep into the ground (with the exception of bent grasses), while crab grass and other weeds are comparatively shallow rooted. It is therefore obvious that sprinkling the lawn so that the water wets only the upper inch of soil encourages the growth of the undesirable plants while the good grasses do not benefit at all. This is particularly true during the hot, dry midsummer months, during which crab grass and other weeds grow vigorously and such grasses as bluegrass lie dormant. Moreover, if the deep grass roots do not get water, they have to come up to the surface for it. This results in surface rooting, and the first dry spell that comes along can take a heavy toll on the desirable grasses.

The amount of water needed by the lawn will depend on the type of soil in which it is growing. Loam and clay soils retain water well and can keep the grass roots supplied with enough water so that droughts can be withstood. On the other hand, where grass grows in sandy or gravelly soils, the slightest drought dries the soil completely and even weeds may be killed along with the grasses. The exposure of the lawn and the temperature and rainfall also influence the amount of water needed by a lawn. Also closely connected with watering is the height to which the grass is cut. A closely clipped lawn increases the rate of evaporation.

Most home gardeners are apt to resign themselves to a dry, straw-like turf during the hot summer months. It is true that the dormant, dead-looking bluegrass lawn eventually becomes green again when the weather becomes cooler and more moist in the fall. But why not enjoy a lush, green lawn during midsummer when it gives such a soothing effect? This is possible! A glance at golf-course putting greens, which retain their green beauty throughout the summer year after year, is evidence enough that lawns can be kept green even in hot, dry weather. To do this, of course, means a carefully planned watering program plus proper feeding. Don't expect miracles even with the best-planned and most religiously executed watering program *unless* the soil is kept properly fed and healthy. Browning is often caused by diseases, attacks of insect pests and even improper mowing.

In using a sprinkler to water the lawn, allow it to stay in one spot for at least an hour, or even two, or until the water soaks down to a depth of 6 to 8 in. Then move it to another spot for some more sprinkling, and so forth until the entire lawn area is well soaked. This may require as long as four to eight hours and even more, but once the soil is well soaked it won't require another watering for days. The spray of water should be light enough so that the soil can absorb it readily. A fine mist does not work well because it is easily blown by the wind. Neither can a coarse stream be used, because it washes the soil and puts the water

FIG. 31
If you do not keep your garden hose conveniently wound around a reel when not in use, you are probably one of those gardeners whose hose twists annoyingly as it is pulled from the coil. But there is a way that you can prevent this twisting. When you put your hose away, form the hose coils as shown in the detail, laying them one on top of the other. You will note that each succeeding loop is made in the direction opposite that of the previous loop

FIG. 32
To prevent a garden-hose nozzle from becoming nicked or cracked when dropped on a hard surface, pierce a sponge-rubber ball and slip it over the tip of the nozzle. If the hose is accidentally dropped, the ball will keep the nozzle from coming in contact with the concrete pavement

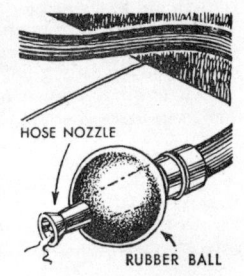

FIG. 33
Repeated kinkings of a garden hose near the faucet connection eventually break the hose and cause a leak. To prevent this, apply a tight spiral wrapping of friction tape around the hose so it extends a foot or so from the connection. Be sure the tape covers the place where the hose kinks

FIG. 34
By supplying water to trees through wooden columns, they will not take as much moisture from other plants

FIG. 35
A length of pipe placed vertically in the ground as illustrated at the left will lead water to tree roots

on too fast. If you can get the sprinkler to provide the effect of a slow, easy rain, the water will soak in as it falls and not flood the ground and run off the slopes. In selecting your sprinkler, choose one which is suitable for the size and shape of your lawn. A sprinkler which does not have to be moved too frequently is, of course, the best kind to have. As for caring for your garden hose, Figs. 31, 32 and 33 make a few helpful suggestions.

An automatic underground sprinkler system is a time and labor-saving device which can water a lawn evenly and efficiently. However, even this convenient method can be abused. Because it is so easy to turn on, there is danger of overwatering. This may not harm lawns growing in sandy soils, but too much water in a compact, clay soil can easily drown the grass roots.

Lawns other than bent-grass lawns (see previous discussions) can be watered at any time of the day. It is true that the high temperature of midday causes higher evaporation, thus resulting in waste of water to some extent. Therefore, most people prefer to do their watering in the morning, late afternoon or evening.

There are special problems that arise in lawn watering. The watering of a newly seeded lawn bed requires special attention. Droughts do not harm seed, but germination is hastened by keeping the soil moist.

A problem arises when there is too much water in a compact soil. This often happens when the spring has been too wet. The result is a waterlogged soil in which the deeper roots die because of poor aeration. When this happens the remaining grass depends on the upper surface of soil for its food and water. This activity in the surface soil causes it to dry rapidly, and moisture must be replaced constantly to keep the grass alive. If you nurse the grass along by daily watering with a fine spray, you can urge the roots to go down deeper and the watering can become less frequent. Where grass roots have been weakened by insect attacks, a similar program can be launched to bring the grass back to health.

Tree-shaded lawns also present a problem in watering. Being protected from the sun, these lawns lose less moisture through surface evaporation. But the tree roots draw water from the undersoil, leaving little moisture in it for the grass roots. It's

FIG. 36
The height to which you mow the lawn is very important. As shown above, the grass should be kept at a 1½-in. height during cool weather and increased to 2 in. in hot weather. Always keep shady lawns 2½ in. high

a good idea to examine the soil under trees from time to time. If the undersoil is found to be dry, replace the water by slow, penetrating applications. You can also cut down the competition between the trees and the grass for water by providing special means to get a sufficient supply of water to the trees. When a young tree is planted, its roots can be assured of a plentiful supply of water if a length of 6-in. pipe is placed in the hole near the roots, as shown in Fig. 35. The pipe is filled with water, which will be fed directly to the roots instead of seeping through the ground to be absorbed by the soil. Another way to provide a water supply for trees is to sink two wooden columns about 3 ft. into the ground on each side of a small tree, as pictured in Fig. 34. Water is run into the columns and passes into the ground around the tree roots through holes drilled in one side of each column. The holes are covered with screen wire to keep the dirt from falling into the columns.

Mowing: One of the most common mistakes made in mowing is to cut the lawn too short. It must be realized that the grass blades play a vital part in transforming food into energy for growth. By continually cutting off these blades, this function is severely crippled, and the grass has a hard time thriving. On the other hand, if you allow the blades to grow too long, they become spindly and form a thin, weak lawn. There is a happy medium between these two extremes where mowing causes the least injury and yet gives a neat appearance to the lawn. You can attain this medium by cutting the grass to 1 in. in cool weather and raising it to 1½ or 2 in. when the atmosphere becomes warmer and drier. If the turf is kept at this height, frequent cutting will not harm it, and the lawn can be kept uniformly trimmed to give it a well-groomed appearance.

The attractiveness of the lawn is the direct result of high cutting, as it is the grass blades that give the green color to the lawn rather than the yellowish stems which would be exposed were the grass clipped closely. High cutting also has a lasting influence on the general health of the grass. Longer top growth (within reason, of course) means a more extensive root system, which in turn can reach more food and moisture. The blades of grass kept over an inch tall also provide shade to reduce the evaporation of soil moisture.

High cutting is especially important in the case of shaded lawns, where sunlight is limited and the longer growth is needed to help combine air, light and soil nutrients necessary for plant growth. Densely shaded lawns should not be cut shorter than 2½ in., with 3 in. being an even more desirable

FIG. 37
If you have difficulty cutting grass along flower or hedge borders with a regular lawn mower, an edge clipper, like the one at the right, will come in handy. An implement of this sort will enable you to trim the edges neatly without damaging any of the plants in the border

FIG. 38
Measure from the cutting edge of the bedknife to the ground to find out at what height the mower will cut

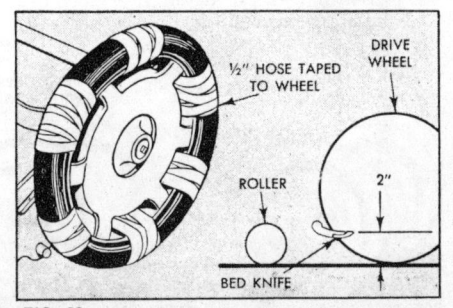

FIG. 39
Wrap a length of hose around the mower wheel, as shown above, if the bedknife cannot be adjusted

FIG. 40
If you cannot adjust the height of the cutter bar on your lawn mower, fit the roller mounting brackets with flat-iron extensions. First, slot and drill two pieces of flat iron as shown in the detail. Then ream the threads out of a small nut and weld it over the hole to serve as a bushing for the roller shaft. The offset in the bracket extensions is bent to suit the length of the roller. Bolting the slotted portions of the extension to the original roller brackets permits the adjustment to be varied to fit seasonal needs

FIG. 41
A broom rake, like the one above, can be used to rake grass clippings off the lawn. In using this kind of a rake, it is often necessary to stop and clean the teeth. This nuisance is overcome and the work speeded if a rubber guard is slipped over the teeth. It is cut from a rubber inner tube and the holes are spaced so they will not spread or pull the teeth together

height. Fig. 36 shows average seasonal mower adjustments.

It often happens that well-fed and well-watered lawns get too thick after a few years of moderately high cutting. In such a situation, mow the lawn fairly close in cool weather, but continue to use the high cut in periods of extreme heat. Bent-grass lawns are the exception to this high-cutting rule. They should be kept at about 1½ in., or better, a 1¾-in. height.

Adjusting the mower so it will cut at the proper height is illustrated in Fig. 38. First place the mower on a flat, hard surface, such as a walk or a garage or basement floor. The next step is to loosen the side brackets at the ends of the roller. As you extend the roller bracket downward, the roller is lowered and the bedknife in front is tilted up. When the cutting edge of the bedknife measures 1½ to 2 in. (or whatever length you are going to cut the grass), tighten the bracket bolts that were previously loosened.

If your mower cannot be adjusted to cut high enough, you will have to obtain longer brackets or extensions (see Fig. 40), from the manufacturer or have them made at a machine shop. Another alternative is to lift the mower wheels by wrapping rubber hose or a ½-in. rope around each wheel rim of the mower (Fig. 39). You can also replace the old wheels with two that have larger diameters.

Naturally, the blades of the mower should be sharp. If not, the grass will be crushed and torn instead of being severed cleanly. Bruising the grass in such a way leaves a brownish cast to the lawn. The time and frequency of mowing are also important. Whether you are to mow twice a week or once a month depends on the condition of the grass and the previous mowing schedule, the rapidity of growth and the weather. If you have mowed to 1½ in., let the grass grow to at least 3 in. before you mow again. If the grass has been cut to ¾ or 1 in., it is better to raise the mower by stages, that is, raise it to 1½ in. and clip the grass a couple of times when it is 2 in. high, then raise to 2 in. and clip when the grass is 3 in. high. Coming into the hot, dry summer months, the frequency of mowings should be determined by the height and condition of the grass and not by intervals of time. In general, it is better to cut too often rather than let the grass get so high that bleached stems are evident after the cutting.

Whether the grass clippings should be removed or not depends on several factors, including your own personal preference. It is interesting to consider the abundant crop produced by grass plants. If you were to save or keep track of all the grass clippings which come from your lawn in one season, you would find the yield to be something like 36 in. deep. Now this is an immense crop, and one can easily imagine the huge amount of plant food that was taken from the soil by the grass roots and up into the blades to produce this crop. The soil cannot continue to yield so bountifully unless plant food is restored to it.

Grass clippings left on the lawn provide, to a small extent, the needed plant food. These clippings dry out, decay and work their way down into the soil. Besides contributing plant food to the soil, clippings also act as mulch to reduce surface evaporation, thus keeping the soil cooler and more moist.

If the clippings are long, say over an inch, by all means remove them. This can be done either by raking (Fig. 41) or by using a grass catcher when mowing. When the clippings are over an inch long, they do not work their way into the soil readily, and as they dry out they leave a brownish cast to the lawn. Also, if the weather is damp or the outlook is for rainy days ahead, grass clippings should be removed. Damp, heavy clippings will mat down and may cause mold and disease to develop in your lawn.

All things considered, removal of clippings from the lawn is of minor importance. Should the clippings lie in strips and make the lawn unsightly, then remove them. Whenever the sight of the clippings is annoying, you can remove them if just for appearance' sake. It is easier to rake away the clippings after they have been on the lawn for a day or two, as they will have decreased in bulk by drying.

Raking and mulching: Leaves should be raked off the lawn, especially if they tend to form a heavy mat. This is so the grass can be encouraged to grow freely in the fall, and a mat of leaves would hinder growth. Some of the leaves can be raked around shrubbery to act as mulch. The remainder can be piled in some corner of the yard to decompose and form leaf mold, which will be useful later on to improve the garden soil. The excess leaves should, of course, be collected (see Figs. 42 and 43) and destroyed.

A winter mulch on the lawn is unnecessary, and its unsightliness is further reason why it should not be used. However, a light covering of straw can be used to cover a new bed of grass seedlings. This will protect them from being heaved out of the soil by the alternate freezing and thawing that take place in the winter. It is a waste of time to put manure on the lawn during the fall or winter, as the soluble fertilizer elements are leached out and washed away before they can do any good for the soil.

FIG. 42
A gunny sack hung from a circular frame, as illustrated above, is a convenient place to keep leaves until you are ready to burn them. The top of the sack is impaled on six hooks evenly spaced around the rack. When the sack is filled, it can be easily unhooked and carried away to dump its contents

FIG. 43
The wheel frame shown above is the answer to your leaf-hauling problem. The frame is 5 x 7 ft., has wire mesh tacked to the underside of it and two wheels attached to it. Leaves are raked onto the frame in the same manner that dirt is swept onto a dustpan

LAWN FURNITURE

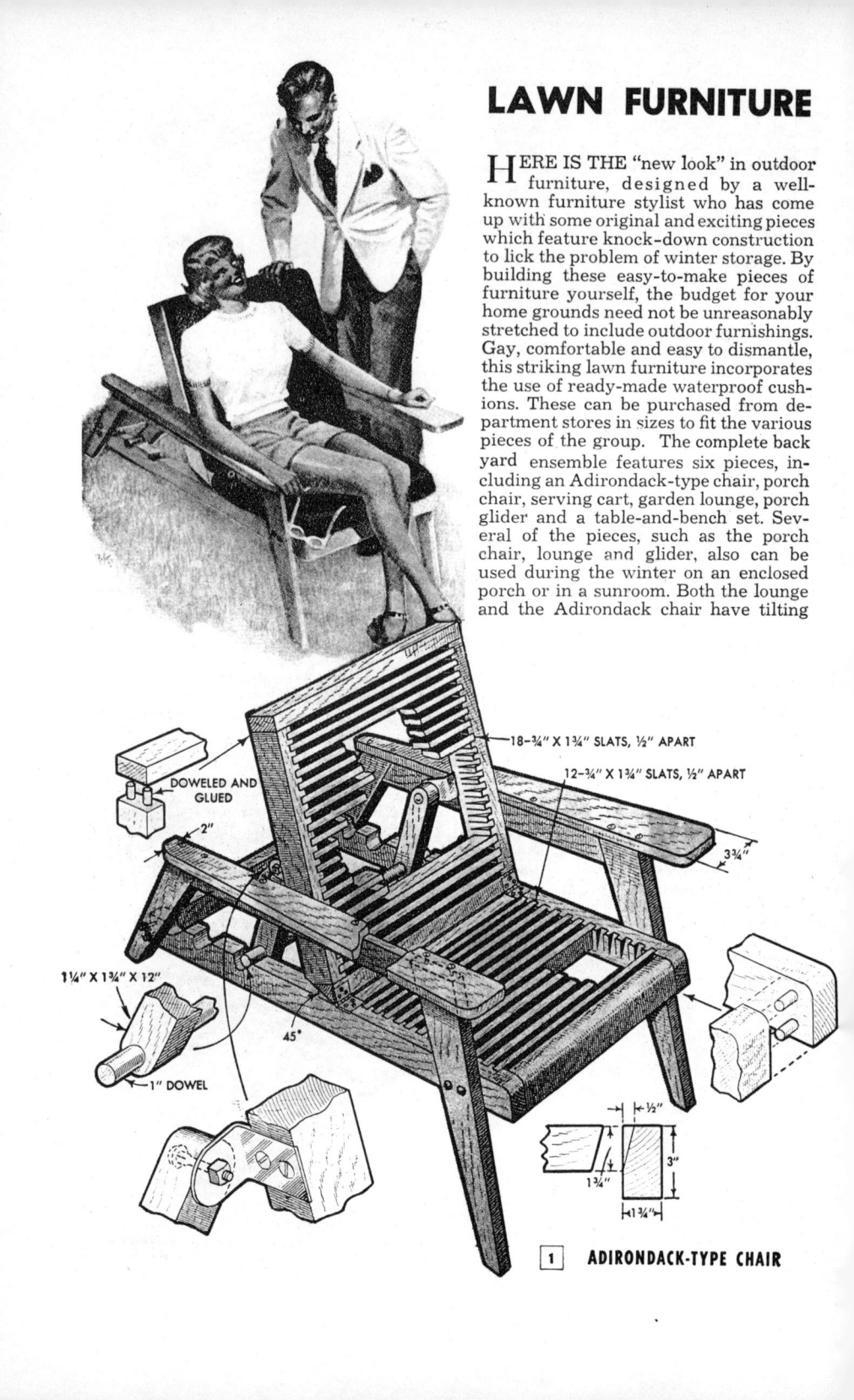

Here is the "new look" in outdoor furniture, designed by a well-known furniture stylist who has come up with some original and exciting pieces which feature knock-down construction to lick the problem of winter storage. By building these easy-to-make pieces of furniture yourself, the budget for your home grounds need not be unreasonably stretched to include outdoor furnishings. Gay, comfortable and easy to dismantle, this striking lawn furniture incorporates the use of ready-made waterproof cushions. These can be purchased from department stores in sizes to fit the various pieces of the group. The complete back yard ensemble features six pieces, including an Adirondack-type chair, porch chair, serving cart, garden lounge, porch glider and a table-and-bench set. Several of the pieces, such as the porch chair, lounge and glider, also can be used during the winter on an enclosed porch or in a sunroom. Both the lounge and the Adirondack chair have tilting

1 ADIRONDACK-TYPE CHAIR

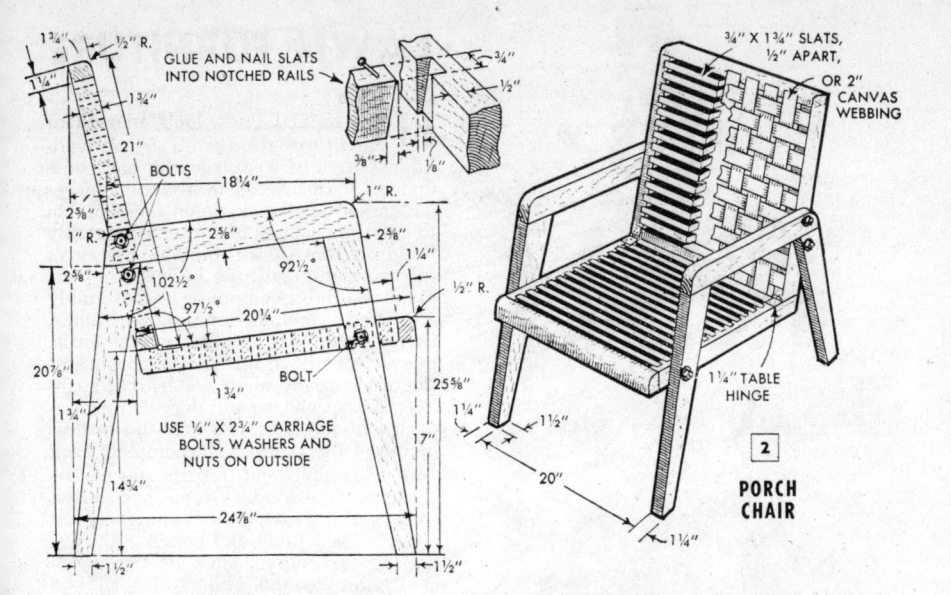

backs that can be adjusted to suit the comfort of the individual. Except in the case of the garden table-and-bench set, a choice of slats or webbing is given in constructing the seats and backs of the furniture. Both types are partially indicated in most of the drawings, and this should not be confused with actual construction details.

While redwood and cypress are two of the most durable woods, especially suitable for outdoor furniture, common lumberyard stock, such as yellow pine or fir, is perfectly satisfactory if the pieces are kept well painted.

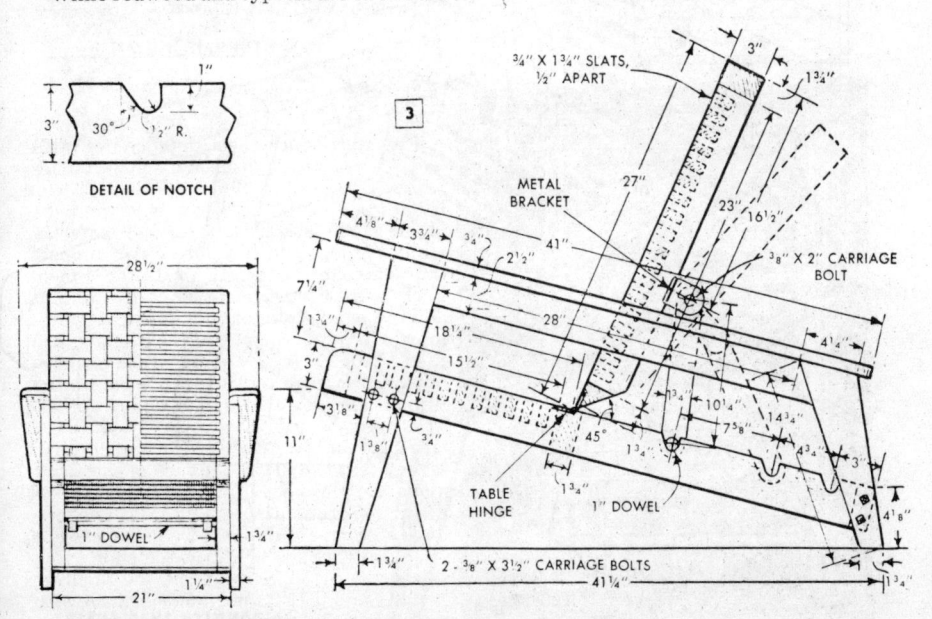

4 SERVING CART

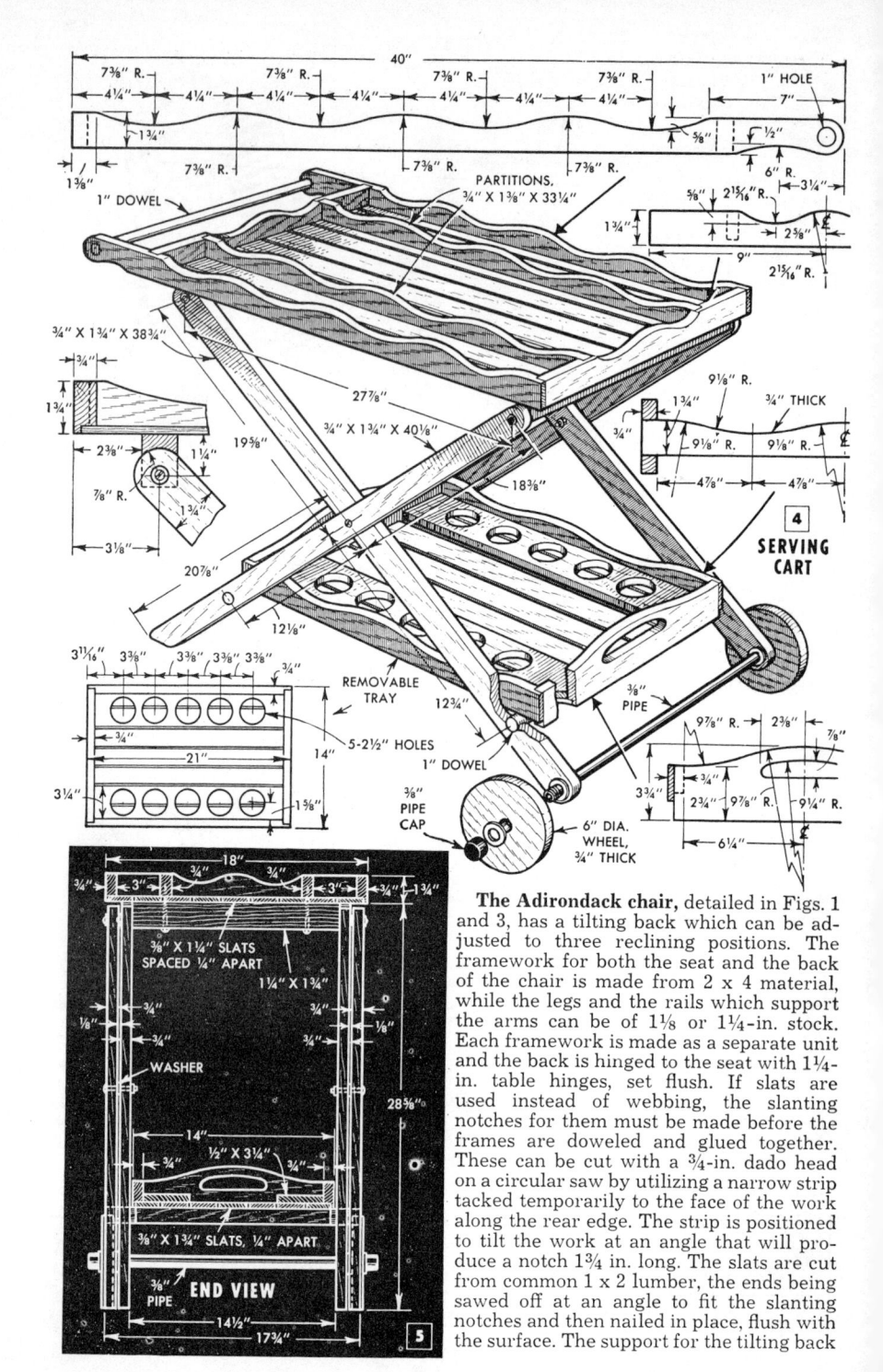

5

The Adirondack chair, detailed in Figs. 1 and 3, has a tilting back which can be adjusted to three reclining positions. The framework for both the seat and the back of the chair is made from 2 x 4 material, while the legs and the rails which support the arms can be of 1⅛ or 1¼-in. stock. Each framework is made as a separate unit and the back is hinged to the seat with 1¼-in. table hinges, set flush. If slats are used instead of webbing, the slanting notches for them must be made before the frames are doweled and glued together. These can be cut with a ¾-in. dado head on a circular saw by utilizing a narrow strip tacked temporarily to the face of the work along the rear edge. The strip is positioned to tilt the work at an angle that will produce a notch 1¾ in. long. The slats are cut from common 1 x 2 lumber, the ends being sawed off at an angle to fit the slanting notches and then nailed in place, flush with the surface. The support for the tilting back is pivoted to it with sheet-metal brackets in the manner shown in Fig. 1. If webbing is preferred to the slats, use either nylon parachute webbing or common canvas webbing and interlace it as shown. The ready-made cushions will hide the tacks used in fastening webbing to the frames.

The porch chair, Fig. 2, is somewhat similar in construction. The seat and back are hinged together to fold flat for storing, while the legs and arms can be taken off as single units by removing only six nuts and washers. Frames for the seat and back are assembled from 1¼ x 1¾-in. stock, notched if slats are used, before doweling and gluing together.

The serving cart, Figs. 4 and 5, can be wheeled about and features a removable beverage tray which rides on two rungs fitted between the crossed legs. Each pair of legs are duplicates, and registering holes for the pipe axle, dowel rungs and screw fastenings are bored at one time through each set. The upper ends of the legs are pivoted with wood screws to 1¼ x 1¾-in. cleats which support the upper tray.

The garden lounge, Figs. 6 and 7, like the Adirondack chair, has a tilting back which is supported in the same manner except that it is adjustable to four positions. The bed frame of the lounge, including the tilting back, is made of 2 x 4 material, with a second frame of lighter stock being bolted to the underside. This second frame carries the wheel axle and also the notches which engage the tilting back support. The end-view detail shows how the pipe axle is bolted in place with lag screws. The arms of the

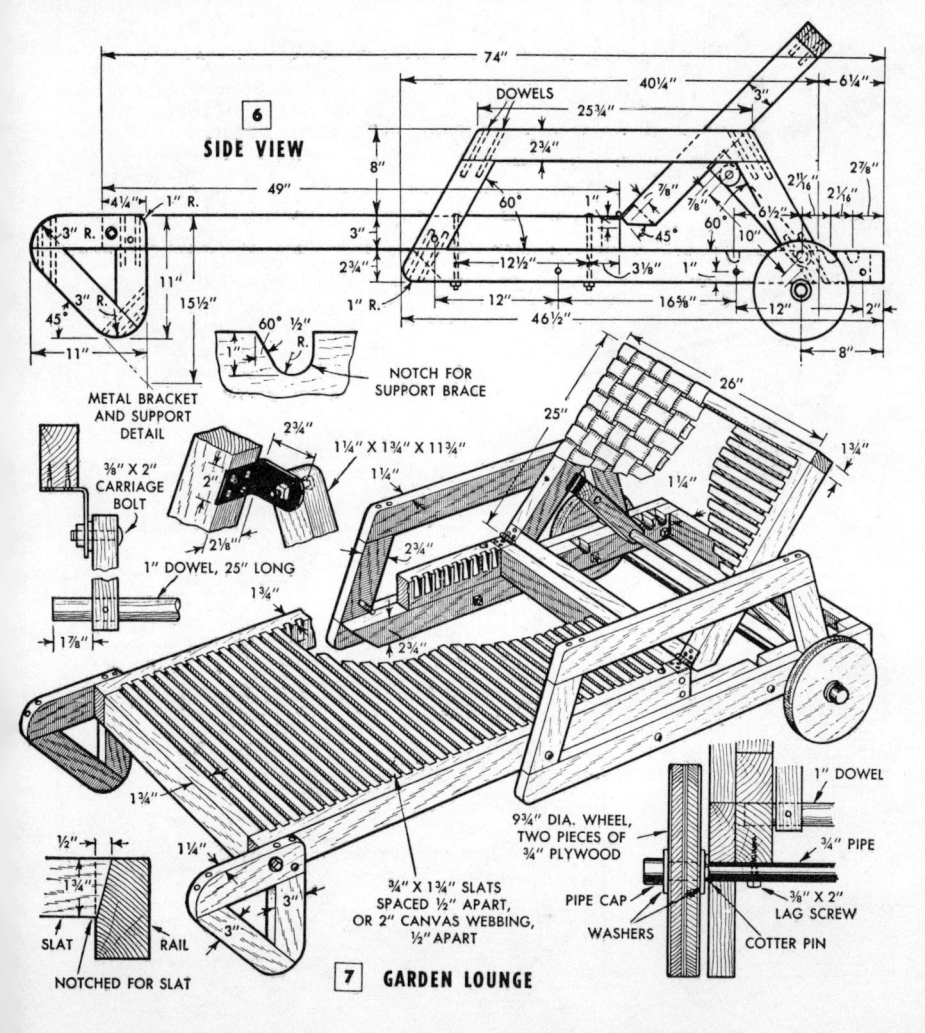

8 PORCH GLIDER

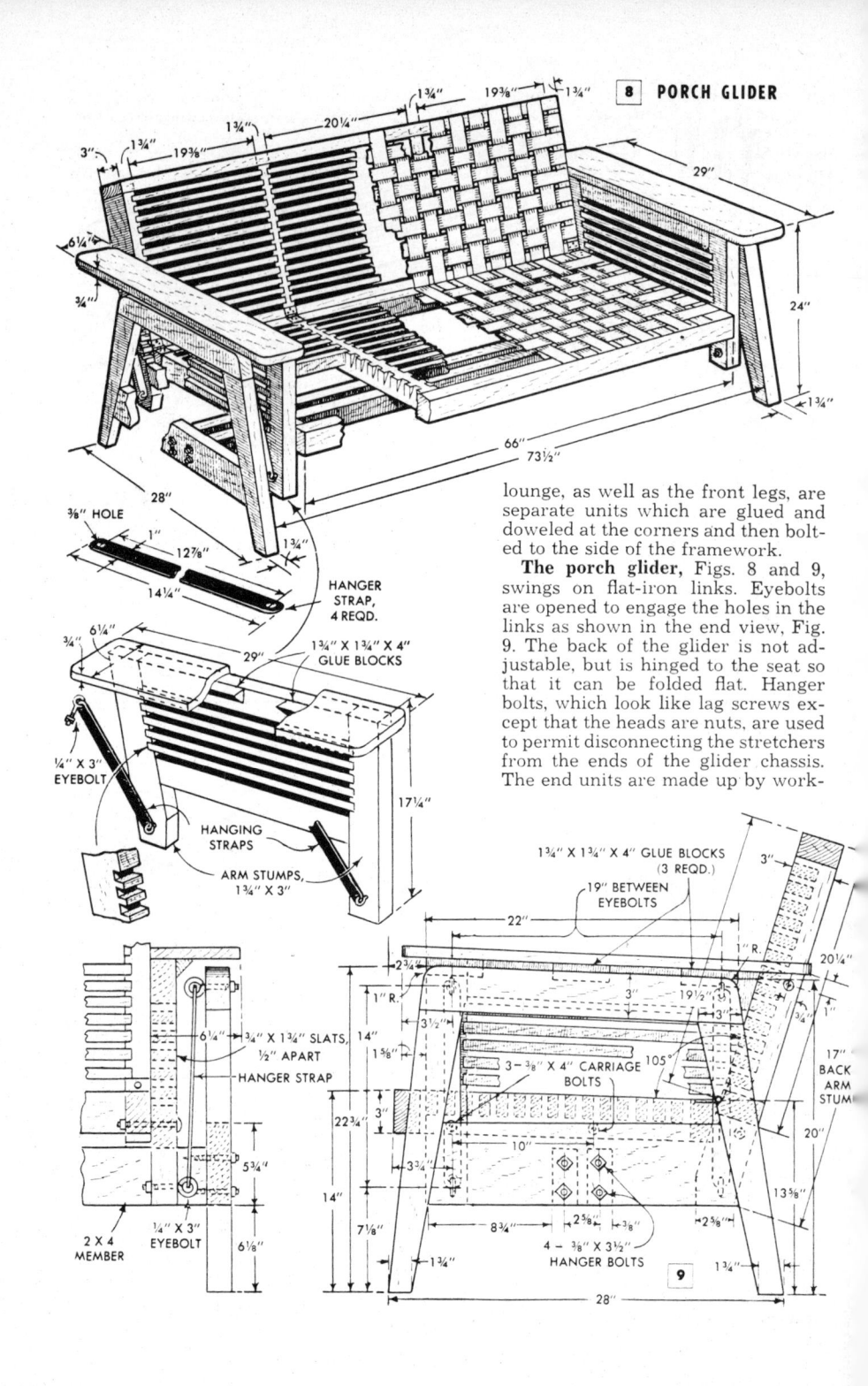

lounge, as well as the front legs, are separate units which are glued and doweled at the corners and then bolted to the side of the framework.

The porch glider, Figs. 8 and 9, swings on flat-iron links. Eyebolts are opened to engage the holes in the links as shown in the end view, Fig. 9. The back of the glider is not adjustable, but is hinged to the seat so that it can be folded flat. Hanger bolts, which look like lag screws except that the heads are nuts, are used to permit disconnecting the stretchers from the ends of the glider chassis. The end units are made up by work-

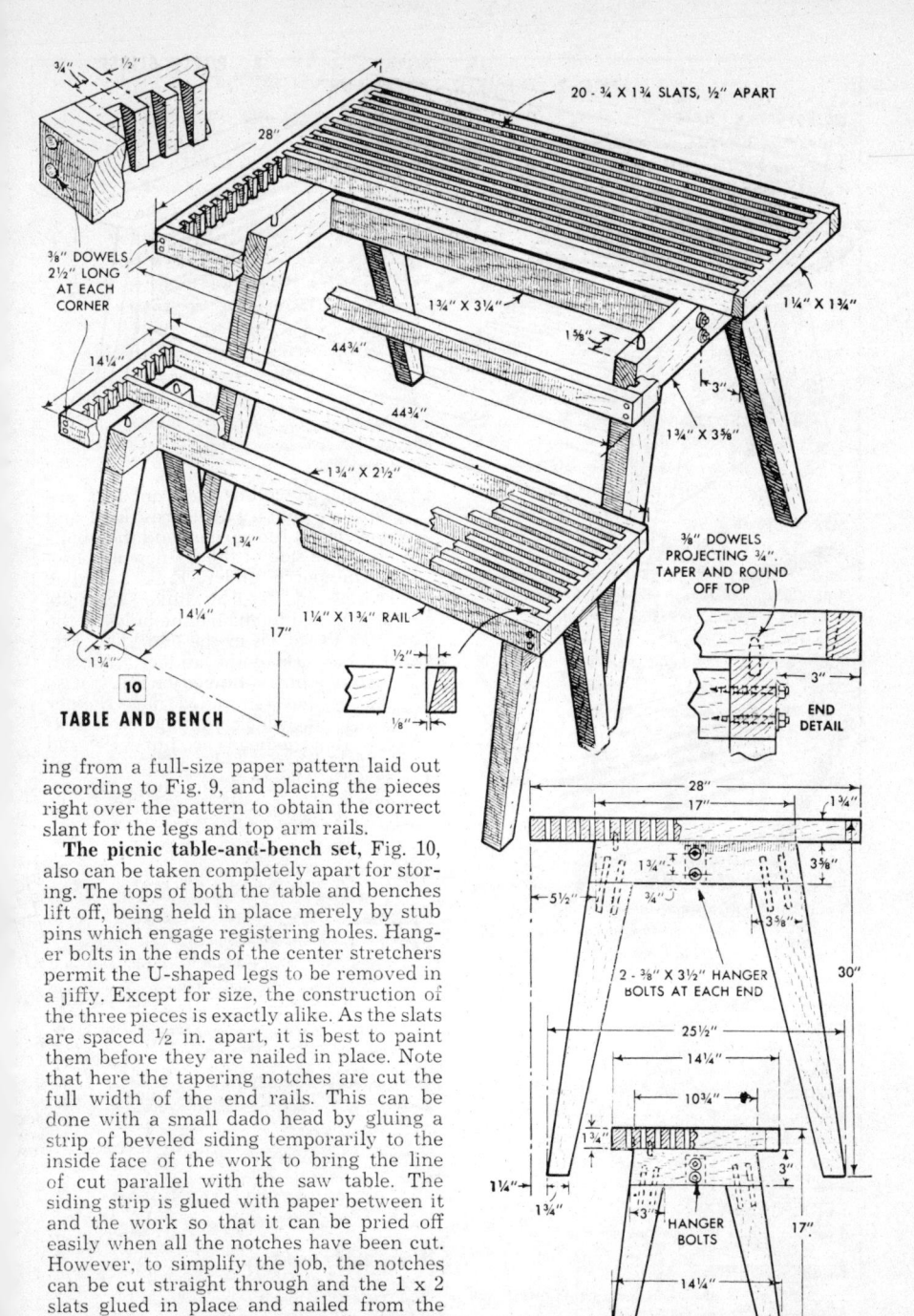

TABLE AND BENCH (Fig. 10)

ing from a full-size paper pattern laid out according to Fig. 9, and placing the pieces right over the pattern to obtain the correct slant for the legs and top arm rails.

The picnic table-and-bench set, Fig. 10, also can be taken completely apart for storing. The tops of both the table and benches lift off, being held in place merely by stub pins which engage registering holes. Hanger bolts in the ends of the center stretchers permit the U-shaped legs to be removed in a jiffy. Except for size, the construction of the three pieces is exactly alike. As the slats are spaced ½ in. apart, it is best to paint them before they are nailed in place. Note that here the tapering notches are cut the full width of the end rails. This can be done with a small dado head by gluing a strip of beveled siding temporarily to the inside face of the work to bring the line of cut parallel with the saw table. The siding strip is glued with paper between it and the work so that it can be pried off easily when all the notches have been cut. However, to simplify the job, the notches can be cut straight through and the 1 x 2 slats glued in place and nailed from the outside face. The legs are cut from 2 x 4 material and glued and doweled to the apron pieces. Screws can be used instead of dowels at the corners of the bench tops.

END VIEWS OF TABLE AND BENCH

LUMBER LIST FOR LAWN FURNITURE

ADIRONDACK CHAIR

2 pcs. 1¾ x 3 x 42 in.—Side seat rails
1 pc. 1¾ x 3 x 21 in.—Front seat rail
1 pc. 1¾ x 3 x 17½ in.—Center seat rail
1 pc. 1¼ x 3 x 17½ in.—Back seat rail
2 pcs. 1¾ x 3 x 27 in.—Side back rails
1 pc. 1¾ x 3 x 21 in.—Top back rail
1 pc. 1¾ x 2¼ x 17½ in.—Bottom back rail
30 pcs. ¾ x 1¾ x 18½ in.—Seat and back slats
2 pcs. 1¼ x 3¾ x 22 in.—Front legs
2 pcs. 1¼ x 3¾ x 14⅝ in.—Back legs
2 pcs. 1¼ x 2½ x 29⅞ in.—Arm aprons
2 pcs. ¾ x 3¾ x 41 in.—Tops of arms
2 pcs. 1¼ x 1¾ x 12 in.—Back stays
1 pc. 1-in. dia. x 20 in. long—Dowel rung
33 yds. 2-in. canvas webbing—Seat and back covering

PORCH CHAIR

2 pcs. 1¼ x 1¾ x 20¼ in.—Side seat rails
1 pc. 1¼ x 1¾ x 20 in.—Front seat rail
1 pc. 1¼ x 1¾ x 17½ in.—Back seat rail
2 pcs. 1¼ x 1¾ x 21 in.—Side back rails
1 pc. 1¼ x 1¾ x 20 in.—Top back rail
1 pc. 1¼ x 1¾ x 17½ in.—Bottom back rail
27 pcs. ¾ x 1¾ x 18½ in.—Seat and back slats
2 pcs. 1¼ x 2⅝ x 23½ in.—Front legs
2 pcs. 1¼ x 2⅝ x 21¾ in.—Back legs
2 pcs. 1¼ x 2⅝ x 19 in.—Arms
20 yds. 2-in. canvas webbing—Seat and back covering

SERVING CART

2 pcs. ¾ x 1¾ x 40 in.—Top side rails
2 pcs. ¾ x 1¾ x 17 in.—Top end rails
2 pcs. ¾ x 1⅜ x 33¼ in.—Partitions
7 pcs. ⅜ x 1¾ x 33¾ in.—Bottom slats
2 pcs. ⅜ x 1⅝ x 33¾ in.—Bottom slats
1 pc. 1¼ x 1¾ x 14½ in.—Brace
1 pc. 1¼ x 1¾ x 16½ in.—Brace
2 pcs. ¾ x 1¾ x 38¾ in.—Legs
2 pcs. ¾ x 1¾ x 40⅛ in.—Legs
4 pcs. ¾ x 6 x 6 in.—Wheels
2 pcs. ¾ x 1¾ x 21 in.—Tray sides
2 pcs. ¾ x 3¾ x 13½ in.—Tray ends
7 pcs. ⅜ x 1¾ x 20 in.—Tray bottom
2 pcs. ½ x 3¼ x 19½ in.—Glass holders
1 pc. 1-in. dia. x 17½ in.—Dowel
1 pc. 1-in. dia. x 17 in.—Dowel
1 pc. 1-in. dia. x 17¾ in.—Dowel

GARDEN LOUNGE

2 pcs. 1¾ x 3 x 49 in.—Seat side rails (front half)
2 pcs. 1¾ x 3 x 22½ in.—Seat end rails (front half)
1 pc. 1¼ x 3 x 22½ in.—Seat end rail (back half)
2 pcs. 1¾ x 3 x 23¼ in.—Seat side rails (back half)
1 pc. 1¾ x 3 x 26 in.—Top rail for back
1 pc. 1¾ x 2½ x 22½ in.—Bottom rail for back
2 pcs. 1¾ x 2¾ x 44¾ in.—Back side rails
2 pcs. 1¼ x 2¾ x 29 in.—Tops of arms
2 pcs. 1¼ x 2¾ x 46⅜ in.—Bottoms of arms
2 pcs. 1¼ x 2¾ x 14¼ in.—Front arm stumps
2 pcs. 1¼ x 2¾ x 11⅛ in.—Back arm stumps
2 pcs. 1¼ x 1¾ x 11¾ in.—Back supports
2 pcs. 1¼ x 3 x 15½ in.—Front leg horizontals
2 pcs. 1¼ x 3 x 8 in.—Front leg uprights
2 pcs. 1¼ x 3 x 17½ in.—Front leg diagonals
53 pcs. ¾ x 1¾ x 23½ in.—Seat and back slats
4 pcs. ¾ x 9¾ x 9¾ in.—Wheels
1 pc. 1-in. dia. x 25 in. long—Dowel rung
60 yds. 2-in. canvas webbing—Seat and back covering

PORCH GLIDER

4 pcs. 1¾ x 3 x 20⅜ in.—Legs
2 pcs. 1¾ x 3 x 22¾ in.—Top leg rails
2 pcs. 1¾ x 5¾ x 22 in.—Lower leg rails
2 pcs. 2 x 4 x 73½ in.—Leg stretchers
2 pcs. ¾ x 6¼ x 29 in.—Arms
2 pcs. 1¾ x 2½ x 20¾ in.—Top arm rails
2 pcs. 1¾ x 3 x 18 in.—Lower arm rails
2 pcs. 1¾ x 3 x 17¼ in.—Front arm stumps
2 pcs. 1¾ x 3 x 17¾ in.—Rear arm stumps
1 pc. ¾ x 1¾ x 108 in.—Arm slats
1 pc. 1¾ x 1¾ x 4 in.—Arm glue blocks
1 pc. 1¾ x 3 x 66 in.—Front seat rail
1 pc. 1¾ x 3 x 62½ in.—Rear seat rail
2 pcs. 1¾ x 3 x 23 in.—Side seat rails
2 pcs. 1¾ x 3 x 19½ in.—Center seat rails
1 pc. 1¾ x 3 x 66 in.—Top rail for back
1 pc. 1¾ x 3 x 62½ in.—Lower rail for back
2 pcs. 1¾ x 3 x 22⅝ in.—Side rails for back
2 pcs. 1¾ x 3 x 16¾ in.—Center rails for back
56 pcs. ¾ x 1¾ x 20⅜ in.—Side back and seat slats
28 pcs. ¾ x 1¾ x 21¼ in.—Center back and seat slats
105 yds. 2-in. canvas webbing—Seat and back covering

TABLE-AND-BENCH SET (table)

2 pcs. 1¼ x 1¾ x 44¾ in.—Top side rails
2 pcs. 1¼ x 1¾ x 25½ in.—Top end rails
20 pcs. ¾ x 1¾ x 43¼ in.—Slats
4 pcs. 1¾ x 3⅝ x 25⅜ in.—Legs
2 pcs. 1¾ x 3⅝ x 18-1/16 in.—Leg rails
1 pc. 1¾ x 3¼ x 35¼ in.—Brace

(Material per bench)

2 pcs. 1¼ x 1¾ x 44¾ in.—Top side rails
2 pcs. 1¼ x 1¾ x 11¾ in.—Top end rails
9 pcs. ¾ x 1¾ x 43¼ in.—Slats
4 pcs. 1¾ x 3 x 11¾ in.—Legs
2 pcs. 1¾ x 3 x 11½ in.—Leg rails
1 pc. 1¾ x 2½ x 35¾ in.—Brace

LAWN FURNITURE

the furnace room in winter or a sunny spot in summer. Also, it's better to use screws than nails as screws are less likely to loosen and pull out. This is especially true in regard to slats used as seats and backs where constant flexing of the slats tends to loosen nails.

Fig. 1 shows a chair of easy but substantial construction. Two short legs are mortised into saplings that form the seat frame and serve as a rear support. To assure rigidity and tight joints, hardwood wedges are inserted into saw cuts in the tenons, as shown in the circular detail, to spread the tenons when

Chairs, Benches and Tables

WITH the simple ruggedness of the tree trunks and branches of which it is made, rustic furniture is appropriate for lawn, porch or cottage, and will withstand the weather indefinitely. Additional protection can be provided by applying spar varnish not only over the raw wood but also over the bark to prevent moisture from seeping between it and the wood. As green wood shrinks considerably, which results in loosening of the joints, only seasoned wood should be used except where sticks or branches are to be bent. These should be curved around a form and left to season for a few weeks in a warm, dry place such as

they are forced into holes drilled in the frame. After the legs have been mortised, nails or screws can be driven through the frame and into the tenons. Seat and back are split saplings flattened at the ends to fit snugly on the frame where they are screwed in place. To provide concave surfaces for greater comfort, these pieces may be shaped with a drawknife before fastening in place, and left slightly roughened rather than plane-smooth for rustic effect. Two X-braces, shown in Fig. 2, are used as reinforcements to prevent twisting. These are bound with raffia, rawhide or stout cord at the intersections, and are screwed to the frame. Arms, which serve as additional supports for the back, are fastened with countersunk screws concealed with plugs.

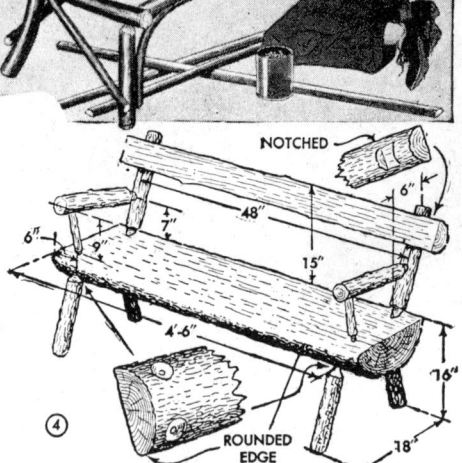

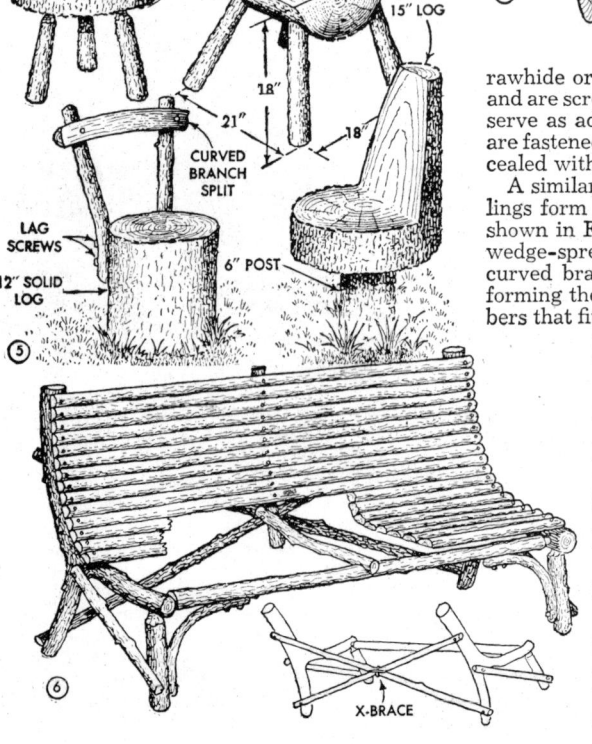

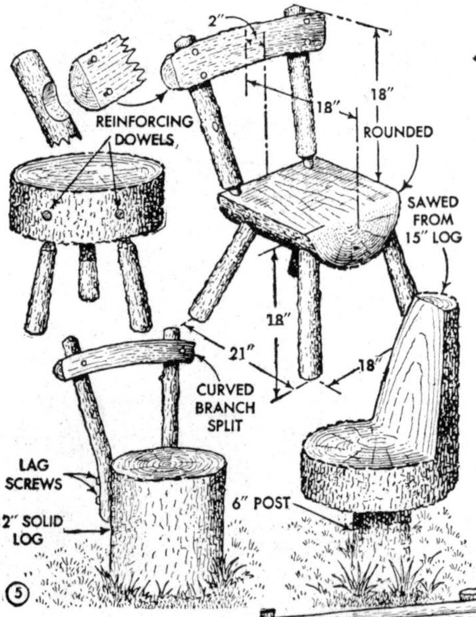

A similar chair is shown in Fig. 3. Saplings form the frame and, as in the chair shown in Fig. 1, two legs are attached by wedge-spread tenons and reinforced by curved braces on sides and front. Sticks forming the seat are nailed to cross members that fit notches in the frame, and split molding covers the ends. On this chair, arms are omitted, a brace being substituted on each side to support a fan-shaped back. This is made of narrow pieces bound with raffia or stout cord to cross members screwed on curved uprights. The chair shown in Fig. 11 has willow whips for the seat and back to provide resiliency. Net sizes of all parts are shown in the cross-hatched drawing, Fig. 10. Ends of the stretchers are cut concave to fit the round contour of the legs against which they are butted and screwed, as shown in the detail

screws driven into counterbored holes, which are plugged. Curved braces may be added for additional support if desired. A back rest is notched as shown and screwed to upright supports, which are drilled for tenons on the arms. For variation, the back and arms may be omitted to make a bench.

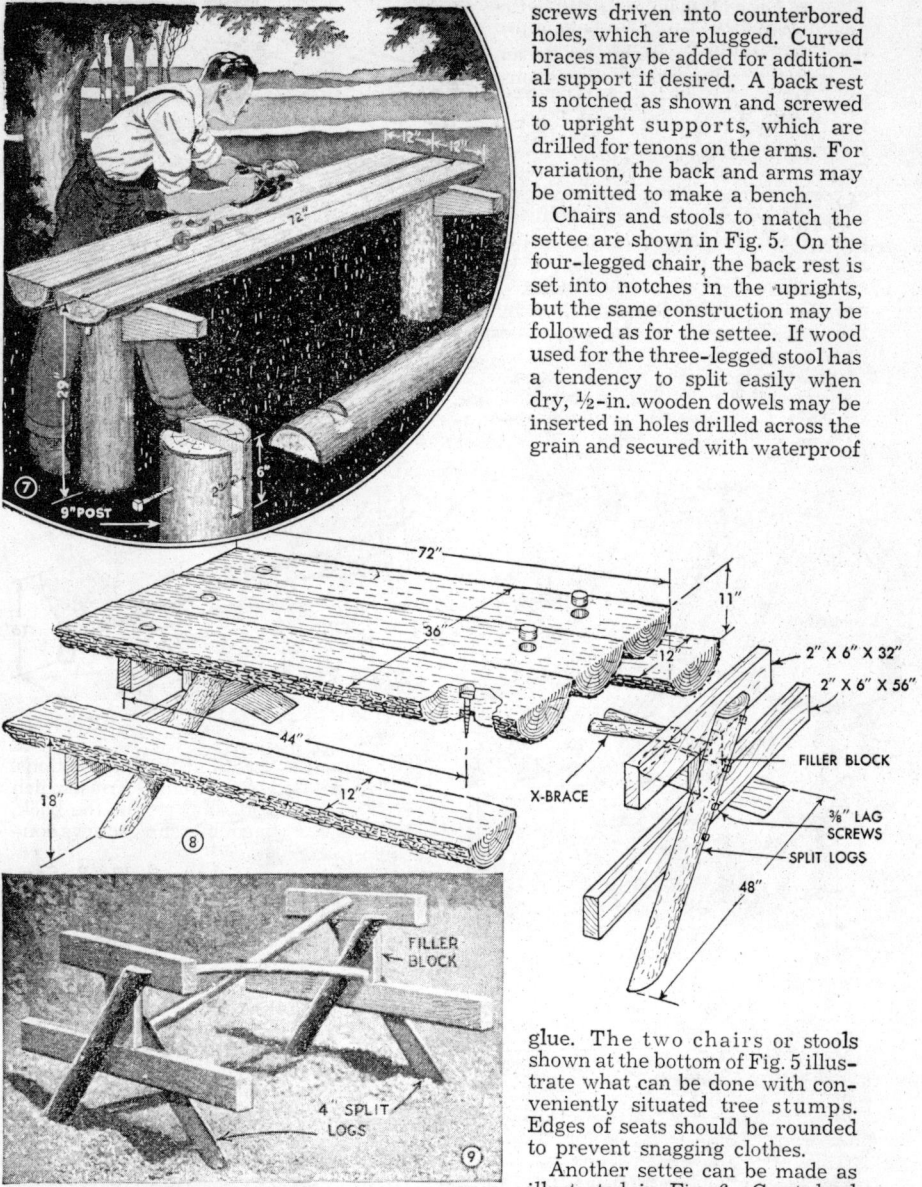

Chairs and stools to match the settee are shown in Fig. 5. On the four-legged chair, the back rest is set into notches in the uprights, but the same construction may be followed as for the settee. If wood used for the three-legged stool has a tendency to split easily when dry, ½-in. wooden dowels may be inserted in holes drilled across the grain and secured with waterproof glue. The two chairs or stools shown at the bottom of Fig. 5 illustrate what can be done with conveniently situated tree stumps. Edges of seats should be rounded to prevent snagging clothes.

Another settee can be made as illustrated in Fig. 6. Crotched limbs, as nearly alike as possible, are cut for back supports and rear legs, and set aside to season. When the wood is thoroughly dried, holes are drilled for front legs, which are secured by wedges and screws as suggested for the chair shown in Fig. 1, and a suitable piece is inserted to form a front rail. Curved braces are screwed to front legs and rail, and an X-brace is attached as shown in the detail.

of Fig. 11. A diagonal brace, attached to front and rear legs, wedges a center rail securely for attachment of willow whips. These are nailed to inside surfaces of the frame as indicated in the circular detail.

For the settee, Fig. 4, a large split log about 9 or 10 in. in diameter is rounded at the edge and smoothed for a seat. Legs are set tightly in holes drilled at the desired angle in the underside, and secured with

Before the seat slats are nailed on, holes should be drilled for the nails to prevent splitting the dried frame.

Split-log slabs are excellent for a fixed, outdoor table that will survive years of hard use. Large posts, well creosoted to prevent decay, are sunk into the ground below the frost line so that they will not heave. They are slotted as shown in Fig. 7 to take 2 by 6-in. crosspieces, which support half logs that form the top. Shallow logs or slabs are best for this purpose, and should be sawed or planed straight on each edge to fit together closely. Screws driven through the slabs into the crosspieces should be countersunk and plugged so as not to be noticeable. A similar table with built-on benches is shown in Fig. 8. The crosspieces that support the table and benches are mounted on split-log legs with a filler block secured at the intersection by a bolt. An X-brace, consisting of 3-in. saplings, is screwed to the crosspieces as shown in the photograph, Fig. 9. Another table, of lighter construction, is shown in Fig. 12. As finished lumber is used for the top, this table is suitable for indoor use in cottages and camps. Legs are glued into holes drilled in a 1 by 3-in. frame. Then the top is screwed to the frame, holes for the screws being counterbored and plugged, and bark-covered, split pieces are nailed along the edge for a molding. Curved braces support legs and frame, and four short spreaders join the stretchers.

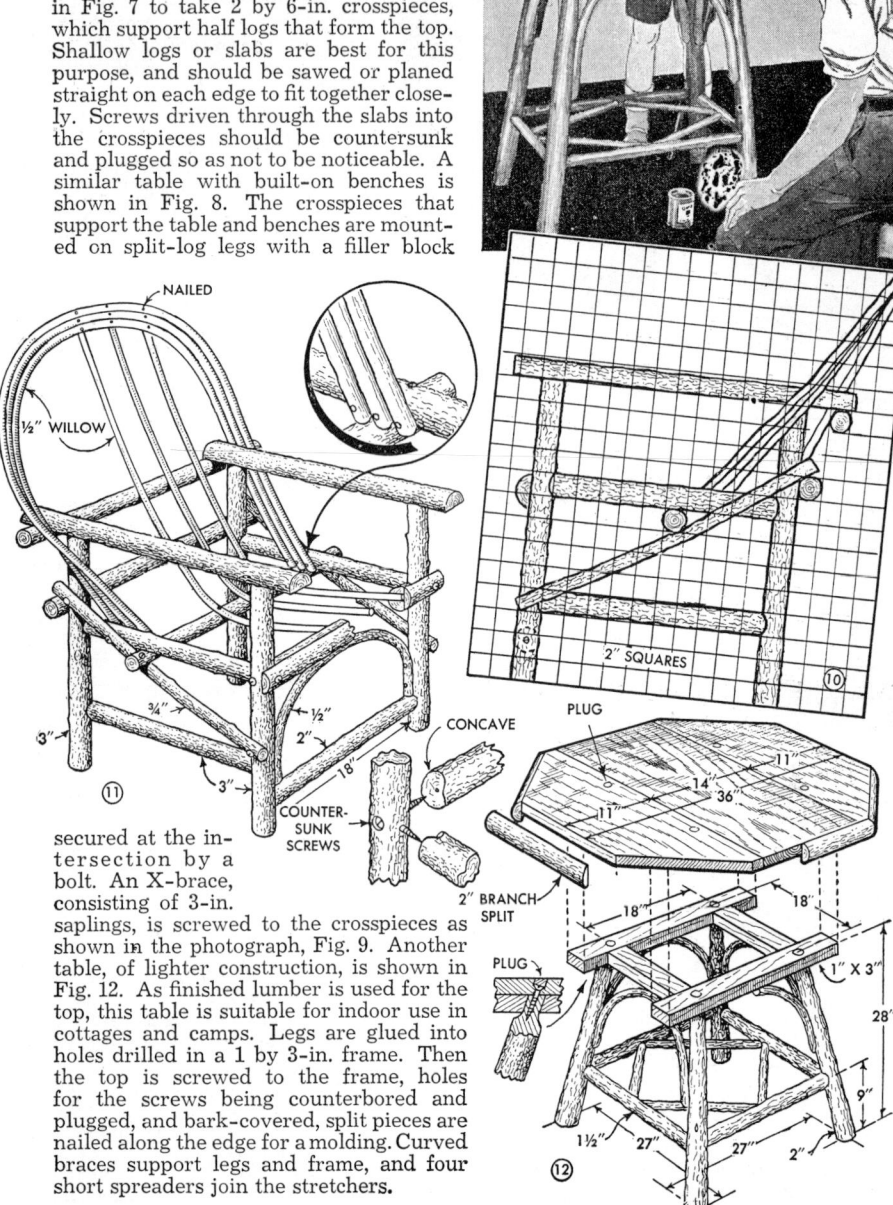

LAWN FURNITURE

JUST the thing for outdoor dining in your back yard or on a picnic, this novel plywood table-and-chair set comes apart so that it can be transported easily or tucked away in the fall in a minimum of space. Each chair consists of four parts, the sides being exactly alike, except for the slots. Fig. 3 gives the patterns for the parts which are cut out of 3/8-in. waterproof plywood. The sides of the chair are mortise-hinged together at the front, while its seat is surface-hinged to the back to fold downward. The slots must be cut to slant inward and accommodate the chair back. Cut them slightly oversize to allow for coats of paint or varnish. To set up a chair, you merely open its sides, Fig. 1, lift up the seat, engage the back piece in the slots and then push down on the back to interlock all of the parts as in Fig. 2.

The top of the table is a piece of plywood, 36 in. square, supported by "frame" legs, one frame being made in two separate parts which slide together in a dovetail groove on the underside of the top, Fig. 4, and interlock the complete assembly. The only fastening used is a removable dowel which pins the sliding legs together. The frames are assembled with open-mortise corner joints, as detailed in Fig. 5. Note that the crossrails of the solid frame are notched to take the ends of the sliding frame. The sliding frame has a dovetail tenon cut in the top edge to match a dovetail groove in the table top. The best way to cut the groove is with a portable router, equipped with a dovetail cutter, guiding it along a straightedge placed diagonally across the plywood and clamped. After the groove is made, the top is cut exactly in half and surface-hinged. When the table is assembled, a hole is drilled through the cross joint formed by the two frames and a dowel is inserted in the hole to lock the parts together. The dowel can be tied to the legs for safekeeping.

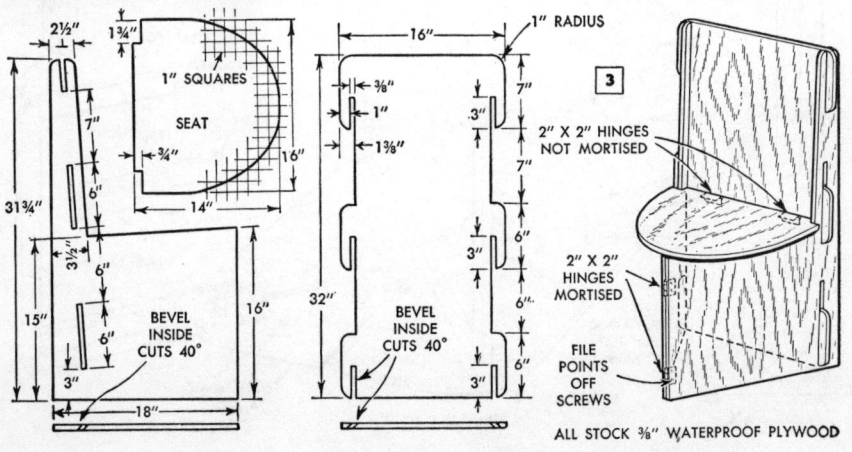

The chair goes together in a jiffy. Just open up the sides, raise the seat, insert the back in the slots and push down to interlock the parts

ALL STOCK 3/8" WATERPROOF PLYWOOD

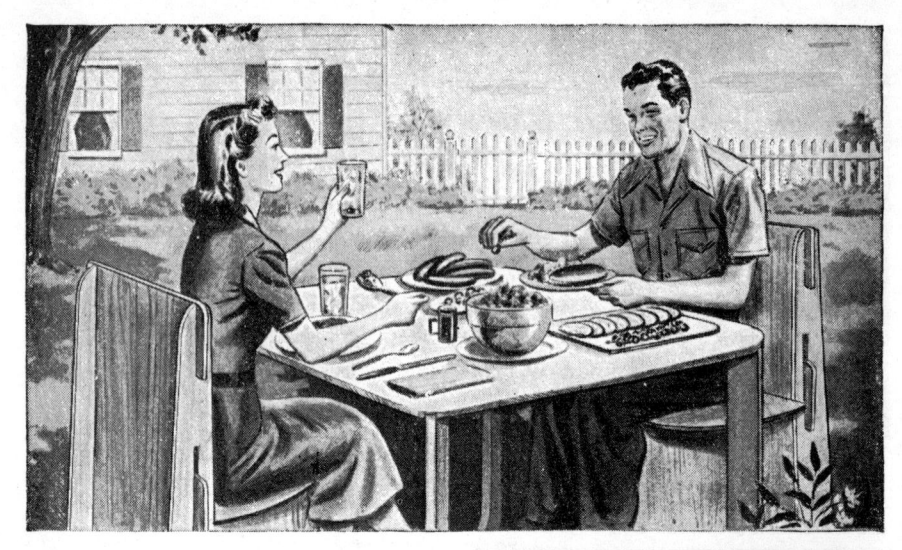

Although outdoor plywood is recommended for this furniture, it should be protected against moisture by finishing both sides and all edges of the parts with two coats of weather-resistant spar varnish. Sand the parts first, and as a touch of decoration, decals can be applied when the first varnish coat is dry. Then the second coat will protect the decals and keep them looking bright. Avoid making the sliding joints too snug as they should be a loose fit. Waxing will help if the parts do not slide freely

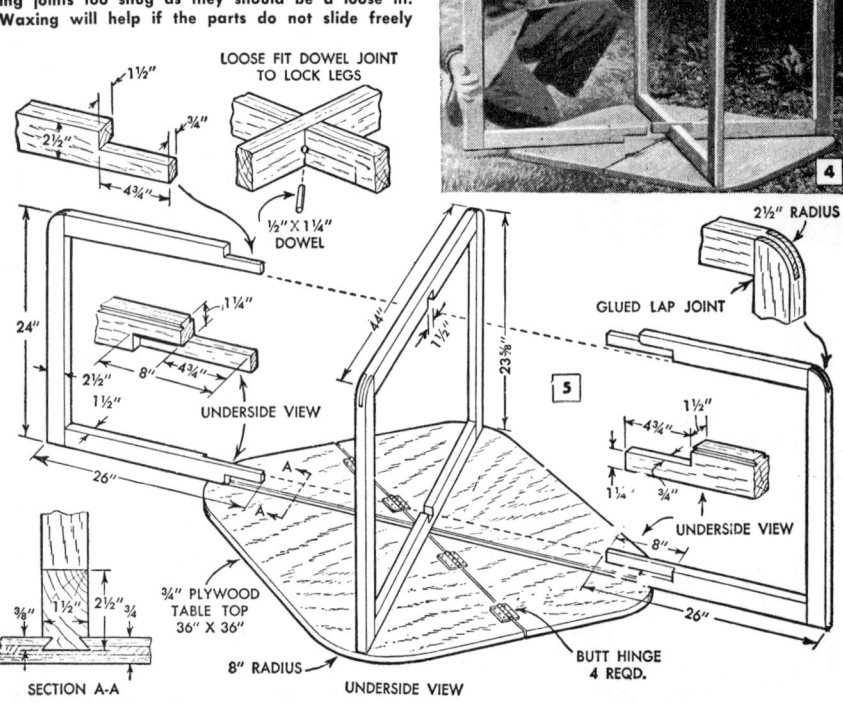

LAWN

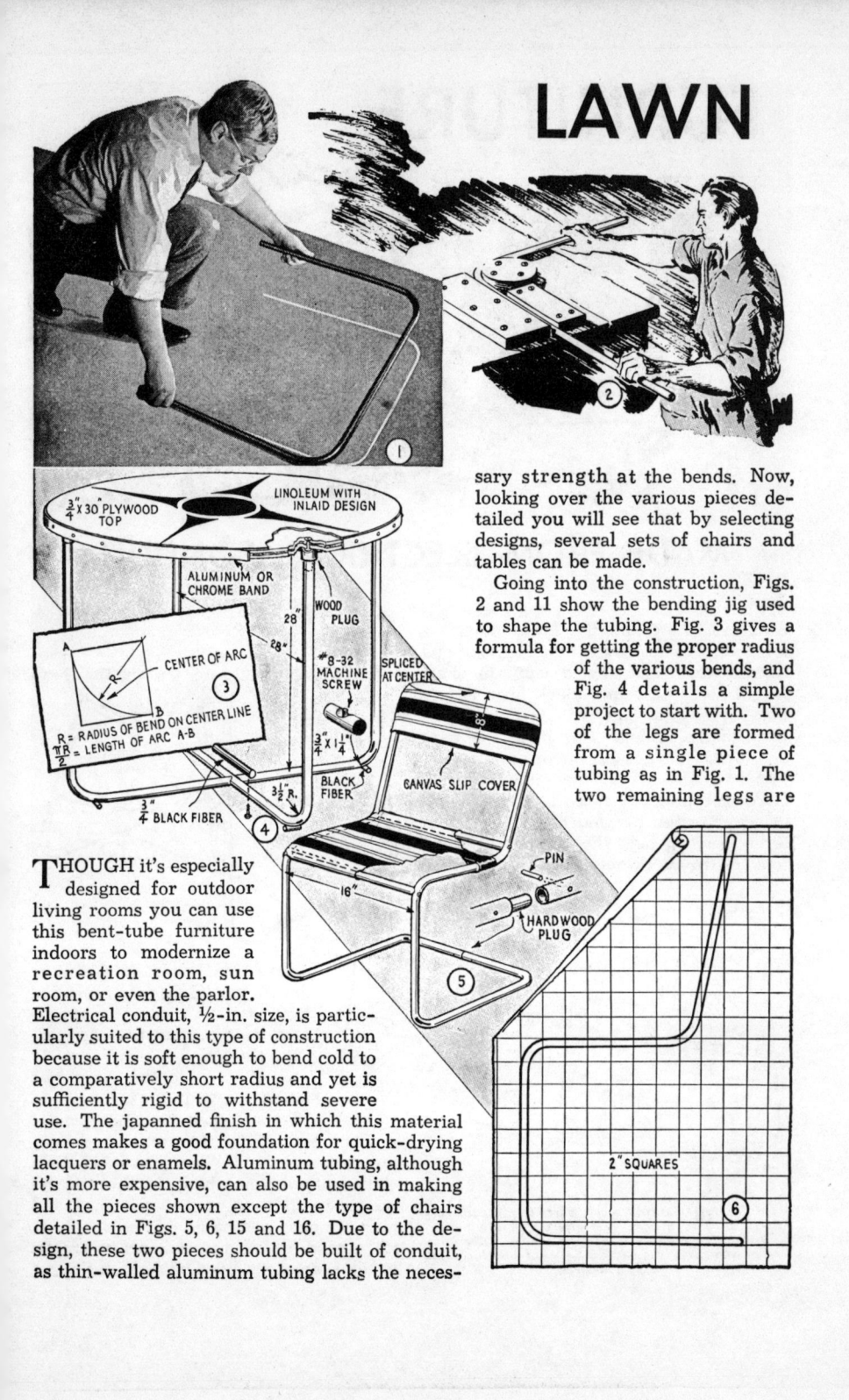

sary strength at the bends. Now, looking over the various pieces detailed you will see that by selecting designs, several sets of chairs and tables can be made.

Going into the construction, Figs. 2 and 11 show the bending jig used to shape the tubing. Fig. 3 gives a formula for getting the proper radius of the various bends, and Fig. 4 details a simple project to start with. Two of the legs are formed from a single piece of tubing as in Fig. 1. The two remaining legs are

THOUGH it's especially designed for outdoor living rooms you can use this bent-tube furniture indoors to modernize a recreation room, sun room, or even the parlor. Electrical conduit, ½-in. size, is particularly suited to this type of construction because it is soft enough to bend cold to a comparatively short radius and yet is sufficiently rigid to withstand severe use. The japanned finish in which this material comes makes a good foundation for quick-drying lacquers or enamels. Aluminum tubing, although it's more expensive, can also be used in making all the pieces shown except the type of chairs detailed in Figs. 5, 6, 15 and 16. Due to the design, these two pieces should be built of conduit, as thin-walled aluminum tubing lacks the neces-

FURNITURE

MADE FROM ELECTRICAL CONDUIT

bent separately and the lower ends are filed concave to fit the first member and form a neat right-angled joint. A short length of ¾-in. round black fiber is fastened over the joint with screws driven through the tubing. This holds the joint and further carries out the modern design. The four feet are of the same material. Hardwood plugs, turned to a tight fit, are driven into the top end of each leg, and screws, which hold the circular plywood top, are driven through the top into these plugs. Linoleum, of whatever design you choose, is cemented to the top. A chrome or aluminum band around the edge finishes the job.

Now, to build the other two tables, shown in Figs. 7, 8, 9 and 19, you follow the same general procedure in bending the tubing and joining parts together. When you bend thin-walled aluminum tubing, it's best to fill the tube with sand, ramming it hard, and plug the ends as in Figs. 17 and 18. Also, it's a good idea to turn a concave groove in the edge of the bending disk on the jig,

A table just suited to use in the outdoor living room for refreshments, card parties, etc. It's arranged to hold a lawn umbrella which can be anchored with a pin driven into the ground. The inlay design shown in Fig. 4 can be used on this table top, if desired

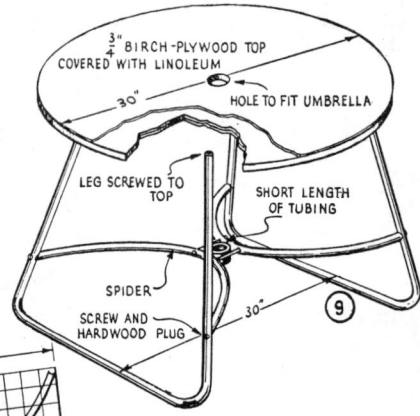

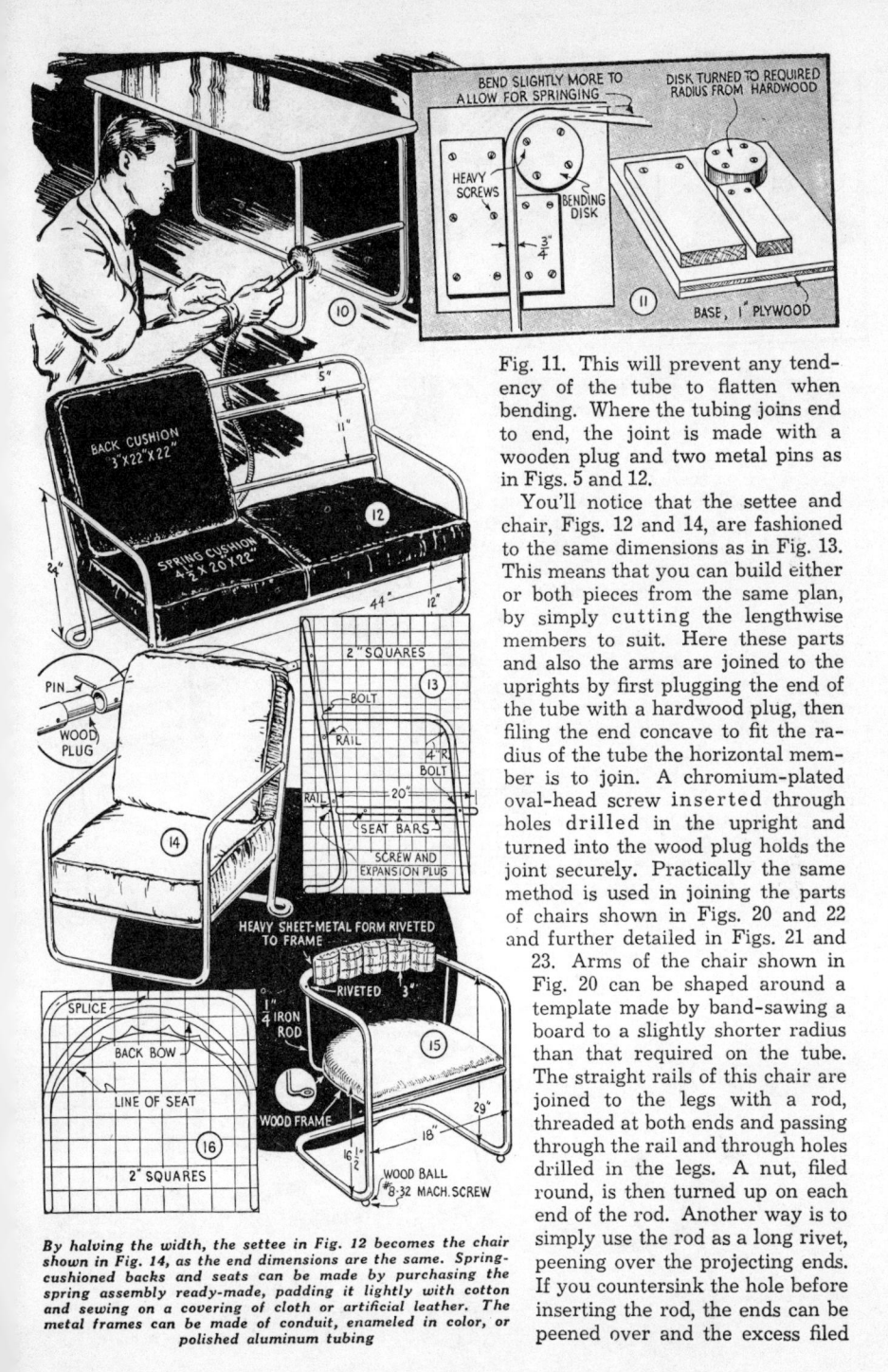

Fig. 11. This will prevent any tendency of the tube to flatten when bending. Where the tubing joins end to end, the joint is made with a wooden plug and two metal pins as in Figs. 5 and 12.

You'll notice that the settee and chair, Figs. 12 and 14, are fashioned to the same dimensions as in Fig. 13. This means that you can build either or both pieces from the same plan, by simply cutting the lengthwise members to suit. Here these parts and also the arms are joined to the uprights by first plugging the end of the tube with a hardwood plug, then filing the end concave to fit the radius of the tube the horizontal member is to join. A chromium-plated oval-head screw inserted through holes drilled in the upright and turned into the wood plug holds the joint securely. Practically the same method is used in joining the parts of chairs shown in Figs. 20 and 22 and further detailed in Figs. 21 and 23. Arms of the chair shown in Fig. 20 can be shaped around a template made by band-sawing a board to a slightly shorter radius than that required on the tube. The straight rails of this chair are joined to the legs with a rod, threaded at both ends and passing through the rail and through holes drilled in the legs. A nut, filed round, is then turned up on each end of the rod. Another way is to simply use the rod as a long rivet, peening over the projecting ends. If you countersink the hole before inserting the rod, the ends can be peened over and the excess filed

By halving the width, the settee in Fig. 12 becomes the chair shown in Fig. 14, as the end dimensions are the same. Spring-cushioned backs and seats can be made by purchasing the spring assembly ready-made, padding it lightly with cotton and sewing on a covering of cloth or artificial leather. The metal frames can be made of conduit, enameled in color, or polished aluminum tubing

away to produce a neat job. In either case, the ends of the rail are filed concave to fit the tubular legs.

Now about finishing. Conduit can be lacquered or enameled with excellent results. Any quick-drying brush lacquer or enamel will do, the latter perhaps being preferable because it does not set so quickly. First, sand the conduit lightly to remove any loose particles and smooth up rough spots. Then brush on the first coat of enamel or lacquer and allow to dry thoroughly. Before applying a second coat, go over the first lightly with fine sandpaper. Then follow with the finish coat, carefully brushed out to avoid sagging on the rounded surfaces. Where aluminum tubing is used it may be polished highly with a buffing wheel driven by a flexible shaft as in Fig. 10. A coat of clear metal lacquer will help to preserve the high polish.

Metal frames for any of these three pieces can be made of either polished aluminum tubing or electrical conduit finished in quick-drying colored lacquer or enamel. By combining these and other designs shown, several sets of attractive porch and garden furniture can be made

LAWN TABLE and CHAIRS

Wrought-iron furniture showing the influence of modern design is now the height of fashion, so designer John Bergen has followed this trend in creating this attractive garden furniture. Parts of these tables and chairs can be welded, riveted or bolted together. A bending jig should be made as shown so all curved parts will conform. Before finishing, the metal should be washed with a strong solution of caustic soda. Use rubber gloves when applying this solution as it will cause severe skin burns. Rinse thoroughly with clear water to remove all traces of caustic soda, then undercoat with a primer of zinc chromite or red lead. Final coat can be either glossy enamel or flat paint, depending on personal preference.

MATERIALS NEEDED

Angle steel
2 pcs.—3/4"x3/4"x13'—Apron rails of table
1 pc.—1"x1"x5'4"—Seat frame of chair
1 pc.—1"x1"x5'2"—Back frame of chair

Strap steel
2 pcs.—1/8"x5/8"x4'6"—Table ends
2 pcs.—1/8"x5/8"x6'—Table sides
1 pc.—1/8"x5/8"x3"—Chair back iron
7 pcs.—1/16"x5/8"x14 1/4"—Seat straps

Cold-rolled steel rod
4 pcs.—1/2" dia., 5'6" long—Table legs
2 pcs.—1/2" dia., 3'4" long—Front chair legs
2 pcs.—1/2" dia., 3' long—Back chair legs
2 pcs.—1/2" dia., 5'2" long—Chair arms

Miscellaneous
1 pc.—33 5/8"x43 5/8"—Plate-glass table top
2 pcs.—3/8" i.d. pipe, 7/8" long—Back spacers
1 foam-rubber pad, cored—2"x17"x15"
1 yd. upholstery, 36" wide

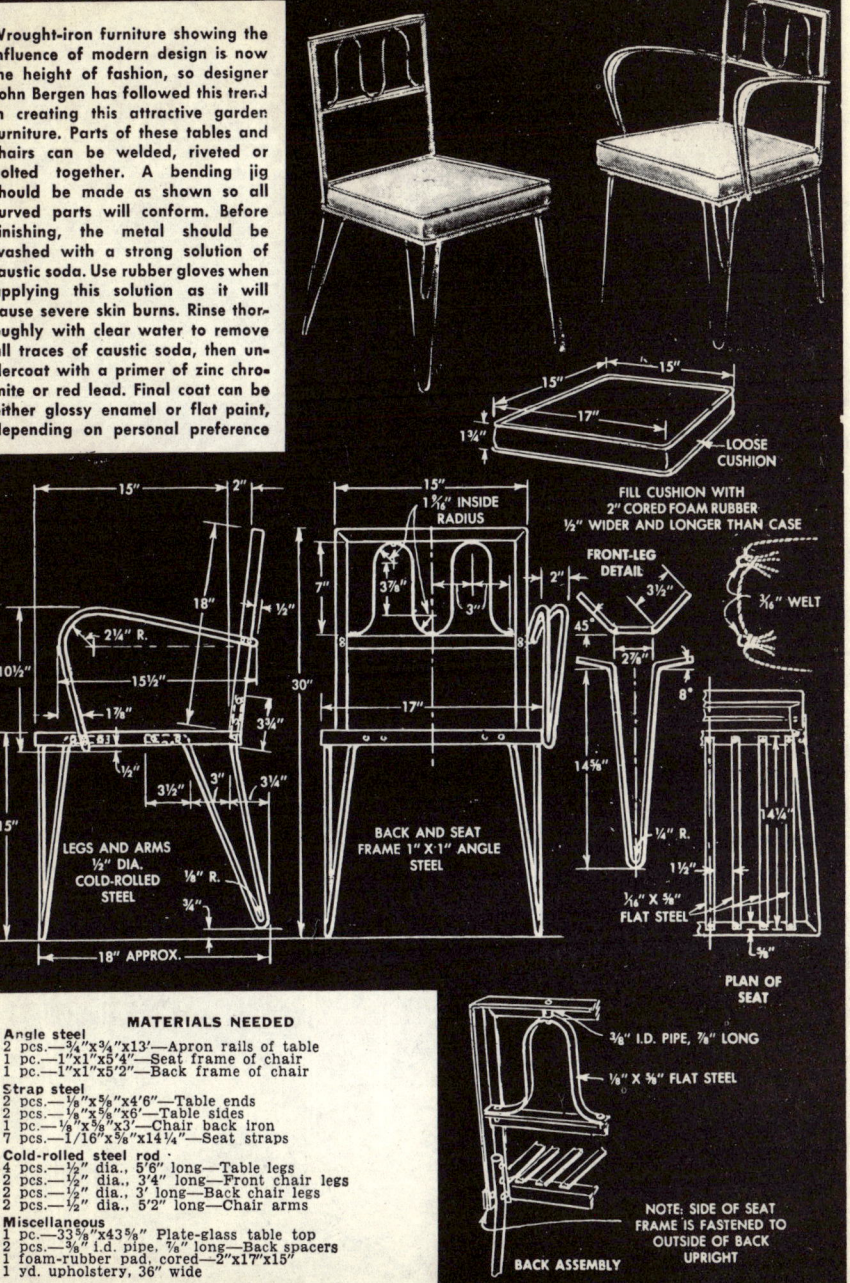

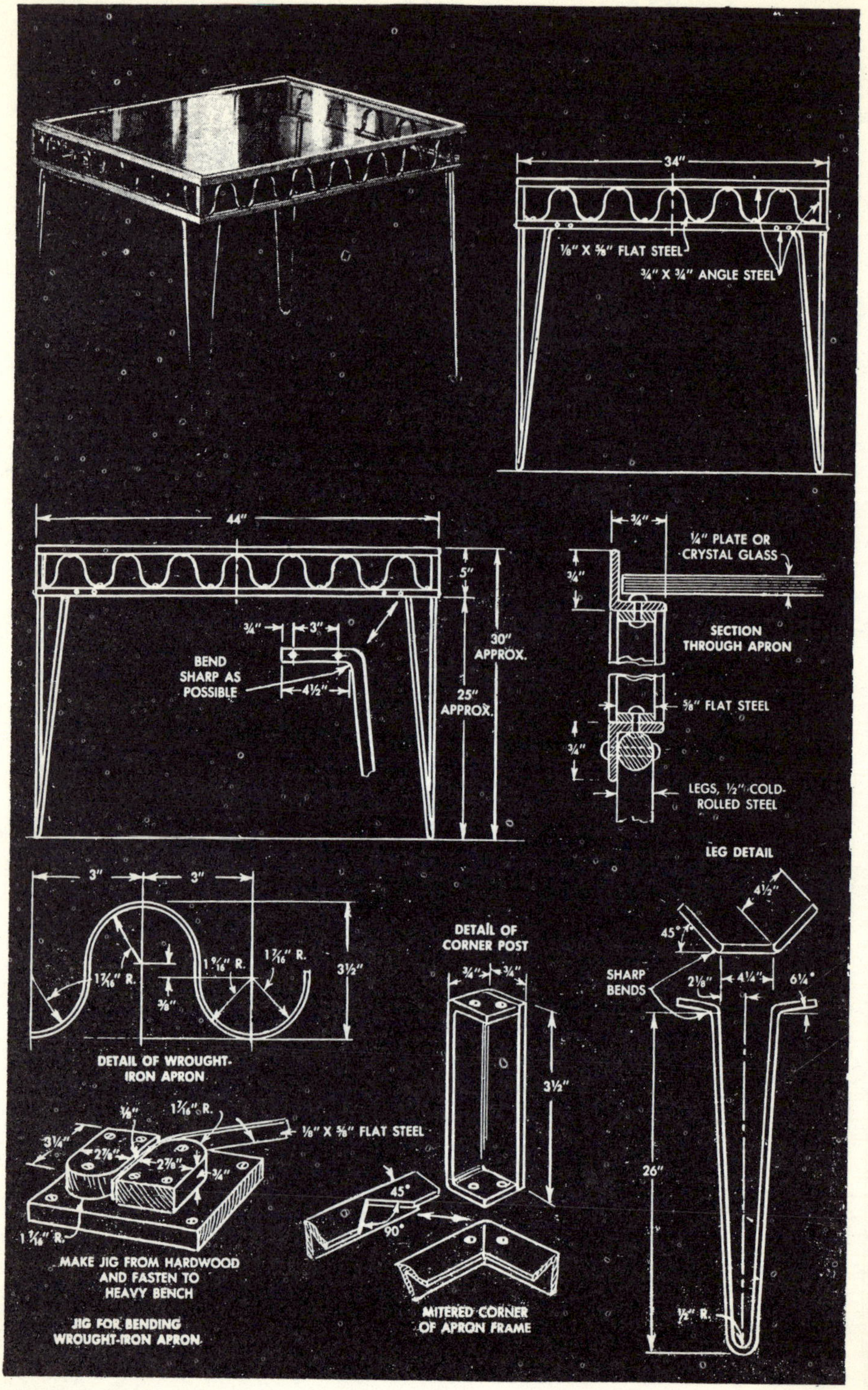

34"
1/8" X 5/8" FLAT STEEL
3/4" X 3/4" ANGLE STEEL
44"
5"
30" APPROX.
25" APPROX.
3/4"
3"
4 1/2"
BEND SHARP AS POSSIBLE
3/4"
1/4" PLATE OR CRYSTAL GLASS
SECTION THROUGH APRON
5/8" FLAT STEEL
LEGS, 1/2" COLD-ROLLED STEEL
LEG DETAIL
3"
3"
1 7/16" R.
1 9/16" R.
1 7/16" R.
3/8"
3 1/2"
DETAIL OF WROUGHT-IRON APRON
DETAIL OF CORNER POST
3/4"
3/4"
3 1/2"
4 1/2"
45°
SHARP BENDS
2 1/8"
4 1/4"
6 1/4°
26"
1/2" R.
3/8"
1 7/16" R.
3 1/4"
2 7/8"
2 7/8"
3/4"
1/8" X 5/8" FLAT STEEL
1 9/16" R.
MAKE JIG FROM HARDWOOD AND FASTEN TO HEAVY BENCH
JIG FOR BENDING WROUGHT-IRON APRON
45°
90°
MITERED CORNER OF APRON FRAME

LAWN SNACK TABLES AND LANTERN

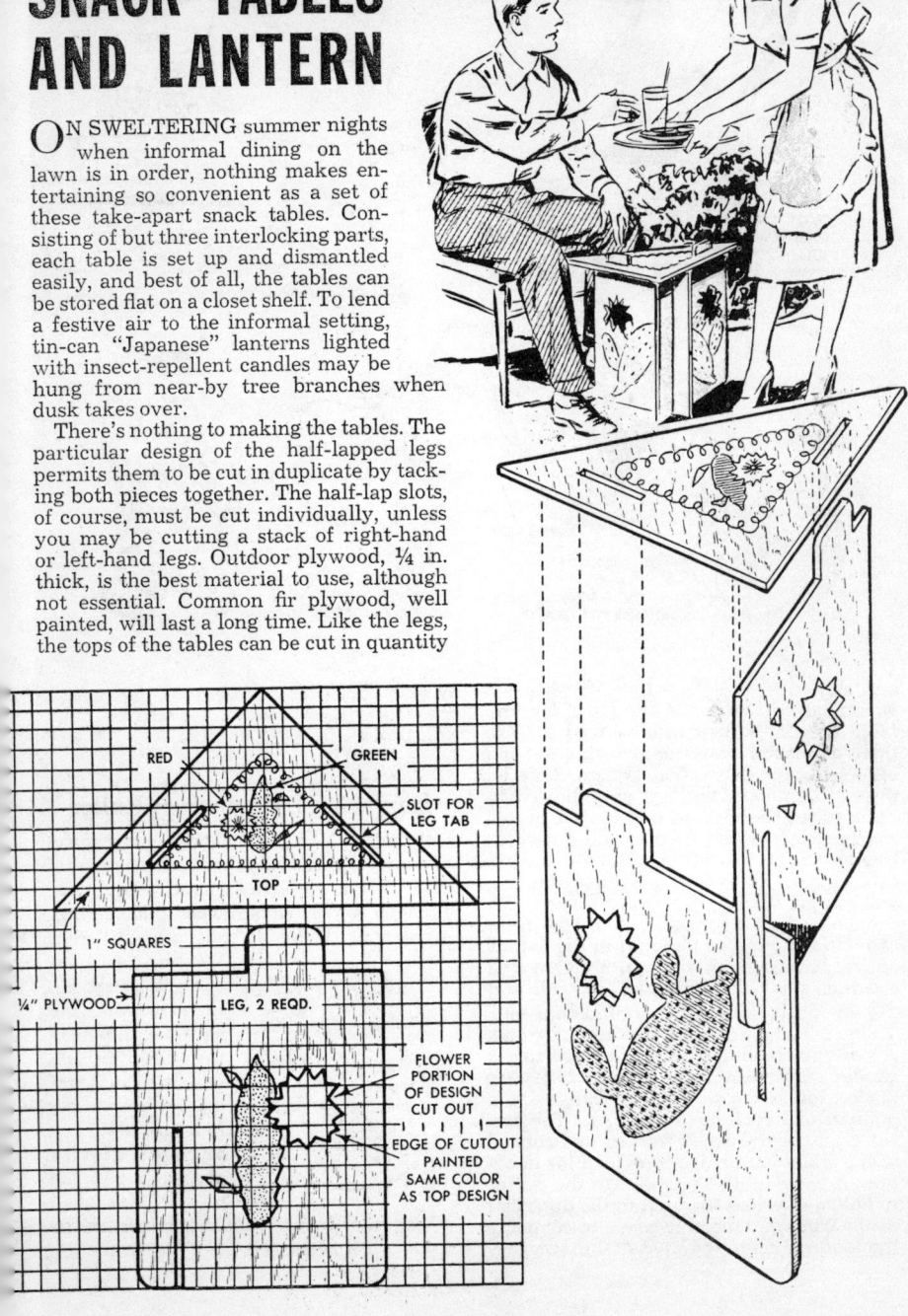

ON SWELTERING summer nights when informal dining on the lawn is in order, nothing makes entertaining so convenient as a set of these take-apart snack tables. Consisting of but three interlocking parts, each table is set up and dismantled easily, and best of all, the tables can be stored flat on a closet shelf. To lend a festive air to the informal setting, tin-can "Japanese" lanterns lighted with insect-repellent candles may be hung from near-by tree branches when dusk takes over.

There's nothing to making the tables. The particular design of the half-lapped legs permits them to be cut in duplicate by tacking both pieces together. The half-lap slots, of course, must be cut individually, unless you may be cutting a stack of right-hand or left-hand legs. Outdoor plywood, ¼ in. thick, is the best material to use, although not essential. Common fir plywood, well painted, will last a long time. Like the legs, the tops of the tables can be cut in quantity

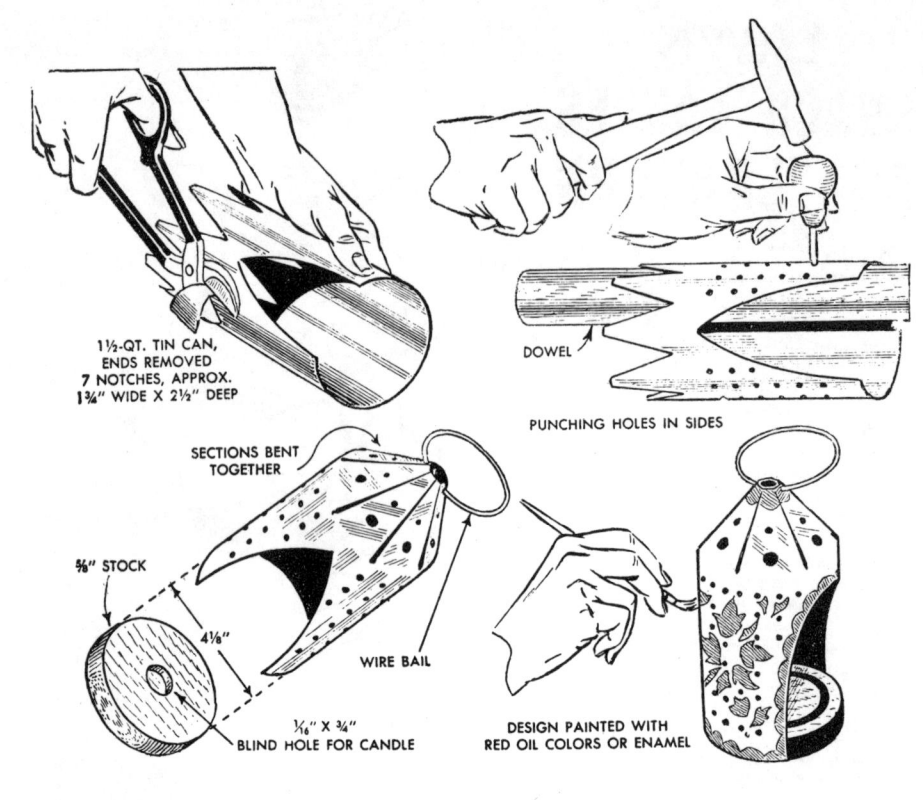

1½-QT. TIN CAN, ENDS REMOVED
7 NOTCHES, APPROX.
1¾" WIDE X 2½" DEEP

DOWEL

PUNCHING HOLES IN SIDES

SECTIONS BENT TOGETHER

⅝" STOCK

4⅛"

WIRE BAIL

1/16" X ¾" BLIND HOLE FOR CANDLE

DESIGN PAINTED WITH RED OIL COLORS OR ENAMEL

and slotted at the same time. Perhaps, it would be best to make the parts for one table first, and then after a trial fit, use them as master patterns in laying out the others. As for decoration, designs may be jigsawed right in the legs as indicated in the squared drawing, or appropriate decals can be selected and used to add a decorative touch. A coat of varnish over the decals, as well as over any painted designs, will protect them.

Each lantern is formed from a 1½-qt. tin can. Starting out as pictured in the details above, both ends of the can are removed and then seven serrations, 1¾ in. wide and 2½ in. deep, are made around one end. Then, cutting up from the opposite end, a V-shape opening is made for inserting a candle. After this, rows of ⅛-in. holes are made around the sides of the can, using a pointed tool such as an ice pick. The serrated end of the can is formed to a conical shape and a wire bail is provided for hanging. A wooden disk recessed in the center to hold a candle is inserted in the open end of the can and tacked in place to complete the lantern. Either painted designs or gay-ly colored decals can be used to decorate the lanterns. A coat of varnish or paint will protect the wooden disk. ★ ★ ★

Replacing Lawnmower Roller

If the roller on your lawnmower is worn, it is a simple matter to make a new one of pipe. Get a piece the correct length and diameter and drive hardwood plugs tightly in the ends. If the old roller was drilled in the ends to provide bearings, do the same with the new one. If pins in the ends of the old one fitted in bearings in the mower frame, use lag screws of the proper size in the new roller as a substitute. The heads of the screws can be sawed off if necessary.

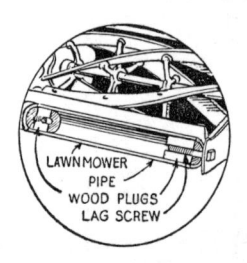

LAWNMOWER PIPE
WOOD PLUGS
LAG SCREW

Lawnmower Held Against Wall By Strapping It to Stud

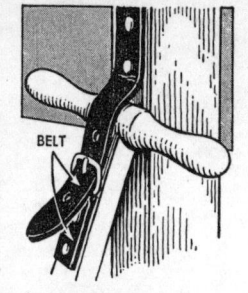

To keep small children from playing with his lawnmower, one home owner strapped it to his garage wall with a discarded trouser belt as shown. The buckle portion of the belt was tacked to the handle of the mower and the strap end to the stud of the wall. By holding the handle of the mower against the stud and inserting the strap through the buckle, the mower may be fastened securely.

LAWN MOWER SHARPENING

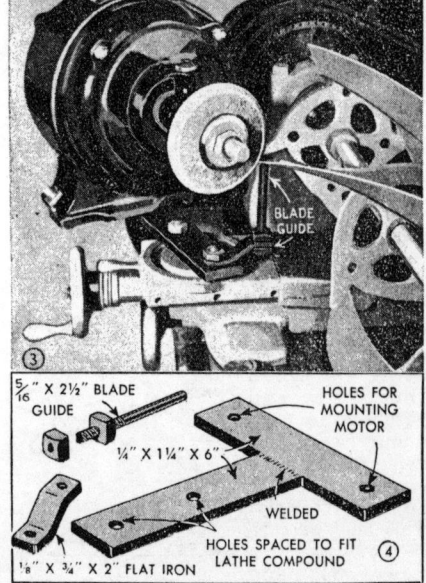

LAWN-MOWER sharpening can be done profitably in any small shop or garage if a metal-turning lathe and tool-post grinder are available. The latter may be improvised from a motor on which a grinding wheel can be attached plus a few fittings to permit it to be bolted in an adjustable manner on the lathe carriage as shown in Figs. 1, 2 and 3. The fittings used are shown in Fig. 4. Remove the lathe centers and insert the shaft ends of the lawn-mower blade reel in the center holes of the lathe head and tailstock so that it floats between them. Next, adjust the grinder so that the latter contacts the cutting edge of blade at right angles, and set the blade guide on the T-shaped iron as in Fig. 3 to hold the blade in this position. Now, start the grinder and the lathe and put the carriage in gear. As the grinding wheel travels along, it grinds the blade quickly and accurately, the guide keeping the blade in the correct position in relation to the grinding wheel. To sharpen the shear plate of the mower, it is mounted in the lathe and the grinding wheel is run over the blade by manual operation of the lathe carriage, see Fig. 1.

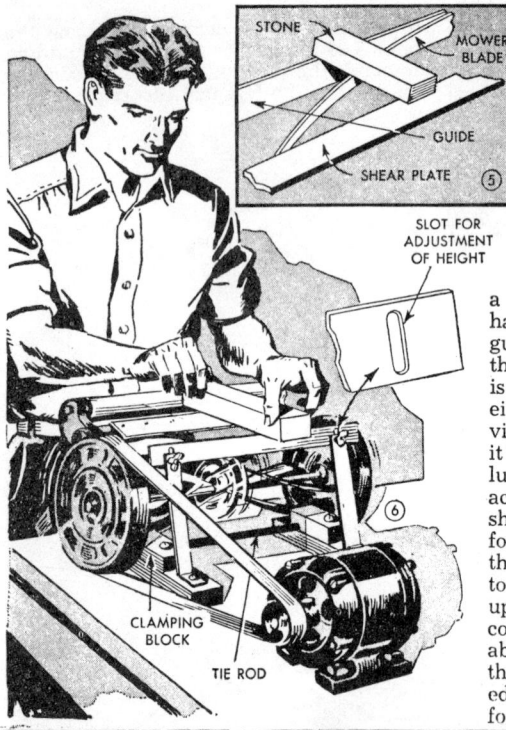

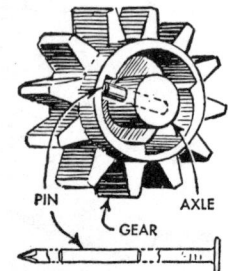

When a lathe is unavailable, you can do a fairly accurate job of sharpening a mower by belting it to an electric motor and using a hand stone as shown in Fig. 6. Clamping-blocks to take the mower tie rod are screwed to a bench so that the end of the mower handle will rest on the floor. Then, a guide bar for the stone is screwed to the blocks, and a small electric motor is belted to one of the wheels. It is best either to gear down the motor or provide a resistance in the circuit so that it rotates about a hundred or so revolutions a minute. In use, hold a stone across the guide bar and the mower shear plate as in Fig. 5, adjusting the former so that the stone just touches the cutting blades, then, start the motor. After the blades have been trued up, adjust the shear plate against them correctly and rotate the mower for about 10 min., applying plenty of oil to the plate. This will smooth the cutting edges, after which the mower is ready for use.

Light Attached to Lawn Mower Powered by Small Generator

When rainy weather and overtime hours kept one man from getting his lawn cut in full daylight he added a generator-powered bicycle light to his power lawn mower. He clamped the headlight to the handle so the beam of light would shine over the motor and onto the lawn ahead of the mower. The bicycle generator used is the type having a friction drive wheel that runs against the bicycle tire. The generator was mounted so the drive wheel would bear against the drive belt of the mower.

Repairing Lawn-Mower Gears

Some lawn-mower drives are fitted with round pawls, and when the pawls and ratchets become unduly worn, the drive will slip or engage and release intermittently. This may be corrected by inserting new pins made from pieces of nail. These should be about 1/16 in. longer than the old pins and the edges should be chamfered slightly instead of rounded, so the pins will not slip over the ratchet teeth. These pins will remain serviceable for about one season and they will not cause additional wear on the gears. You may have to file the pins lightly and check several times before you obtain the proper length.

Bed Knife of Mower Held Down While Cutting Tough Grass

A simple arrangement to hold the roller and bed knife of your lawnmower to the ground when cutting tough grass can be made with a length of No. 6 wire and a screen-door spring. Hook each end of the wire into a hole in the castings that support the roller and extend it upward behind the handle brackets so that it forms a triangle with its upper corner 6 in. directly above the lower end of the handle. Now attach one end of the spring to this upper corner, which has been indented for the purpose, and extend it downward over the end and along the underside of the handle. Stretch the spring reasonably tight and fasten the end to the handle with a small screw eye.

Mower Sharpened Easily
With This Gadget

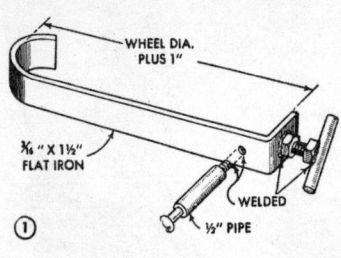

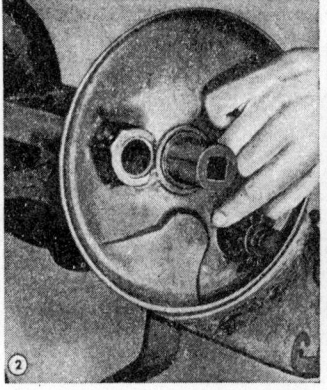

Velvet-smooth cutting edges on lawn-mower blades are something you think about when pushing a dull mower through heavy grass. A professional job of sharpening is quite simple as you can see from Figs. 1, 2 and 3 and the upper right-hand photo. First, make a wheel crank as in Fig. 1 and attach it to one drive wheel with the clamping screw shown in the detail, Fig. 1. Then clamp the mower in a vise, remove the drive wheels and interchange the reel-driving pinions as in Fig. 2. Replace the drive wheel to which the crank is attached. Then set the bed knife, Fig. 3, so that the blades "drag" very lightly when the reel is turned backwards. Mix medium-grade emery powder with crankcase oil to the consistency of thick cream and drop or paint the mixture on the edges of the blades. Turn the crank, spinning the reel backwards at moderate speed. Reset the bed knife and continue the operation until the blades are uniformly sharp. Finally, clean, oil, and reassemble the mower.

Rubber-Headed Tacks in Tires Give Mower Better Traction

When rubber tires on a lawn mower wear smooth, traction can be gained by driving rubber-headed tacks in the tires as indicated. The tacks should be long enough to clinch when they penetrate the tire.

Full-Size Mower Cut in Half Provides Lawn Trimmer

A regular lawn mower cut in half does an excellent job of trimming and mowing along fences, around flower beds and other places where a full-size mower would be awkward to use. After cutting the mower in half, a welded bracket of flat and angle steel is made to support the reel, cutter bar, roller and cross bar. The flat-steel fork of the handle is reshaped so the handle is moved to one side and closer to the wheel.

MOTORIZED HAND LAWN MOWER

¼" X 1" X 1" STEEL ANGLE, 4 OR 5 AS REQUIRED

ENDS OF KNIVES CUT FLUSH WITH SPIDER

HUB SAWED OUT

6" DIA. V-PULLEY

THESE RIVETS REMOVED

RATCHET PAWLS REVERSED

①

IT'S easy to take the "push" out of a hand-power lawn mower with a ⅝ or ¾-hp. gasoline engine and a few pieces of flat iron. These small, compact engines can be mounted on almost all hand lawn mowers of 18-in. cut or wider. Due to the relatively higher speed and steady forward movement over the ground the grass is cut more cleanly and uniformly. Moreover, the mower will handle taller and heavier growth than can be cut by hand power alone. In this particular type of drive the motor is belted directly to the cutting reel so there's no need to bother about "differential" or a clutch with a complicated throw-out mechanism. Needed differential is taken care of by slippage of the drive wheels induced by a light downward pressure on the handle as you steer the mower in the direction desired. Declutching is accomplished simply by bearing down on the handle when you want to stop or make a sharp turn. This, in turn, raises the drive wheels and shifts the weight to the roller, allowing the mower to be turned in any direction or pulled backward to clear a fence, shrub or other obstruction.

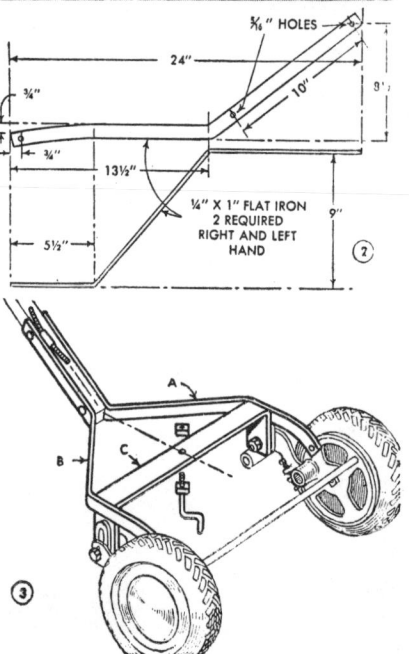

3/16" HOLES

24"

¾"

10"

9'

¾"

13½"

¼" X 1" FLAT IRON 2 REQUIRED RIGHT AND LEFT HAND

5½"

9"

②

A

B C

③

intended merely as a rough guide to the size. Different types of mowers will require some variation. Measure the mower at hand to determine the dimensions required before cutting material.

Connect Fig. 3 with the details in Figs. 4 and 5 and you will readily see how the additional parts of the engine support are fitted on the mower frame. Note that the engine sub-base, made from a steel plate, is hinged on the mower shrub bar as in Fig. 5, and rests on the end of the belt-tightener screw, Fig. 4. This simple arrangement allows easy adjustment of the belt tension. With these parts made and fitted, reassemble the mower with the ratchet pawls reversed in the drive pinions. Be sure to slip the V-belt over the reel pulley first. Bolt the engine and fuel tank in position as in Fig. 4, making sure that the pulleys align properly. Size of the engine drive pulley will depend on the average speed of the engine and you can easily calculate the required diameter of the pulley once the speed of the engine is known. By installing a throttle control on the handle as in the photo, Fig. 1, the engine speed usually can be controlled so that the mower is driven at a comfortable walking pace.

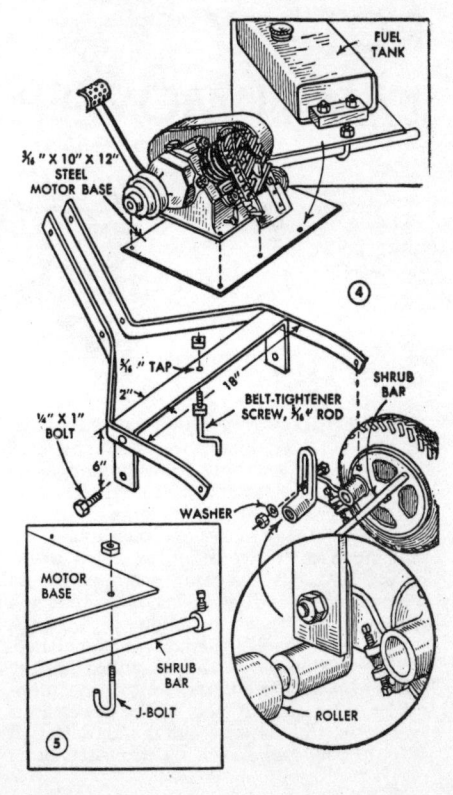

The first thing to do when installing the power drive is to disassemble the mower, removing the reel and the handle. Save the wood section of the handle assembly. Usually, because of the nature of the installation, it will be necessary to make a new yoke, Fig. 2. When removing the reel be careful not to lose or damage any part of the bearings. Fig. 1 shows the method of attaching a pulley to the end of the reel. Diameter of the pulley may have to be varied to suit the mower at hand. The four-blade reel shown in Fig. 1 is not standard as some mowers are fitted with five blades. Ends of the blades projecting beyond the spider are cut off flush and the rivets removed as shown. Steel angles are attached to the spider with stove bolts which pass through the rivet holes. Although a standard metal V-pulley can be used as in Fig. 1, it may be necessary to cut the hub to obtain a free fit between the spider and the inner end of the reel-shaft bearing housing. A pulley also can be turned from ¾-in. hardwood or waterproof plywood of the same thickness. Use lock washers on all bolts in this assembly.

Fig. 3 shows the flat-iron parts which are made and attached to the mower frame. The two parts of the handle yoke, A and B in Fig. 3, are detailed in Fig. 2. However, the dimensions and the bends indicated are

ROTARY LAWN MOWER

POWERED by a 1½-hp. air-cooled engine, the rotating blades on this powerful mower cut everything from dandelions to the tallest and toughest weeds. An outstanding feature of this machine is that the front wheel may be set to either side of the cutting path to avoid rolling down uncut grass. This insures clean cutting with no ridges left standing. Also note that the center of gravity of the engine is well back of the rear wheels to secure a nice balance.

The engine platform and guard for the cutter blades are cut and bent from a piece of No. 10-ga. sheet steel to the dimensions given in the upper right-hand detail of Fig. 2. Then a band of flat iron is welded around the edge of the guard. Two lengths of 1½ by 1½-in. angle iron are welded to the top of the guard and the bottom of the platform to give additional strength and furnish mounting surfaces for the handle and the rear-wheel sleeve bearing. The cutter assembly consists of a cut-down spindle and hub from the front wheel of an auto. A short piece of rod is welded to the spindle and threaded for a castle nut. This rod also serves as a shaft for the driven pulley, center detail. As shown in the lower left-hand detail, the cutter blades are made from two corn-binder knives and are bolted to a 12-in. disk of No. 10-ga. sheet metal. After these parts are completed, the disk is

2 EXPLODED VIEW

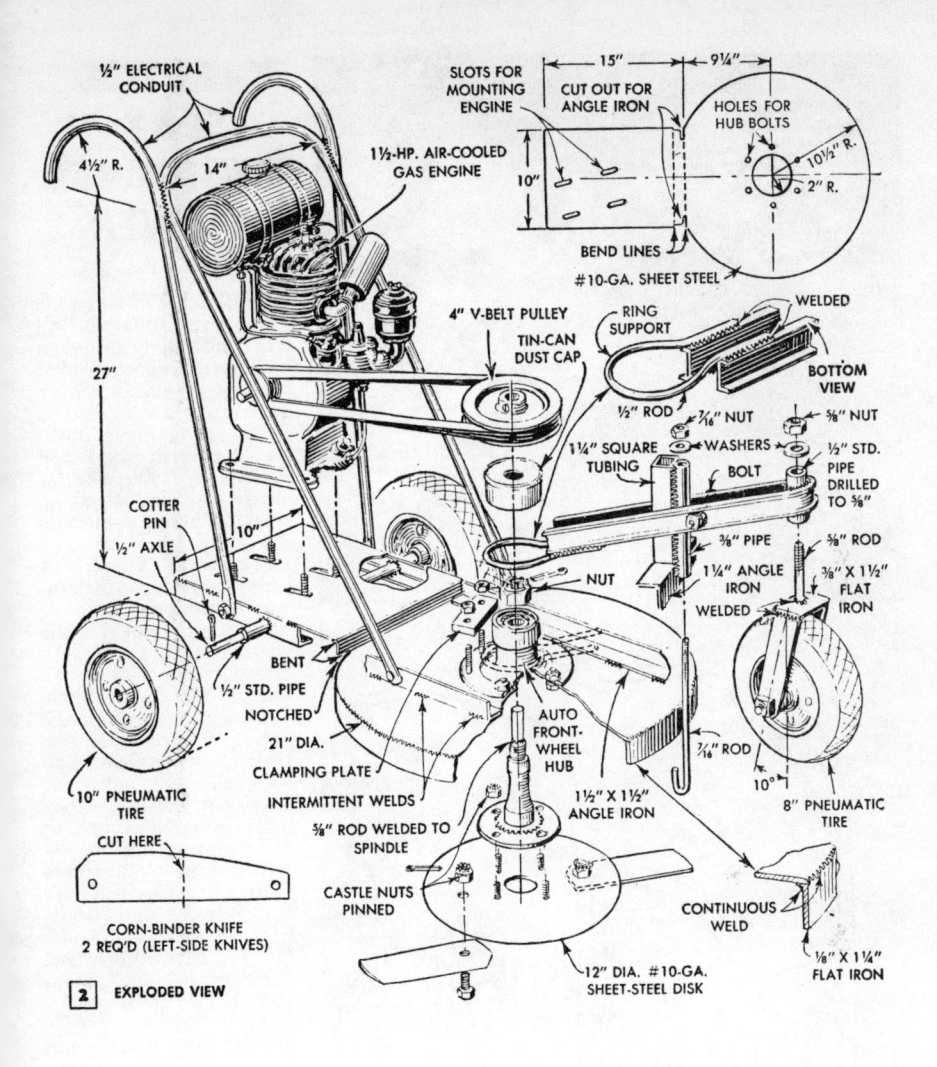

bolted to the spindle, the hub is bolted to the guard and a V-belt pulley and tin-can dust cap are fitted for a trial assembly. The parts should turn freely and be in line, Fig. 1.

The front-wheel assembly, which consists of an adjustable arm, a clamping plate that bolts to the auto hub, and a fork, are illustrated in the right-hand center detail of Fig. 2. The height of the arm, and consequently the height of the cutting blades from the grass, is adjusted by means of a J-bolt and sleeve that fit between the two sides of the arm. In addition, the arm furnishes a means of lateral adjustment so the wheel can be set to either side of the cutting path. The rear wheels ride on an axle that turns in a pipe-sleeve bearing welded to the platform angle iron. When slotting holes for the engine mounting, notice that these holes are off-center and at an angle so the centers on the driving and driven pulleys will line up correctly. To complete the mower, form the handle and brace of ½-in. rigid electrical conduit.

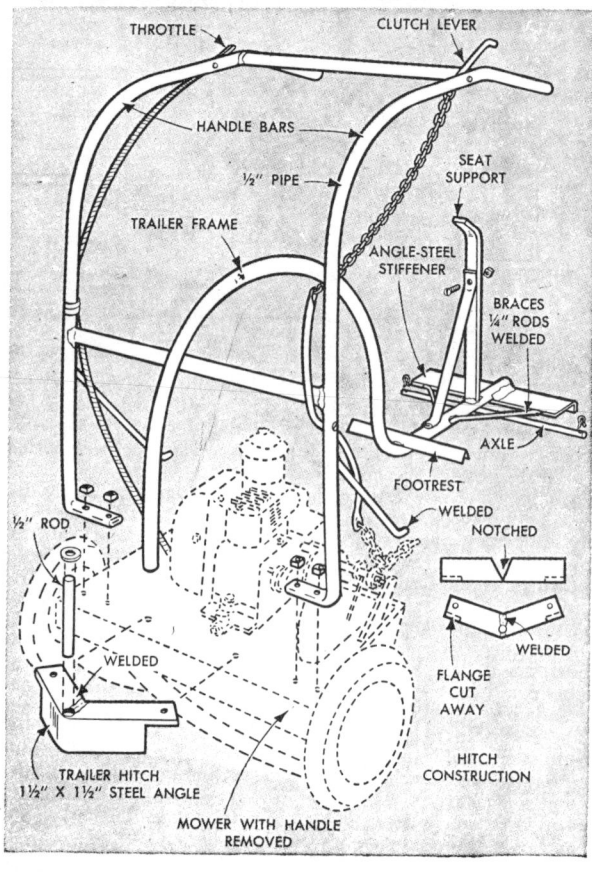

TRACTOR SEAT FOR POWER MOWER

A BIG LAWN looks a lot smaller when you are comfortably seated on this power-mower trailer. Use of a riding trailer such as this one reduces grasscutting time as much as 20 percent since you can make faster, shorter turns and also speed up mowing above a normal walking pace on the straightaway. The trailer unit requires no alteration of the power mower other than removal of the regular handle and substitution of special handle bars and a trailer hitch.

Dimensions given in the detail drawings of the trailer and handle bars on the following page are only approximate. They must be altered to suit different mowers. The detail at the left shows clearly the position of the trailer frame and handle bars in relation to the mower. An improvised extension of the clutch-control mechanism is located on the left handle bar and consists of a chain linkage which works through a bell crank attached to the handle bar below the cross brace. On some mowers the throttle may be utilized but on others an extension may be necessary.

Ordinary black pipe is used for the handle bars and braces. For the curved trailer frame extra-heavy pipe should be used as this unit is subjected to considerable bending stress. If you do not have facilities for making true-radius bends in pipes of the sizes re-

The lawn mower-trailer unit can be made to turn in its own length, an essential feature when mowing around flower beds and in irregularly shaped plots

This photo calls attention to the simplicity of the trailer hitch and manner of assembling the handle bars and controls. The regular mower handle is removed

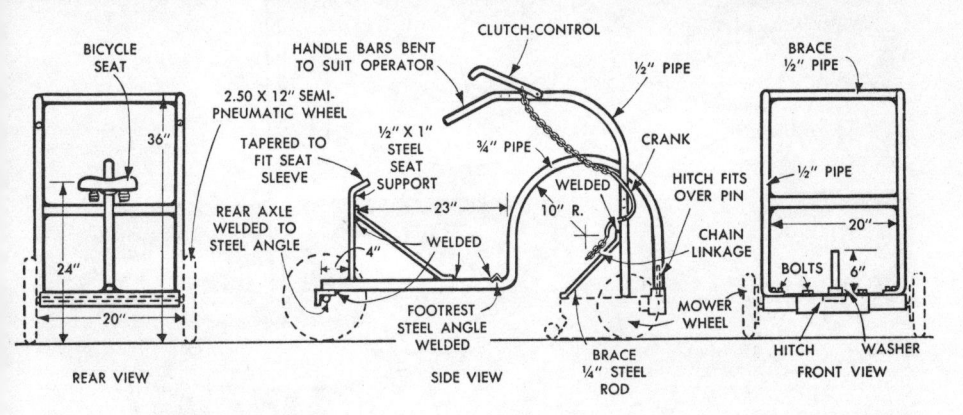

quired take the parts to your plumber. He can do the job at a nominal cost. Note that the lower ends of the handle bars are bent at right angles and flattened. Holes are drilled in the flattened ends for bolts which pass through holes drilled in the mower shield, or guard. This usually is of heavy sheet metal and will require no reinforcing. Handle-bar braces are welded in place as indicated.

The simple drawbar shown can be adapted to nearly all types of power mowers. It is made by V-notching a short length of steel angle so that it can be bent and the joint welded. Flanges at the ends are cut away to clear the mower guard and holes are drilled for bolts and a ½-in. steel rod, 5 in. long, which is welded to the drawbar angle in the vertical position and serves as a pivot for the front end of the trailer frame. It should be noted that the curve of the frame is stopped at a point about 8 in. from the end. The remainder of the pipe is straight so the parts can swivel freely when it is placed over the pivot pin.

A short length of steel angle welded near the center of the frame just below the bend serves as a footrest. At the rear end the frame is welded to an angle-steel stiffener which in turn is welded to the axle. The seat support is welded to the frame and braced as indicated. A bicycle-type saddle seat is used, and some filing may be necessary to fit it securely over the tapered projection of the support. Angle braces projecting to the rear support the handle bars and the lower ends of the braces are welded or bolted to the mower frame. The trailer wheels are 12 in. in diameter and are of the semi-pneumatic type having integral bearings. They are attached to the axle with cotter keys and washers.

LAWN UMBRELLA

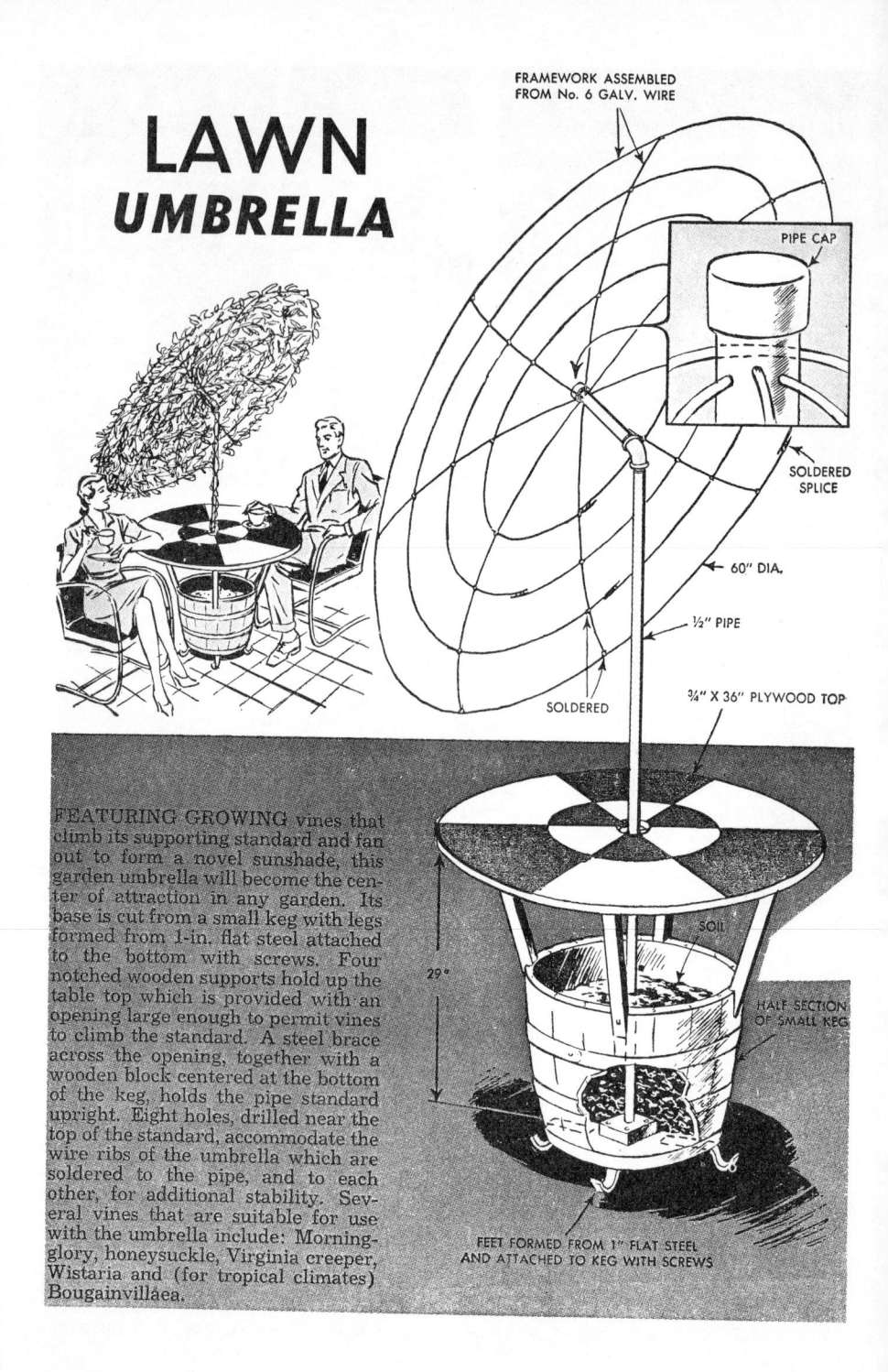

FRAMEWORK ASSEMBLED FROM No. 6 GALV. WIRE

PIPE CAP

SOLDERED SPLICE

60" DIA.

½" PIPE

¾" X 36" PLYWOOD TOP

SOLDERED

SOIL

29"

HALF SECTION OF SMALL KEG

FEET FORMED FROM 1" FLAT STEEL AND ATTACHED TO KEG WITH SCREWS

FEATURING GROWING vines that climb its supporting standard and fan out to form a novel sunshade, this garden umbrella will become the center of attraction in any garden. Its base is cut from a small keg with legs formed from 1-in. flat steel attached to the bottom with screws. Four notched wooden supports hold up the table top which is provided with an opening large enough to permit vines to climb the standard. A steel brace across the opening, together with a wooden block centered at the bottom of the keg, holds the pipe standard upright. Eight holes, drilled near the top of the standard, accommodate the wire ribs of the umbrella which are soldered to the pipe, and to each other, for additional stability. Several vines that are suitable for use with the umbrella include: Morning-glory, honeysuckle, Virginia creeper, Wistaria and (for tropical climates) Bougainvillaea.

LAZY SUSAN'S BIG SISTER

WHEN it comes to serving dinner or clearing the table after a meal, a Lazy Susan set in the wall between the kitchen and dining room is a real step-saver. The unit rotates on a ½-in. rod held to the top of the opening with a bearing bracket and to the bottom with a plate set flush with the framing. A ball thrust bearing carries the weight for smooth operation.

Disks of ½-in. plywood form two shelves which are braced with plywood uprights. Note how the uprights and plywood partition are fastened around the rod, and also how the partition is pinned to the rod. A cleat under the upper shelf adds rigidity to the partition. The unit is locked by a simple latch which engages notches in the bottom of the lower shelf.

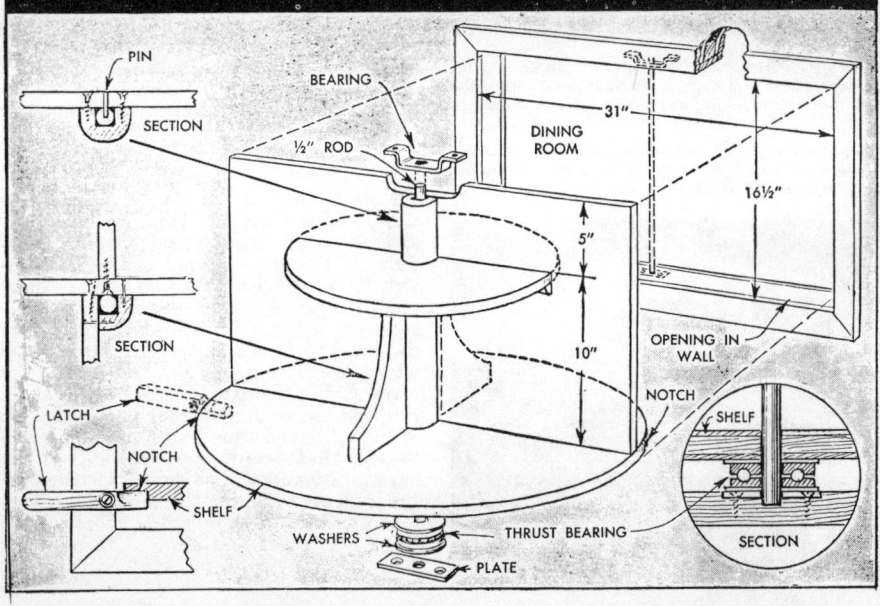

LAZY SUSAN TABLE AND BENCHES

This view shows the table proper completed and the wagon wheel in place ready to be covered with a disk of ½-in. outdoor plywood which later is varnished

The strips to which the table boards are screwed should be a loose fit in the notches so that the entire top can be lifted off easily for winter storage

ADD TO THE enjoyment and convenience of your back-yard picnics by building this rustic Lazy Susan lawn table. The feature of the table is a large rotating tray made from an old wagon wheel which is covered with a plywood disk. Legs and cross braces of both the table and matching benches are sections cut from tree limbs. The framework for the upper portion of the table is made from common 2 x 4 and 1 x 2 stock.

Note in the detail at the right how the tree-limb sections and the framework are assembled with bolts and metal angles. Sides of the four legs are notched to provide flat surfaces for the two pairs of diagonal stringers as well as for the apron. The stringers are half-lapped at the center of the table to form a holder for the axle on which the wagon wheel is mounted. The lower end of the axle is cut off so the skein projects the desired height above the top of the table and then is fastened to one of the cross braces with two metal angles. The aprons are notched, as shown, to take four 1 x 2 cleats to which the table top is fastened with wood screws. The notches hold the top in place and allow it to be lifted from the framework for winter storage.

1" SQUARES

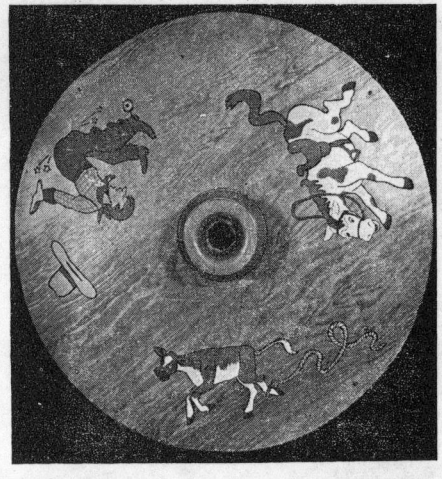

In keeping with the "chuck wagon" theme of the table, the Lazy Susan is decorated with colorful Western figures and the table top with branding-iron motifs

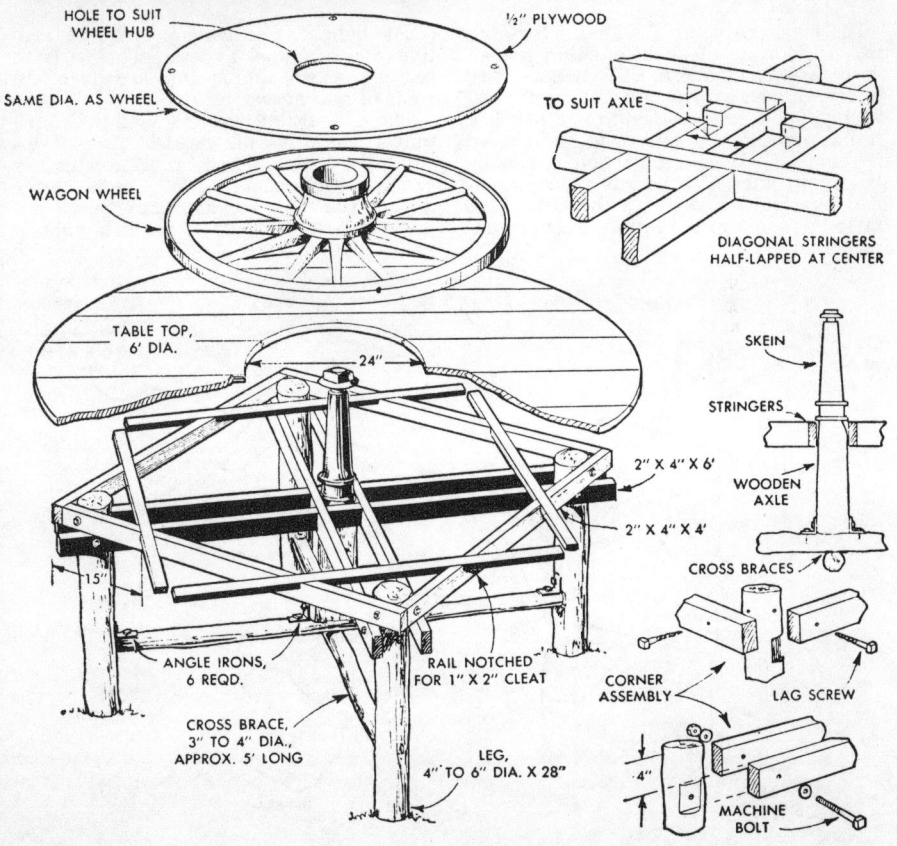

- HOLE TO SUIT WHEEL HUB
- ½" PLYWOOD
- SAME DIA. AS WHEEL
- WAGON WHEEL
- TO SUIT AXLE
- DIAGONAL STRINGERS HALF-LAPPED AT CENTER
- TABLE TOP, 6' DIA.
- 24"
- SKEIN
- STRINGERS
- 2" X 4" X 6'
- WOODEN AXLE
- 2" X 4" X 4'
- CROSS BRACES
- 15"
- ANGLE IRONS, 6 REQD.
- RAIL NOTCHED FOR 1" X 2" CLEAT
- CORNER ASSEMBLY
- LAG SCREW
- CROSS BRACE, 3" TO 4" DIA., APPROX. 5' LONG
- LEG, 4" TO 6" DIA. X 28"
- 4"
- MACHINE BOLT

Above, note how the side braces for the bench legs are nailed in slanting notches cut in the edges of the plank seats and sawed off flush with the top. Left, the 2 x 12 bench tops are nailed to the leg assemblies. The cross members of the leg assemblies are half-lapped and bolted as shown in the detail

The top is built up of wide tongue-and-groove flooring, the boards being temporarily assembled to form a 6-ft. square. Two concentric circles are laid out on the assembly, one 2 ft. in dia. and the other 6 ft. in dia., and then the boards are numbered in sequence, disassembled and cut along the layout lines. The boards are reassembled by screwing them to the cleats, the latter being set in the apron notches during assembly.

The benches are made of tree-limb sections and 2-in. stock as detailed below. Both seat and legs of the benches are given two coats of marine varnish.

The table is finished by giving the entire unit a coat of marine varnish. Then, if desired, the cartoons and brand designs are added with enamel or oil colors. After drying, the table top and Lazy Susan are given two more coats of varnish, rubbed lightly with steel wool and waxed. ★ ★ ★

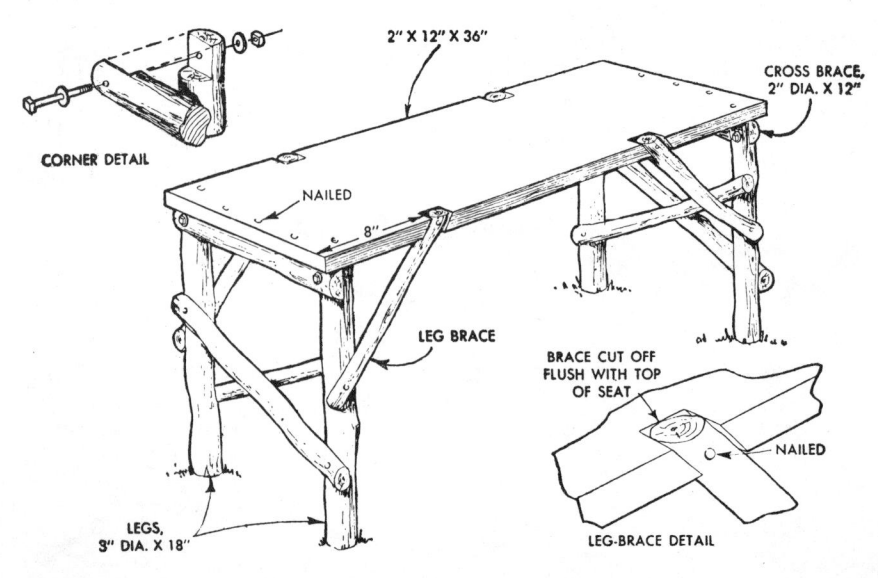

LEATHER BRAIDING

[1]

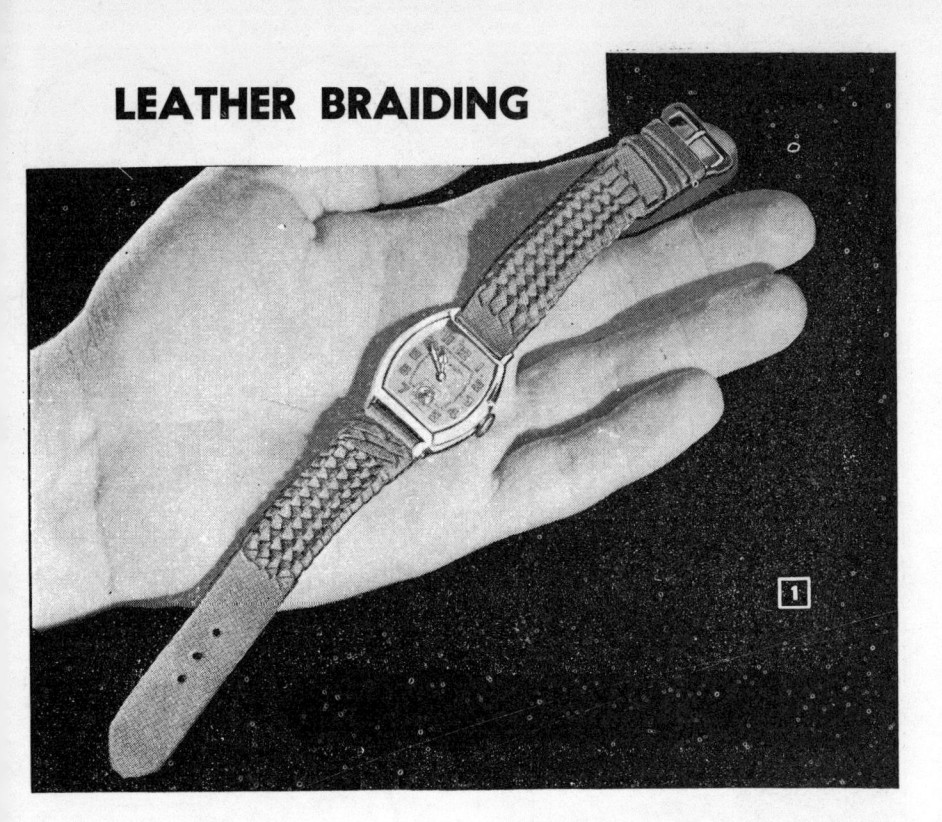

A DAPTED from the ornate braiding adorning the sword belts of early Spanish conquistadores, these two examples of leatherwork, although seemingly complicated, actually are quite simple and fascinating to braid. The wrist-watch band, Fig. 1, and the belt, Fig. 2, which is of the same braid, only made wider, are but two of the many practical accessories to which this attractive braid is suited. In addition to the procedure given for braiding a strap or belt, instructions also explain how to braid a turk's-head knot and how to cover a buckle.

The wrist-watch band requires a piece of calfskin 5/8 in. wide by 8 1/8 in. long, three yards of 1/8-in. beveled goatskin lacing and a buckle with a 5/8-in. opening. For tools, you'll need a knife, a thonging chisel, a leather punch, an awl or fid, a lacing needle and a tube of cellulose cement. The end of the lacing is skived, then inserted in the needle and held with a dab of cement. The calfskin strip is cut into five pieces as indicated in Fig. 3. The piece forming the strap end is rounded at one end, punched for the buckle tongue and perforated with four slits. The slits are made with a thonging chisel held at a 45-deg. angle, 1/16 in. in from the end. The buckle

To braid a belt, you merely weave it twice as wide as the wrist-watch band, as the braiding is the same

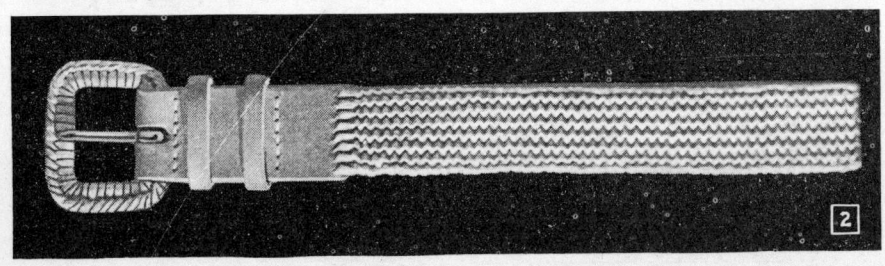

[2]

3 Strap Parts Required

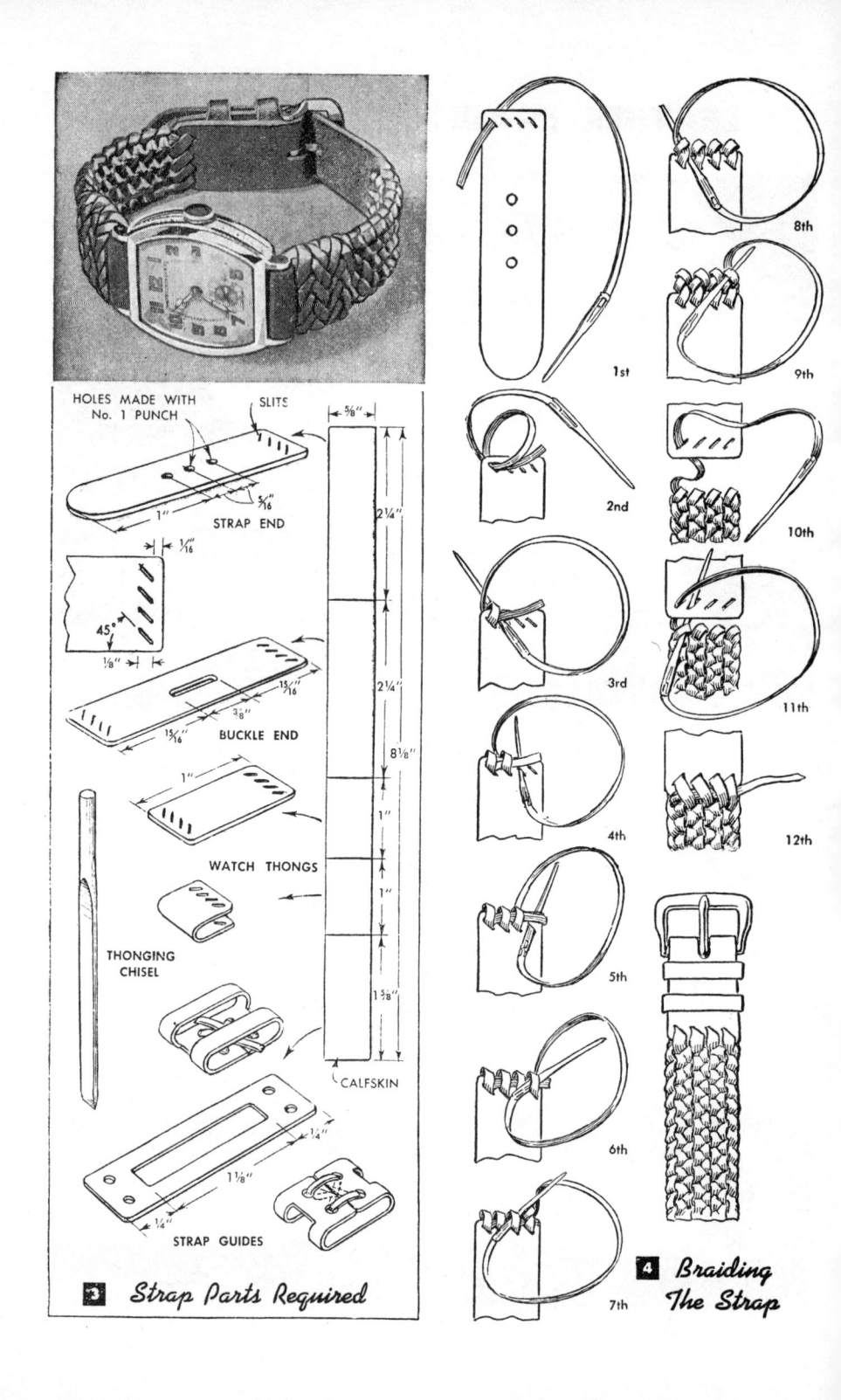

HOLES MADE WITH No. 1 PUNCH
SLITS
STRAP END
5/8"
1"
5/16"
1/16"
45°
1/8"
2 1/4"
2 1/4"
BUCKLE END
15/16"
3/8"
15/16"
8 1/8"
WATCH THONGS
1"
1"
THONGING CHISEL
1 5/8"
CALFSKIN
STRAP GUIDES
1 1/8"
1/4"
1/4"

4 Braiding The Strap

1st, 2nd, 3rd, 4th, 5th, 6th, 7th, 8th, 9th, 10th, 11th, 12th

Braiding Turk's-Head

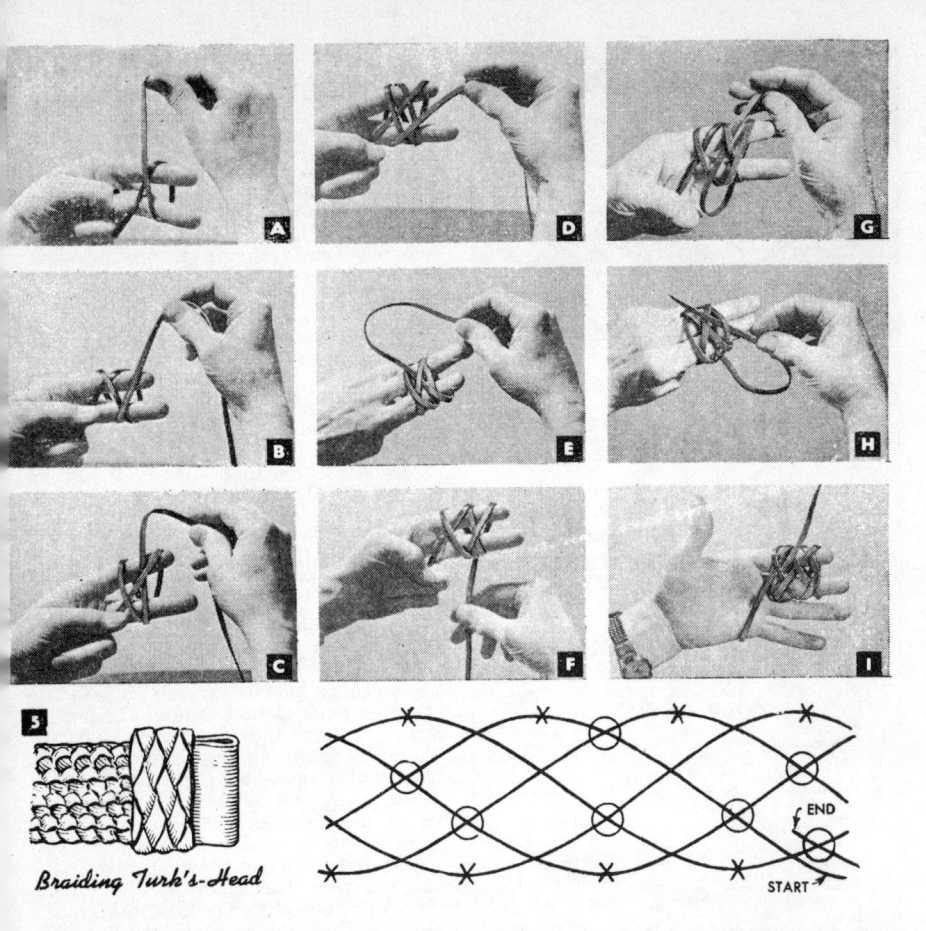

end, which is the same length as the strap end, has a slot cut in it. This is made by punching a hole at each end of the slot and removing the portion between the holes. Aligning slits in each end of the piece are punched at one time while the strip is folded, finish side out. The two pieces used for the thongs are punched in the same manner. The fifth piece is slotted and folded and the ends are laced together, as shown, to form strap guides.

To braid the wrist-watch band, take the strap end and start lacing through the slit at the extreme left as in Fig. 4, step 1. Draw the lacing or thong all the way through except for about an inch. Then come back around and go through the same slit again, step 2, drawing the lacing tight. Continuing, the needle is brought forward and pushed through the next slit, step 3, and drawn tight, the end of the thong being underneath. This is repeated as in steps 4 and 5 and then, being sure that the needle passes to the left of the working strand, bring it to the front and through the right-hand loop, step 6. Pull it through and tighten and then proceed to come around again and pass through the second loop, step 7, pulling it tight. Keep going back and forth as in steps 8 and 9 until you have about twenty rows braided.

Now, take one of the watch-thong pieces and insert the needle through the left-hand slit as in step 10. Then pass the needle through the left-hand loop of the braiding and through the second slit as in step 11. Proceed in the same way, finishing by bringing the lacing under the loops as in step 12. Cut off the excess and secure the end with a dab of cement. The same procedure is followed in joining and braiding the other parts but, before attaching the braiding to the other thong, check to see if the band is the right size. In determining the size, remember that the braid when first woven stretches and, therefore, it should be made short enough so it will be fastened by the first hole in the strap.

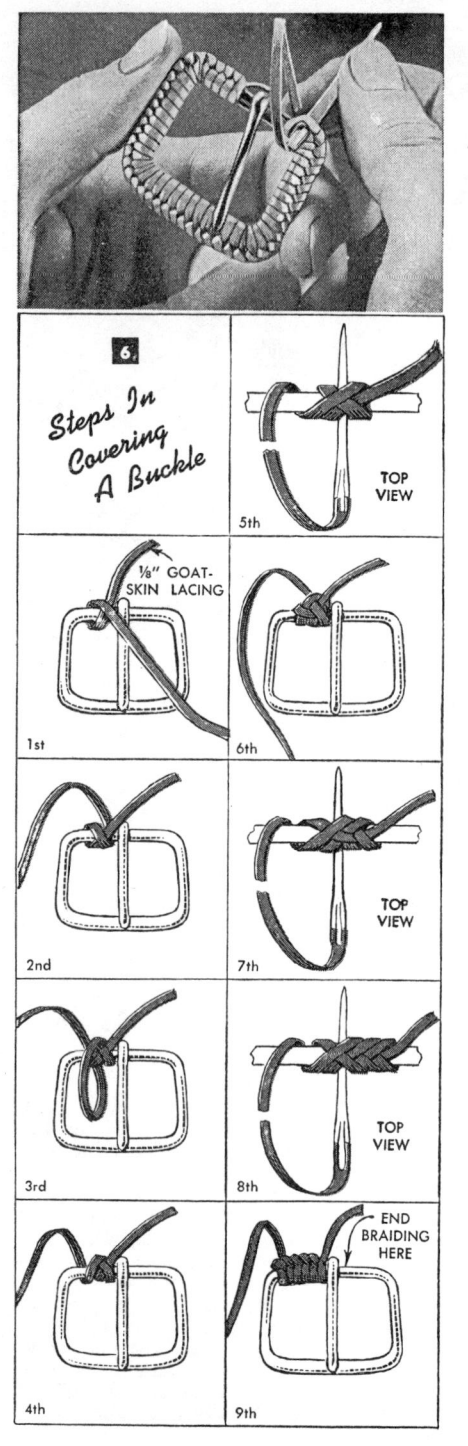

Braiding a turk's-head knot, which is used to add a finish to the watch band, is done as follows: First, make a paper pattern following the diagram in Fig. 5 and form it in a roll so that the ends of the diagram meet and the lines are continuous. Place pins at points marked X. Now, starting where indicated, begin following the lines with the lacing, going around the pins and passing under previous strands at points circled. When finished, remove the work from the pattern and place it over the strap where the braiding joins the watch thongs. Tighten the turk's-head by gradually taking up the slack, going over the knot several times. When tight, place a drop of cellulose cement between the lacing where the ends meet and cut it off. As you become adept at braiding the turk's-head, it can be formed rapidly on the fingers as shown in the series of progressive steps pictured in photos A to I inclusive. To start, wrap the lacing around the first two fingers of the left hand as in photo A. The part held by the left thumb is called the standing end, the other the working end. Pass the working part over the standing part and completely around the fingers as in B. Now the working part is passed beneath the standing part, C, and around the fingers again. Then continue the working part, inclining it to the extreme right, photo D. Note that it passes along the right side of the standing part but, when it reaches the tip of the forefinger, it passes over the standing part and along its left side, passing under the diagonal strand just as the standing part does, photo E. Now, bring the working part to the front once more, photo F. From this point on, the sequence will be over one strand and under one, leaving the lacing fairly loose, G. Working on the back of the hand, photo H, the lacing is passed under one of the scallops and over another. Finally, bring the working part alongside the standing part once more as in photo I.

Covering a belt buckle is shown in Fig. 6. Starting with a yard of 1/8-in. goatskin lacing, and holding the buckle with the heel upward, begin as in step 1. Then pass the lacing around and bring it to the front as in step 2. Go around again as in steps 3 and 4. Now, closely following the top view, step 5, open the lacing with an awl and insert the needle between the first and second strands, pulling the lacing tight, step 6. Repeat the same procedure, inserting the needle this time between the second and third strands, step 7. Continuing as in step 8, pass the lacing around the buckle, pulling each loop tight to produce the braid shown in step 9. The finished buckle is shown in the photo.

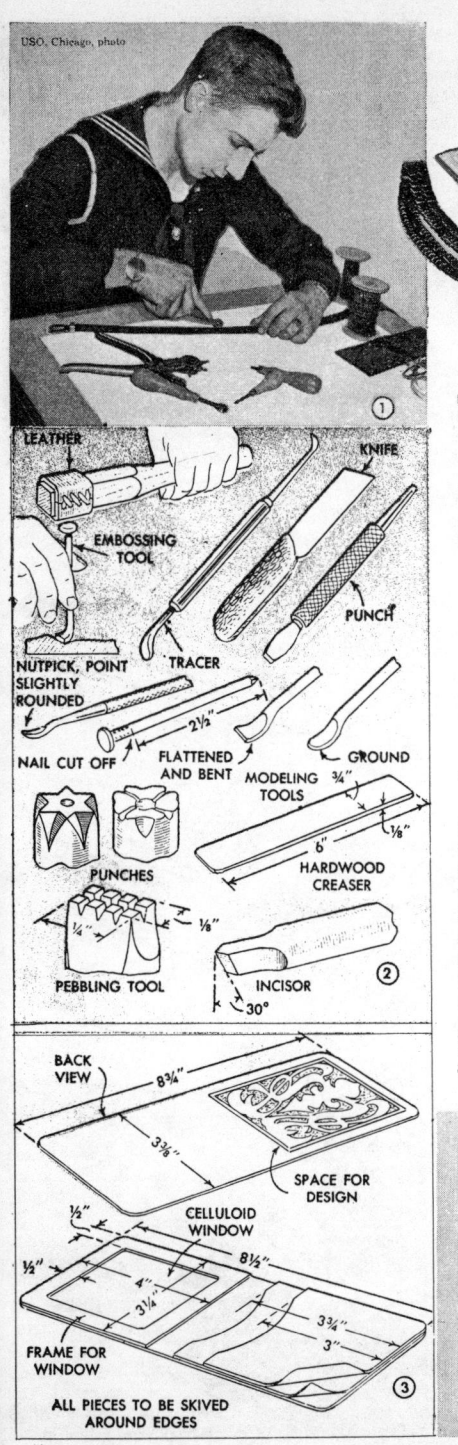

TOOL IN LEATHER
for Fun and Profit

Working with leather is fascinating and profitable. Leather articles attractively embossed and neatly laced make welcome gifts and find a ready sale. Most of the simple tools needed can be made at home

WORKING in leather is so easy that even a beginner can produce pleasing, salable articles. Few tools are needed, and most of these can be made from nails, bolts, nutpicks and other common objects, as shown in Fig. 2. A jackknife will do for cutting, but a leather-cutting knife is better adapted for skiving, or beveling the edges of leather, Fig. 13. Though it is not necessary to have a punch like that shown in Fig. 1, this tool is handier for making lacing holes than a nail, which of course can be used. Or a shoemakers' awl can be used as in Fig. 5, the holes being punched in the leather with the aid of a rawhide or wooden mallet.

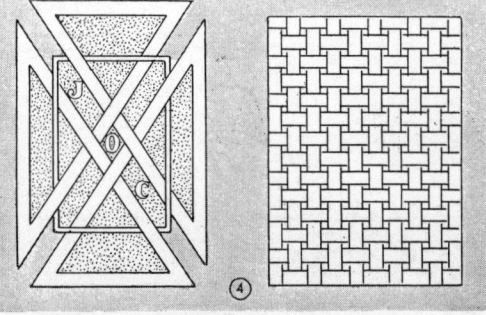

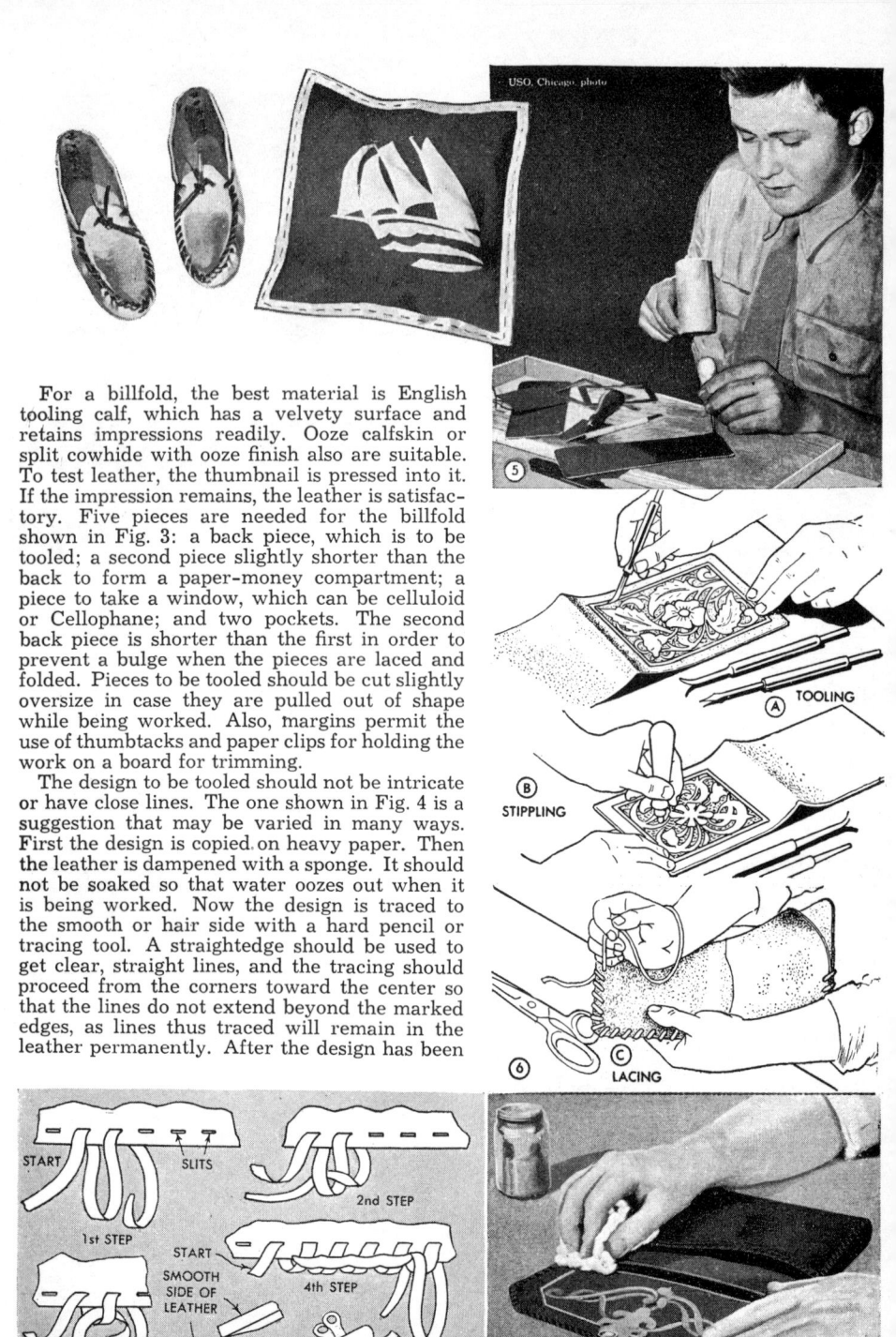

For a billfold, the best material is English tooling calf, which has a velvety surface and retains impressions readily. Ooze calfskin or split cowhide with ooze finish also are suitable. To test leather, the thumbnail is pressed into it. If the impression remains, the leather is satisfactory. Five pieces are needed for the billfold shown in Fig. 3: a back piece, which is to be tooled; a second piece slightly shorter than the back to form a paper-money compartment; a piece to take a window, which can be celluloid or Cellophane; and two pockets. The second back piece is shorter than the first in order to prevent a bulge when the pieces are laced and folded. Pieces to be tooled should be cut slightly oversize in case they are pulled out of shape while being worked. Also, margins permit the use of thumbtacks and paper clips for holding the work on a board for trimming.

The design to be tooled should not be intricate or have close lines. The one shown in Fig. 4 is a suggestion that may be varied in many ways. First the design is copied on heavy paper. Then the leather is dampened with a sponge. It should not be soaked so that water oozes out when it is being worked. Now the design is traced to the smooth or hair side with a hard pencil or tracing tool. A straightedge should be used to get clear, straight lines, and the tracing should proceed from the corners toward the center so that the lines do not extend beyond the marked edges, as lines thus traced will remain in the leather permanently. After the design has been

along the edge of a metal straightedge. If the laces are too short, they can be glued and spliced as in the lower details of Fig. 8, which also shows one method of lacing. After lacing around the article, the ends are pulled through two slits, glue is applied and the ends are tucked under the lacing. Then the billfold is ready to be waxed to protect it from scratches and wear. The work is laid on a hard surface where pressure can be applied, and wax is spread over it with a soft cloth, Fig. 7, after which the billfold is polished with a dry cloth to bring out the luster. Or wax can be applied with a soft brush such as a discarded toothbrush.

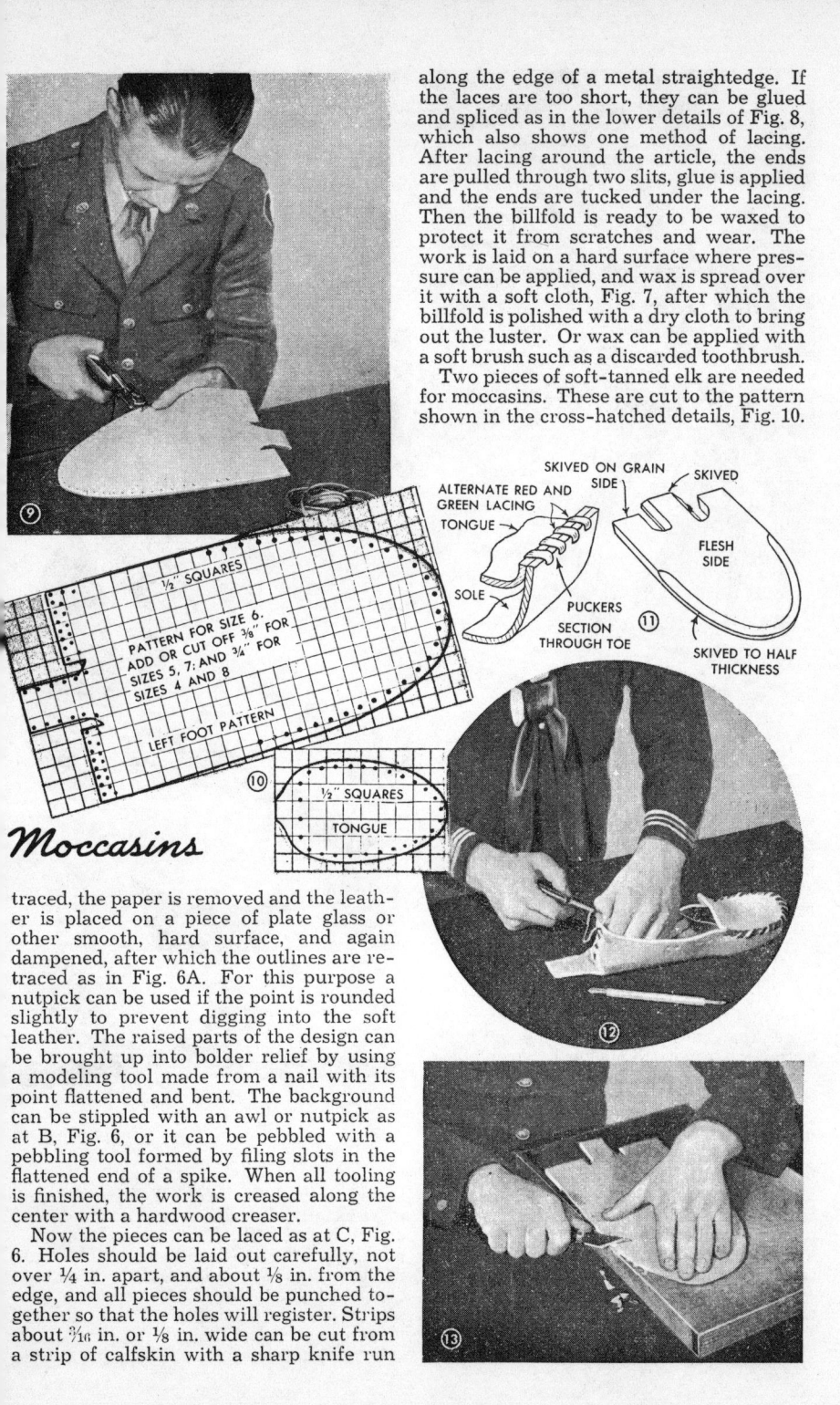

Two pieces of soft-tanned elk are needed for moccasins. These are cut to the pattern shown in the cross-hatched details, Fig. 10.

Moccasins

traced, the paper is removed and the leather is placed on a piece of plate glass or other smooth, hard surface, and again dampened, after which the outlines are retraced as in Fig. 6A. For this purpose a nutpick can be used if the point is rounded slightly to prevent digging into the soft leather. The raised parts of the design can be brought up into bolder relief by using a modeling tool made from a nail with its point flattened and bent. The background can be stippled with an awl or nutpick as at B, Fig. 6, or it can be pebbled with a pebbling tool formed by filing slots in the flattened end of a spike. When all tooling is finished, the work is creased along the center with a hardwood creaser.

Now the pieces can be laced as at C, Fig. 6. Holes should be laid out carefully, not over 1/4 in. apart, and about 1/8 in. from the edge, and all pieces should be punched together so that the holes will register. Strips about 3/16 in. or 1/8 in. wide can be cut from a strip of calfskin with a sharp knife run

Belts

Note that the holes in the tongue are spaced closer than those around the toe and heel, so that the latter will pucker between holes when laced. As elk hide is heavy, the sole should be skived as indicated in Fig. 11 to permit puckering. Skiving is done by laying the piece along the edge of a board and slicing the edge of the leather at a slant or bevel, Fig. 13. Holes can be punched for laces easily with a leather punch, as indicated in Fig. 9, but the heel should not be punched until the moccasin has been fitted to the foot. Lacing is done with pointed laces, as they should fit the holes snugly. Two colors of 1/8-in. lacing may be used, each lace being about 2½ times as long as the distance around the toe. Now the moccasin is put on the foot and the flaps are folded around the heel. If these do not meet exactly, the back should be cut down and reshaped to fit. Then lacing holes are punched in the flaps and laces are inserted in the outer row of holes and the flaps are tied as shown in Fig. 12, after which the stay is brought up and laced through the inner row of holes.

Overlaying designs on a background of

spaced ⅜ in. from the edge and ½ in. apart.

Braided leather belts are popular and easy to make. The ten-strand type, Fig. 16, is most in demand, though belts can be made with as many strands as desired. A piece of leather ⅓ longer than finished size, and 1⅝ in. wide, is needed. The leather is cut into ten strands with a sharp knife. The cuts should start 7 in. from one end, and continue to the other end. Braiding is done as shown in Fig. 18, the uncut end being held by a nail. Strand 1 is laid over strand 2, under 3, over 4, and so on, the other strands being braided in the same way, as indicated in the left-hand detail, Fig. 16. All strands are pulled tight, then

Embossing

flattened at intervals by being hammered on a hard surface while covered with a board, as in Fig. 19. Buckle and loop then are added, as shown in the center detail of Fig. 16. The strands at the ends of the belt are skived and glued between folds of the tongue after the buckle and loop have been inserted. When the glue is dry, holes are punched for the lacing, which goes through the strands and tongue.

leather is another interesting example of leathercrafting. Fig. 17 shows a pillow being decorated in this way. A pattern, Fig. 15, is laid out on heavy paper, then traced to the leather which is to form the overlay. Suede lambskin is best for both overlay and background. The design then is cut out with a sharp knife and placed, smooth side up, on the rough side of the background piece, where it is mounted with ordinary liquid glue that has been thinned slightly by being placed in a pan of warm water. The glue is applied to the rough side of the design, then scraped off with a knife to remove any surplus. If any parts fail to adhere when pressed against the background, additional glue can be applied with a pointed brush, Fig. 17. After the gluing is finished, the design is covered with tissue paper and weighted with a board loaded with a heavy object. A 1½-in. binding strip then is folded over the edges of the cover pieces (front and back of the pillow) and laced together with ⅛-in. lacing, as in Fig. 14, a paper clip being used to hold the pieces together. Holes are

Embossing a belt requires a die in which a design is cut, as in Fig. 20. Of course, a design can be tooled into the leather as previously described, but a die-formed design is more distinct. For the die, a 5-in. length of heavy cowhide about 1¼ in. wide is needed. A design such as the one shown in Fig. 22 is cut into it with an incisor, illustrated in Fig. 2. The tool is held firmly, and almost vertically, and cuts should be ¹⁄₁₆ in. deep. Next, the belt strap is moistened and laid on a hardwood board or slab of marble, the die is laid over it and pounded with a mallet as indicated in Fig. 21. The die then is moved along the belt and hammered again, until the entire belt has been embossed. After this, a buckle and strap are sewed on, and, if desired, the background can be darkened with leather stain to bring out the pattern.

LENS MOUNTINGS

Upper right, a quick, easy way of mounting a lens for experimental work. Two cardboard tubes are selected, one with an inside diameter equal to that of the lens, the other of a size which will telescope firmly inside the first tube. Rings A and B are cut from the smaller and forced into the larger tube with the lens between. Right, a simple semipermanent mounting in a metal tube, making use of a split-wire retainer

Lenses subjected to heat from a nearby light source should be spring-mounted in such a way that the glass is free to expand without danger of cracking. One of the simplest ways to do this is to support the lens in metal holders, or brackets, attached to the end of the tube with screws. In making this mounting, care should be taken to shape the holders in duplicate so that the lens will be held firmly

This mounting is similar to the cardboard mounting described above except that the metal tube is machined with a shoulder to provide a back support for the lens. The friction ring is machined to a press fit, and its inner edge is carefully beveled to fit the curve of the lens. The lens should fit in the shouldered portion of the tube with a slight amount of play to allow for possible expansion of the glass

This is a variation of the friction-ring mounting employing a flat, split ring as a retainer. The ring is machined to an outside diameter slightly greater than the inner diameter of the shouldered section of the tube. The ring is then split and a short section removed so that when compressed it is an easy fit in the shouldered end of the lens tube. Note also that the inner edge of the ring is beveled to fit lens

Another variation of the split-wire retainer especially suitable for holding lenses of large diameter. Three equally spaced bosses are formed in the wire ring with long-nosed pliers. The ring snaps into a semicircular groove machined in the shouldered portion of the tube. Special care must be taken when machining to locate this groove the correct distance from the shoulder so that, with the lens in position in the tube, the ring will bear against it lightly when it is snapped into place in the groove

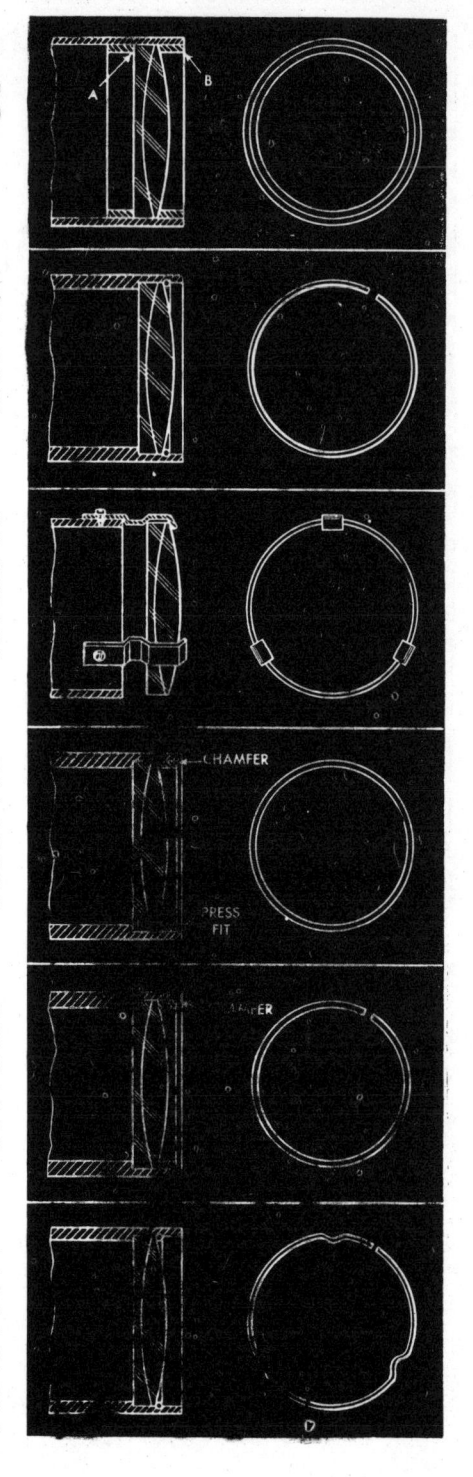

Letter Press of Pipe Fittings for the Home Shop

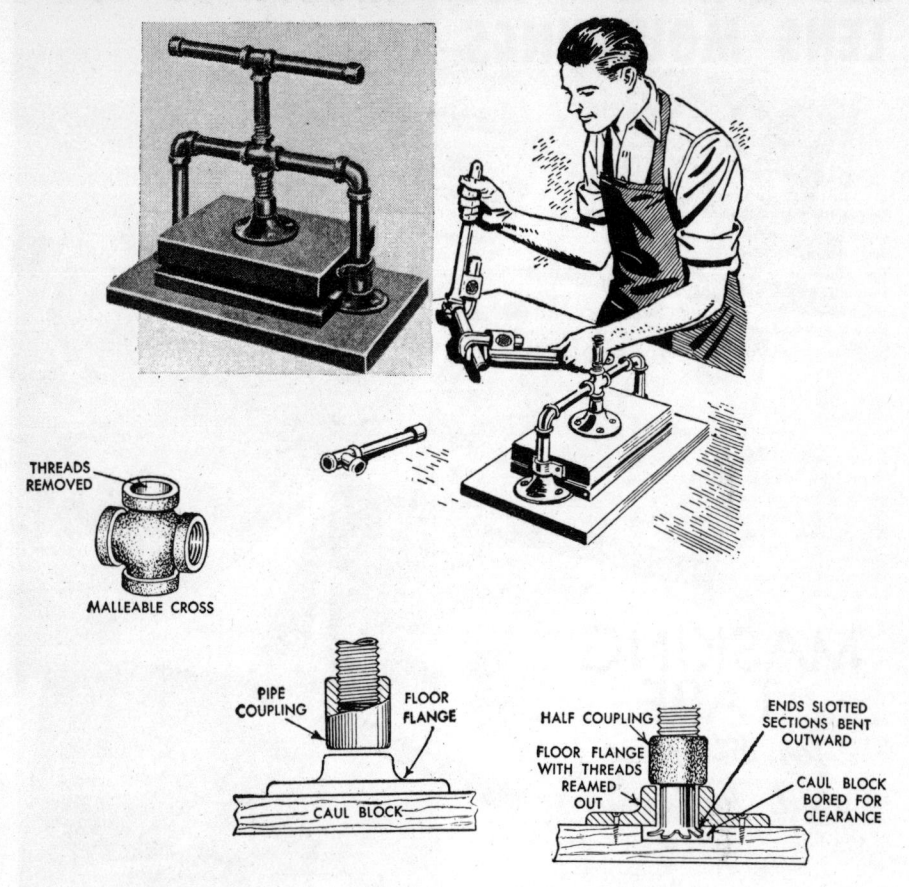

There's nothing quite so handy in the home workshop as a letter press when it comes to veneer work and other small gluing jobs. Commercial presses are expensive, but you can have this sturdy press in your shop just for the cost of a few pipe fittings. The fittings needed are three floor flanges; two elbows; six nipples, or seven if you use the method of mounting the spindle on the caul block shown in the lower left-hand detail; two pipe caps; two tees, or one tee and a cross; and a length of threaded pipe for a spindle. To make the press, fashion the base, caul block and base block of thick hardwood. Screw the flanges and base block to the base and assemble the frame. If a cross is used to take the spindle, the threads must be removed from the top opening. These may be bored out on a lathe, or you can take the cross to a local plumbing shop and have the threads reamed out. If a tee is used instead of a cross, drill a hole through the top of the tee to permit insertion of the spindle. The caul block is guided between the vertical nipples of the frame by means of pipe straps fastened to the ends of the block. A pipe coupling is turned on the lower end of the spindle to press down on a floor flange fastened to the top of the caul block. If this method is used, the caul block will have to be raised by hand. However, if you wish the spindle to raise the caul block, have the threads of the flange reamed out and turn a nipple into the bottom of the spindle coupling. The lower end of the nipple is slotted and the sections bent outward to hold it to the flange. The caul block must be recessed to clear the pipe-nipple sections. The handle is made from two nipples, two pipe caps and a tee.

LETTERING WITH MASKING TAPE

MASKING TAPE for Striping and Letttering

Direct taping

(2) SPRAY STRIPE AREA
(3) APPLY MASKING TAPE
(4) SPRAY GROUND COAT
(5) REMOVE TAPE

MASKING tape is a great help to the finisher with an occasional job of lettering or striping. The most popular tape is the crepe-back paper variety, which has enough flexibility for curves. The adhesive usually is rubber-base cement, which holds well, yet strips clean. However, plain paper or cloth tapes can be used.

Two methods of taping are commonly employed. In one, the tape is applied to the letter or stripe, the background being sprayed on while the stripe or letter is thus protected. In the second method, the background is sprayed first, then the letter is outlined and sprayed. Direct taping as applied

Outline taping

By this method, the work is covered with paper and tape so that only the stripe or lettering is exposed for spraying

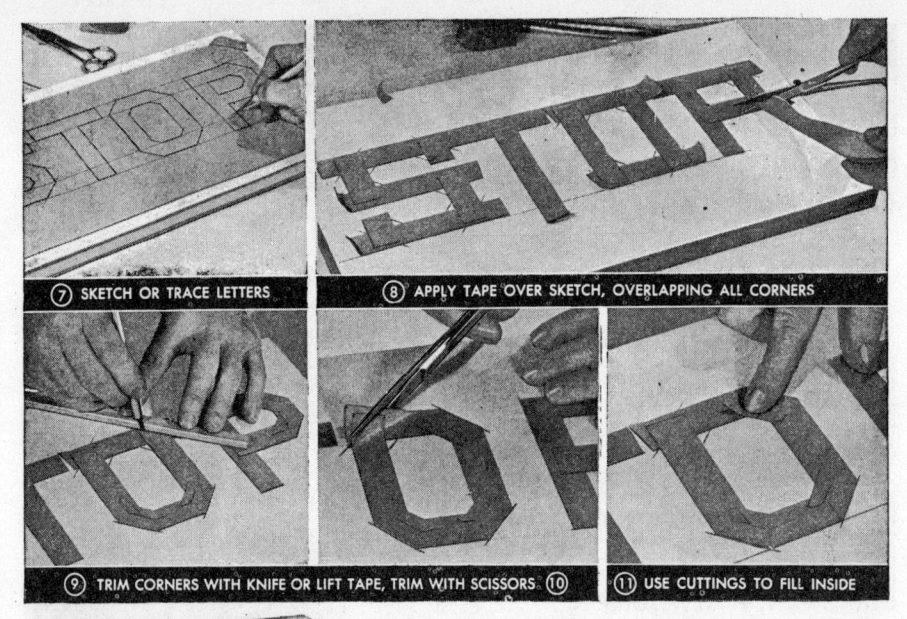

(7) SKETCH OR TRACE LETTERS (8) APPLY TAPE OVER SKETCH, OVERLAPPING ALL CORNERS

(9) TRIM CORNERS WITH KNIFE OR LIFT TAPE, TRIM WITH SCISSORS (10) (11) USE CUTTINGS TO FILL INSIDE

(12) *Direct lettering*

to a simple stripe is shown in Figs. 2 to 5 inclusive. The area comprising the stripe is sprayed first, using a round spray and allowing the edges to fan out. After this partial coat has dried, the tape is applied and pressed or rolled down firmly. The whole panel then is sprayed, after which stripping of the tape reveals the stripe. The direct method of spraying also is illustrated by Fig. 1. In this instance, the black stripe was first roughed in, followed by the white upper portion sprayed right over the tape, and finished by spraying the lower red portion while protecting the upper white with a cardboard mask.

This method used for letters 3 to 10 in. high. Tape gives uniform stems

(13) **TAPE WIDTH TO USE FOR DIRECT LETTERING**

Height of Letter	Width of Tape *		
	Thin	Medium	Heavy
3	.515 (½)	.60 (½ or ¾)	.75 (¾)
3½	.59 (½)	.70 (¾)	.87 (⅞)
4	.69 (½ or ¾)	.80 (¾ or ⅞)	1.00 (1)
4½	.76 (¾)	.90 (⅞)	1.12 (1)
5	.86 (⅞)	1.00 (1)	1.25 (1¼)
6	1.03 (1)	1.20 (1¼)	1.50 (1½)
7	1.20 (1¼)	1.40 (1½)	1.75 (1¾)
8	1.37 (1¼)	1.60 (1½)	2.00 (2)
9	1.54 (1½)	1.80 (1¾)	2.25 (2)
10	1.70 (1¾)	2.00 (2)	2.50 (2½)

* Based on stems ⅙, ⅕ and ¼ the height. First figure is exact. Second figure is nearest standard width. All dimensions in inches

ABCDEFGH
5 4½ 4½ 4½ 4 3¾ 4¾ 4½

JKLMNOP
4½ 4¾ 3¾ 6 4¾ 4½ 4

QRSTUVW
5 4½ 4¾ 4½ 4½ 5 6½

XYZ1234
5 5 4½ 4½ 4½ 5

56789 *Block letters* (14)
4½ 4½ 4¼ 4½ 4½

WIDTHS MARKED IN UNITS 1/6 THE HEIGHT

Outline lettering

This is especially useful for large letters. Tape is applied to the outlines of a paper pattern to make a combined mask and stencil

The second method of masking as used for a stripe is pictured in Fig. 6. The whole area of the work must be protected with tape and paper aprons, leaving the stripe area clear for spraying.

Direct masking of lettering is shown in photos 7 to 12. The inexperienced worker first should draw the lettering on paper, transferring the lines to the work by the familiar method of blackening the back of the paper and then tracing, Fig. 7. However, with very little experience it is quite easy to letter simple block letters, such as shown in Fig. 14, directly on the work. The table, Fig. 13, gives the width of tape to use for thin, medium and heavy stems. The widths of the letters themselves are determined by using a measuring unit one-sixth the height of the lettering. Note when taping, Fig. 8, that all corners are overlapped. This not only stiffens the letter for possible re-use, but also provides a guide when trimming the corners. Corners usually are cut with a stencil knife, as in Fig. 9, but it is also practical to mark corners with a pencil and then lift the tape and cut with a pair of scissors, Fig. 10. In either case, the piece cut off fills the inside of the corner as shown in Fig. 11.

Letters over 10 in. high are done best with the outline method. Fig. 15 is an example. In this setup, the background color is finish-sprayed. A paper pattern of the lettering then is made, after which the lettering is transferred to the work as in Fig. 16. The paper pattern in this case serves a double purpose, since it also acts as a stencil. How this is done can be seen in Figs. 17 and 18, the letters being cut outside the line, and the tape applied to the exact outline of the letter and also overlapping the paper pattern, Fig. 18. Spraying is done as in Fig. 19. This combined mask and stencil can be stripped and used several times if needed.

Letters less than 3 in. high are best worked with a stencil cut directly from a wide tape. Wide tapes also are useful for more ornate lettering or designs. This kind of tape can be cut directly over the work.

(15)

(16) PAPER PATTERN IS FITTED IN PLACE AND LETTERS ARE TRACED

(17) LETTERS ARE CUT OUT

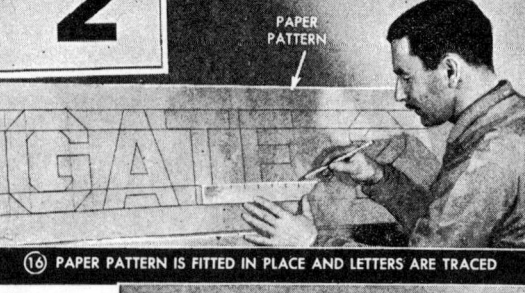

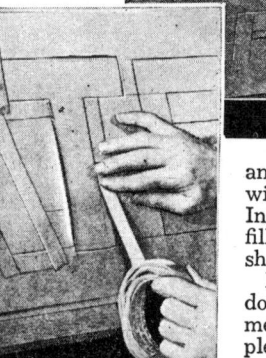

(18) TAPE APPLIED

(19) SPRAYING IS DONE WITH ROUND SPRAY

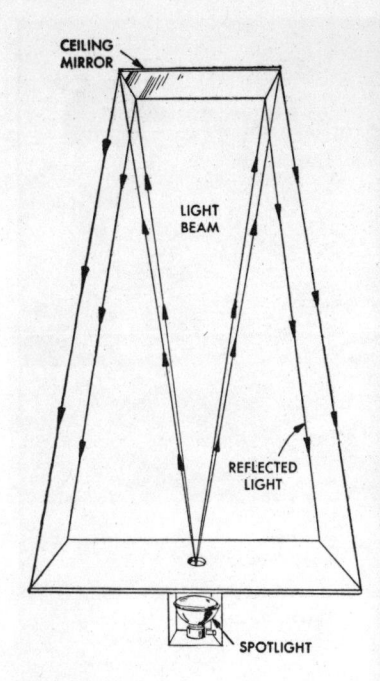

Diagram labels: CEILING MIRROR, LIGHT BEAM, REFLECTED LIGHT, SPOTLIGHT

Photos courtesy General Electric Co.

LIGHTING

ADEQUATE HOME LIGHTING falls into two groups which must be combined for best results. These are over-all room illumination and lighting for specific tasks.

The indirect lighting arrangement shown on this page is one example of lighting for a specific purpose. A spotlight under the table reflects from an overhead mirror and adds sparkle to the formal dinner without the tiring glare of direct lighting or the shadowy uncertainty of candles. The diagram above shows the setup. The spotlight is trained on a mirror attached to the ceiling. The light beam passes through a 2 5/8- in. hole bored in a table leaf. The size of the mirror, which is centered over the table, should be one half that of the table top. Use a projector-type spot lamp of 100 to 150 watts and mount it below the table top in a porcelain socket attached to a bracket. Enclose the light in a metal housing provided with louvres for ventilation and flanges for fastening. The table leaf is protected with asbestos and a glass centerpiece can be used to conceal the lamp.

Other recommended lighting recipes for various specific purposes are given on the following pages. Space does not permit

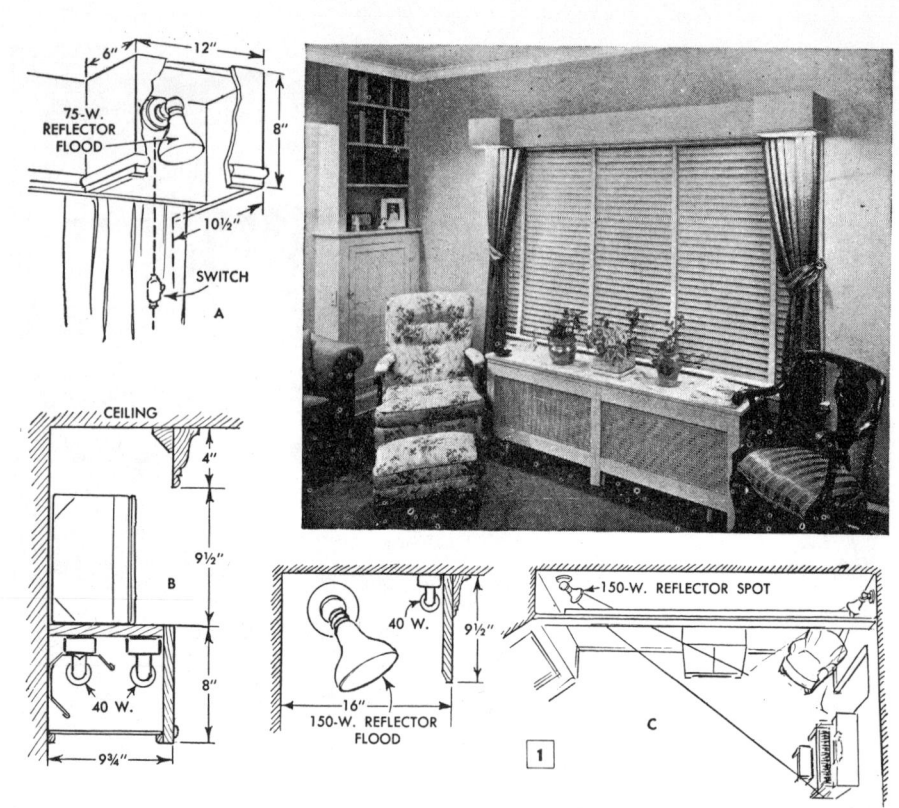

Indirect valance lighting, more efficient portables and a concealed mantel light brighten cozy living room

Kitchen

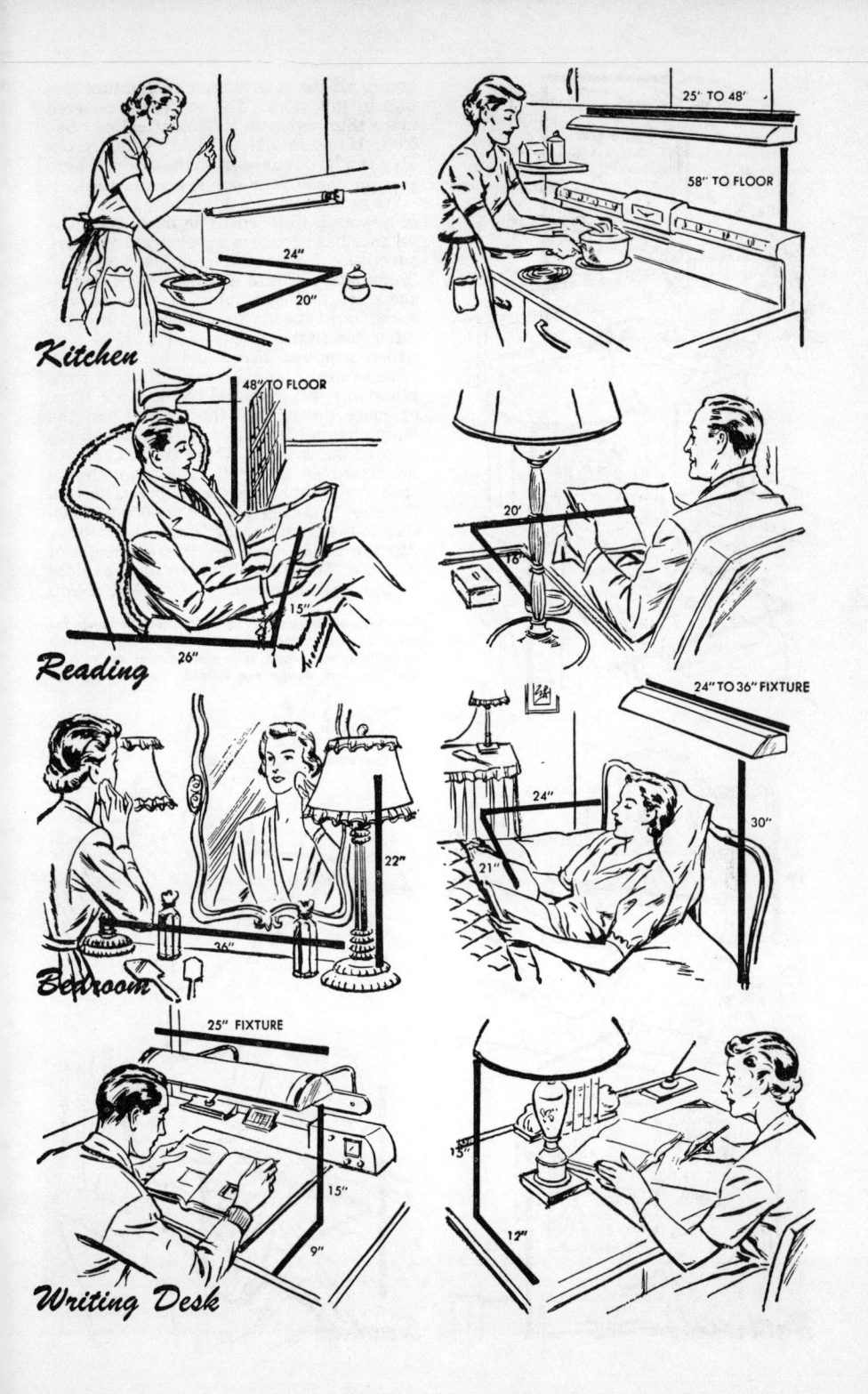

Reading

Bedroom

Writing Desk

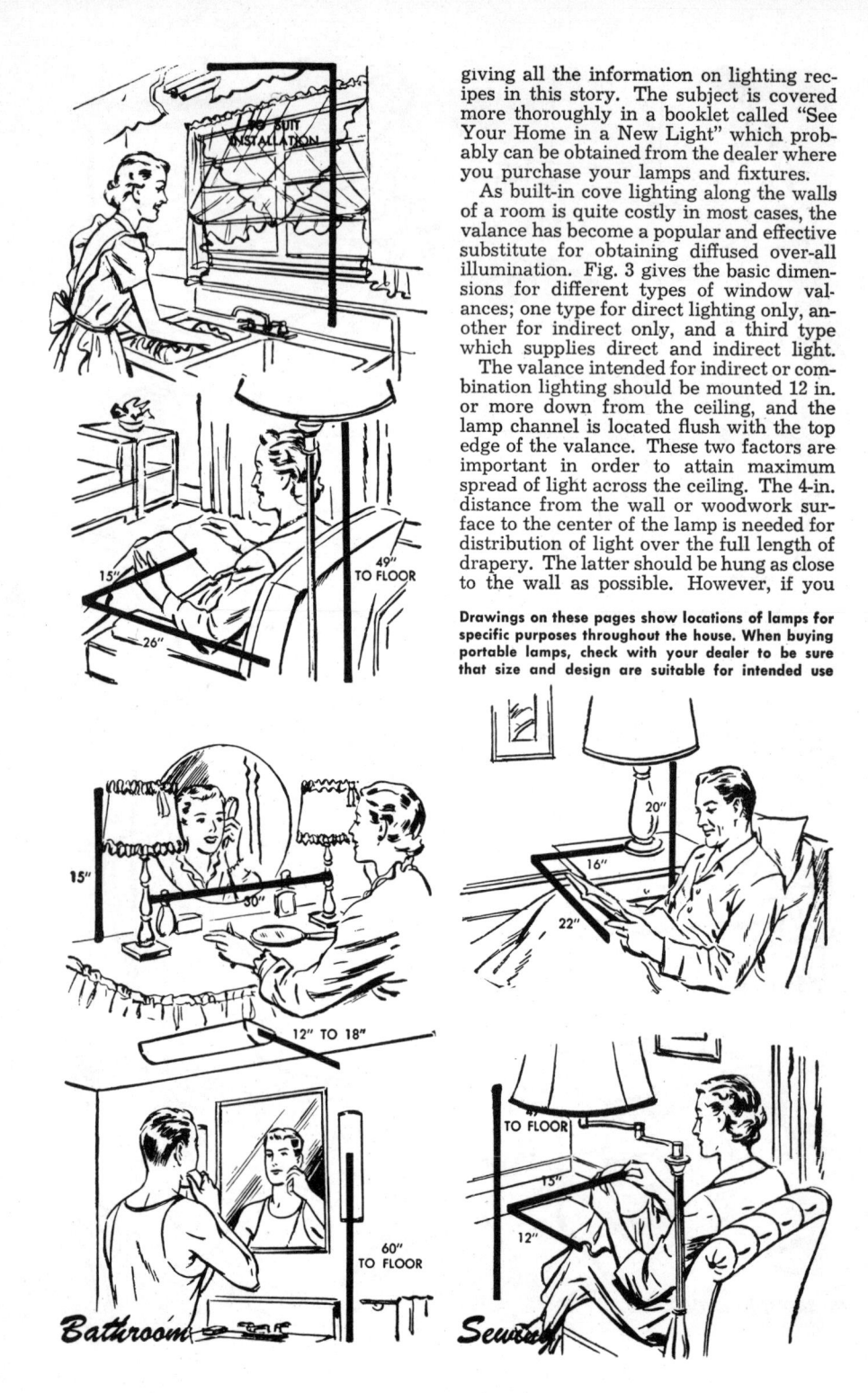

giving all the information on lighting recipes in this story. The subject is covered more thoroughly in a booklet called "See Your Home in a New Light" which probably can be obtained from the dealer where you purchase your lamps and fixtures.

As built-in cove lighting along the walls of a room is quite costly in most cases, the valance has become a popular and effective substitute for obtaining diffused over-all illumination. Fig. 3 gives the basic dimensions for different types of window valances; one type for direct lighting only, another for indirect only, and a third type which supplies direct and indirect light.

The valance intended for indirect or combination lighting should be mounted 12 in. or more down from the ceiling, and the lamp channel is located flush with the top edge of the valance. These two factors are important in order to attain maximum spread of light across the ceiling. The 4-in. distance from the wall or woodwork surface to the center of the lamp is needed for distribution of light over the full length of drapery. The latter should be hung as close to the wall as possible. However, if you

Drawings on these pages show locations of lamps for specific purposes throughout the house. When buying portable lamps, check with your dealer to be sure that size and design are suitable for intended use

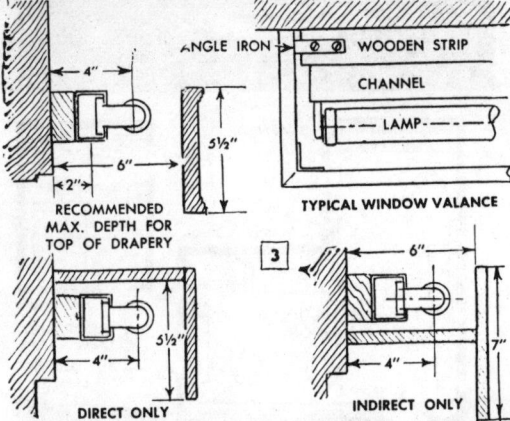

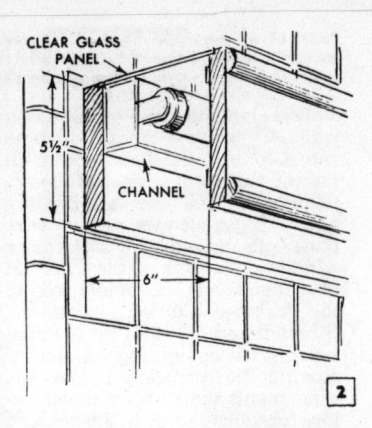

find that with your present traverse rods, the 4-in. dimension does not bring the fluorescent tube beyond the drapery, increase the thickness of the wooden backing strip to bring the channel farther out from the wall. Then increase the distance from the valance to the wall proportionately. Paint all inside surfaces of the valance white to reflect the maximum amount of light.

The length and location of valances as well as the type selected is determined by the arrangement of the room and the effect you have in mind. In one instance a valance may be used to provide lighting along the entire length of a single wall to balance the light from portable lamps. In another, full-length valances may be in-

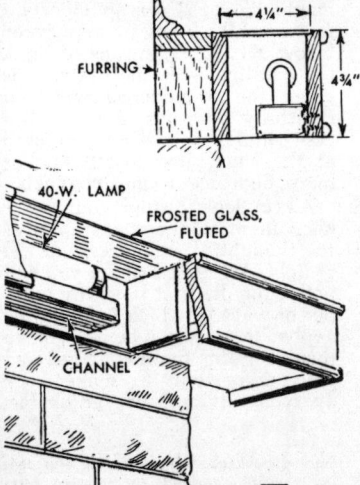

Panel of white-painted fiberboard above student's desk provides reflecting surface to attain maximum light from pin-up lamps

stalled along opposite or adjoining walls, or window-length valances can be used over the windows all around the room so each wall contributes to the over-all lighting effect. Bays, alcoves, picture windows, offsets and wall sections or windows flanking a fireplace all lend themselves extremely well to the valance-lighting treatment. You will find, too, that the installation of valance lighting will permit moving the furniture to new arrangements which were never before practical due to the lack of sufficient light. Fig. 5 offers an example of how furniture and portable lamps can be arranged when the general lighting level has been increased with valance lighting.

A few of many possible lighting installations are pictured and detailed on these pages. Fig. 1 details the drapery brackets, soffit lighting and the wall-length valance with its concealed spot and flood lights which combine to light-condition the redecorated living room shown in the photos. The wall bracket detailed in Fig. 2 not only provides plenty of light for working at the kitchen range, but doubles as a handy and attractive spice shelf. Utilizing a clear-glass panel for the shelf permits a certain amount of indirect illumination to produce an interesting treatment. The indirect fireplace-mantel light, Fig. 4, is an important feature of the living room pictured below Fig. 1. By means of light, the painting over the fireplace is given the prominence which it rightfully deserves. For a bookcase similar to the one pictured at the right, just install one or more fluorescent-lamp channels.

As you look through your home, keep in mind these examples of the thoughtful and successful use of light. Without a doubt you will notice for the first time any number of ways in which light can be applied in your own home to make the decorating more effective, and more important, to make your everyday living more enjoyable.

★ ★ ★

Lighted bookcase with glass top plus ceiling spotlight trained on painting form bright, dramatic corner in small apartment

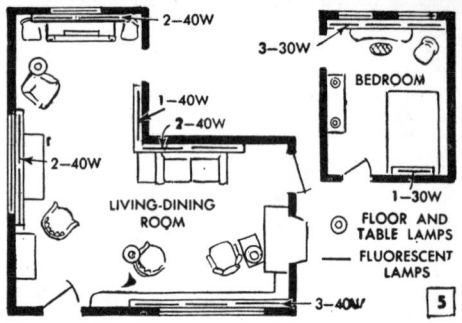

Oxen-Yoke Ceiling Fixture for Dining Room Imparts Pioneer Atmosphere to Ranch Home

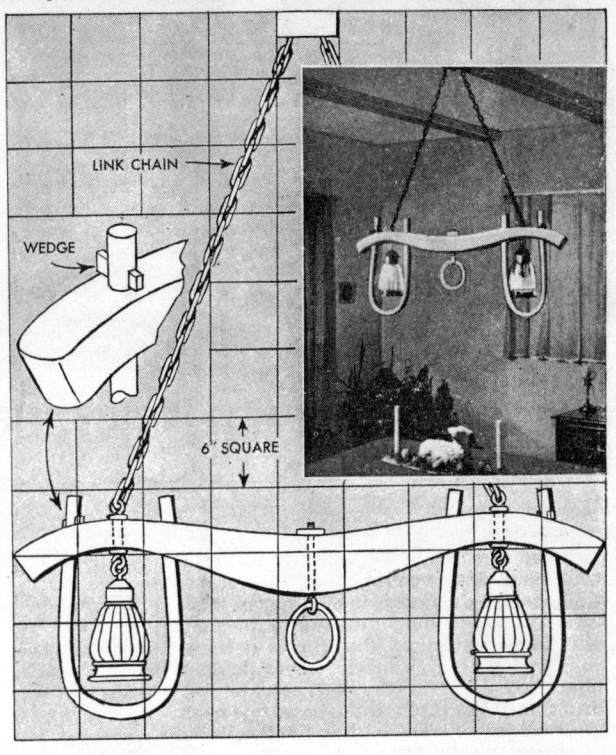

Suspended from the ceiling of a studio living room or informal dining room, this unusual fixture conveys a feeling of pioneer days to your cabin or ranch-type home. It's especially appropriate if western-style architecture and furnishings prevail in the room. The yoke can be bandsawed from a 2 or 3-in.-thick piece of wood, preferably oak, to the size indicated in the squared pattern. The bows may be bent by steaming them and clamping in shape, and they are fastened to the yoke by means of wedges, as shown. Hickory, if available, is excellent wood for the bows. The fixture is held to the ceiling with two lengths of chain which are attached to eyes in the yoke. Although these parts on the original were taken from an old lighting fixture, threaded brass tubing and eyes can be purchased for this purpose. The yoke is drilled to take the tubing and the eyes for the chain and light are turned on each end of this. Then, a lantern-type light, which is attached to the lower eye, is wired through the tubing and along the chain to the ceiling outlet. Finally, a large iron ring is bolted through the center of the yoke.

Recessed Ceiling Light Made From Auto Headlight

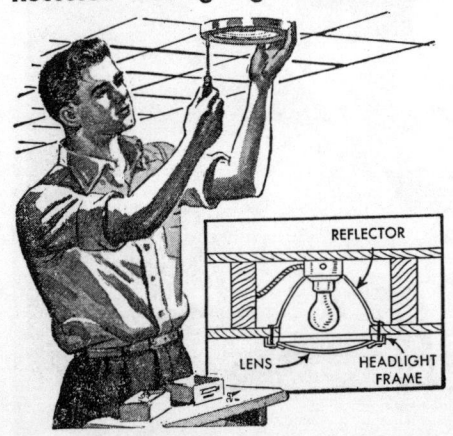

A recessed ceiling light in a playroom or workshop is less likely to become broken than one that is surface mounted. Although such fixtures are fairly expensive, one that will provide excellent illumination can be made from an automobile headlight and installed quite simply. Cut an opening in the ceiling about the same diameter as the front of the reflector. Then slot the edge of this opening so that the reflector can be slipped through it, first cutting an opening in the small end of the reflector for a porcelain socket. Fasten the socket in place and mount the reflector lens and headlight frame as indicated. To remove bulbs from the light, just unscrew the frame holding the lens. A wall switch will be found most desirable and convenient for this type of light fixture, especially in the workshop.

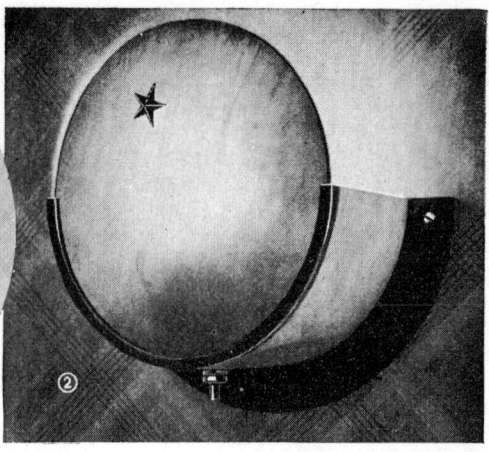

LIGHTING FIXTURES

PLASTIC cylinders, obtainable in various sizes and colors, can be converted readily into charming bracket lights. Figs. 1 and 2 are typical examples. The construction of the first of these is shown in Fig. 3. The coloring should be translucent, such as pink, rose or emerald quartz.. A standard size of plastic rod is used for the spacers and for the two ornamental buttons on the front of the cylinder. The backplate is sheet-plastic stock, although wood, smoothly enameled or lacquered, can be substituted. The lighting arrangement consists of a porcelain ring (sign) receptacle. This type of socket consists of two parts, which are screwed together through a 1/8-in. plastic disk cemented inside the cylinder. A similar disk seals the lower end and also provides a mounting for the small rotary canopy switch. The bulb is of the tubular type, 6 in. long.

Figs. 4 and 5 show the construction of the second bracket light. A half cylinder of plastic is mounted to a plastic wall plate, using cement and drive screws. Fastened to the backplate is a small ledge of crystal plastic, cut out in the center to accommodate the sign receptacle.

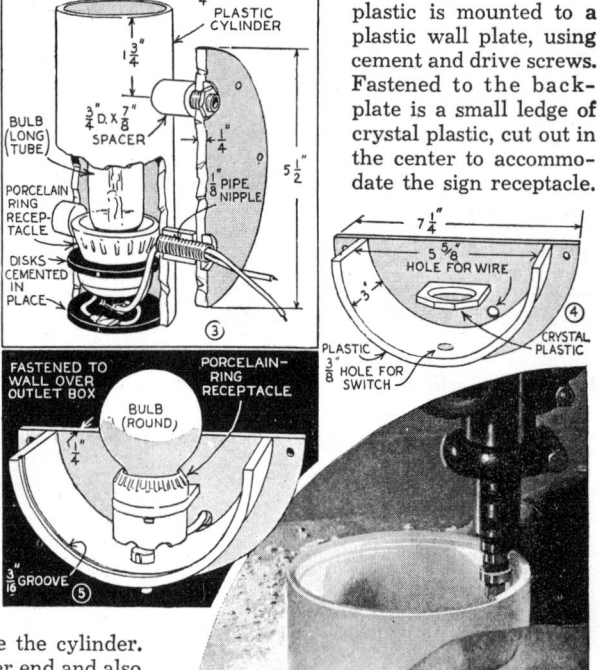

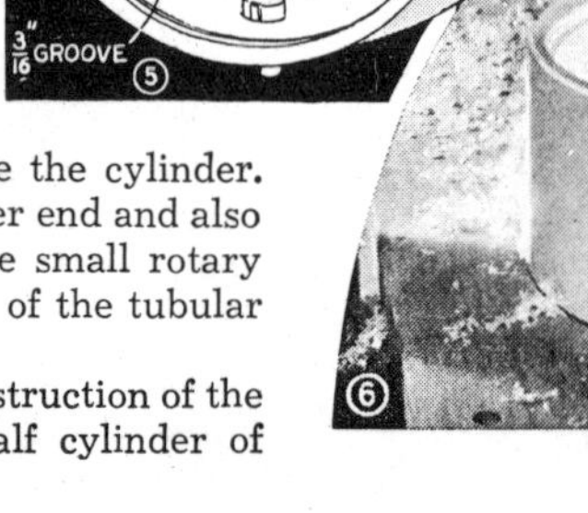

FIXTURES

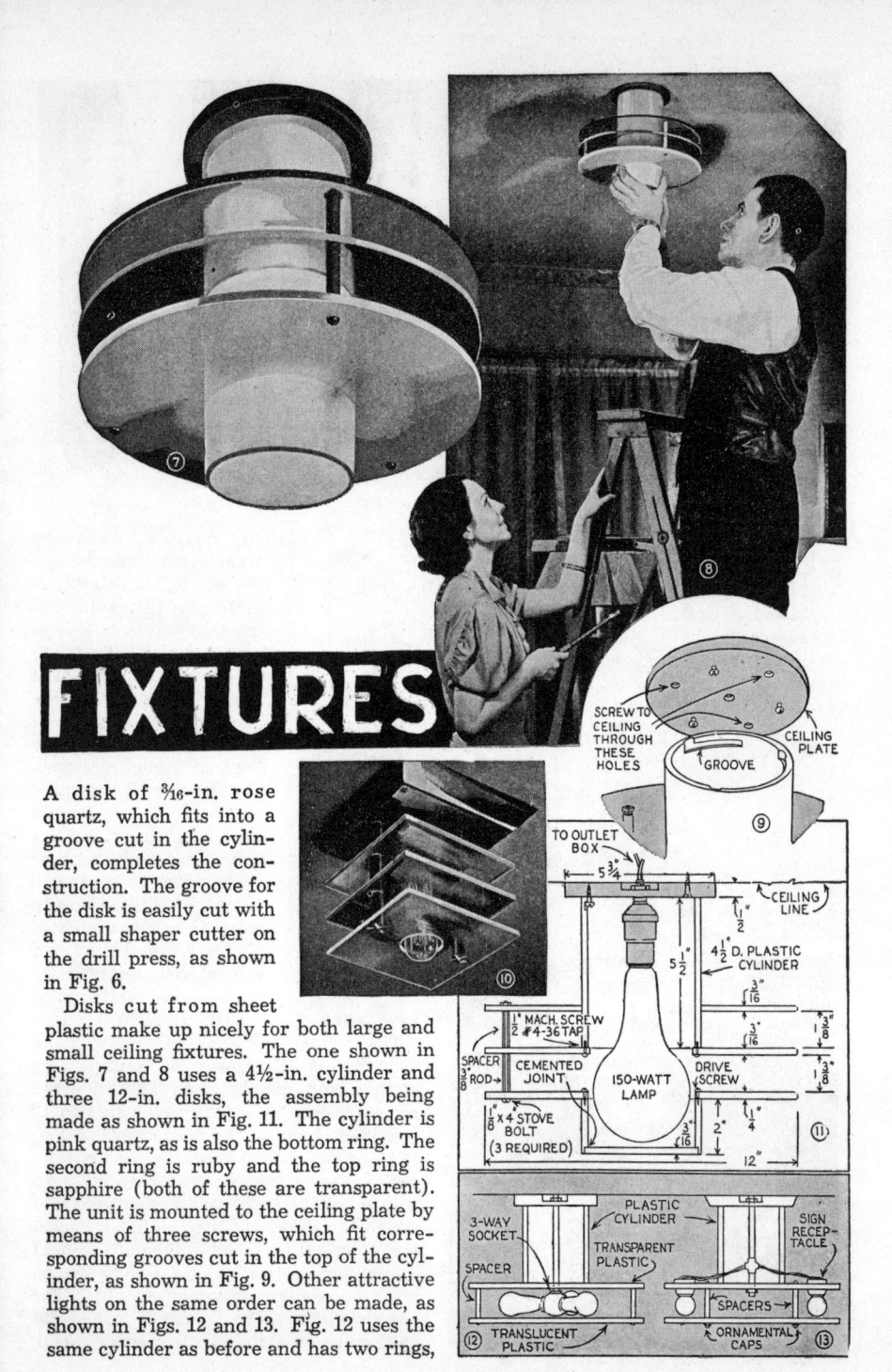

A disk of 3/16-in. rose quartz, which fits into a groove cut in the cylinder, completes the construction. The groove for the disk is easily cut with a small shaper cutter on the drill press, as shown in Fig. 6.

Disks cut from sheet plastic make up nicely for both large and small ceiling fixtures. The one shown in Figs. 7 and 8 uses a 4½-in. cylinder and three 12-in. disks, the assembly being made as shown in Fig. 11. The cylinder is pink quartz, as is also the bottom ring. The second ring is ruby and the top ring is sapphire (both of these are transparent). The unit is mounted to the ceiling plate by means of three screws, which fit corresponding grooves cut in the top of the cylinder, as shown in Fig. 9. Other attractive lights on the same order can be made, as shown in Figs. 12 and 13. Fig. 12 uses the same cylinder as before and has two rings, the bottom one being translucent while the upper one is transparent. A ⅛-in. pipe fitting through the unit serves as a support and also connects to a 3-way socket. The light shown in Fig. 13 can be made with either three or four lights around the edge of the plastic ring. The bottom disk is held in place with ornamental caps as it must be removed to permit changing bulbs. Fig. 10 shows a small hall light made from three squares of plastic—crystal, ruby, and crystal. The ceiling plate is a sheet-metal box made from chromium or stainless-steel stock.

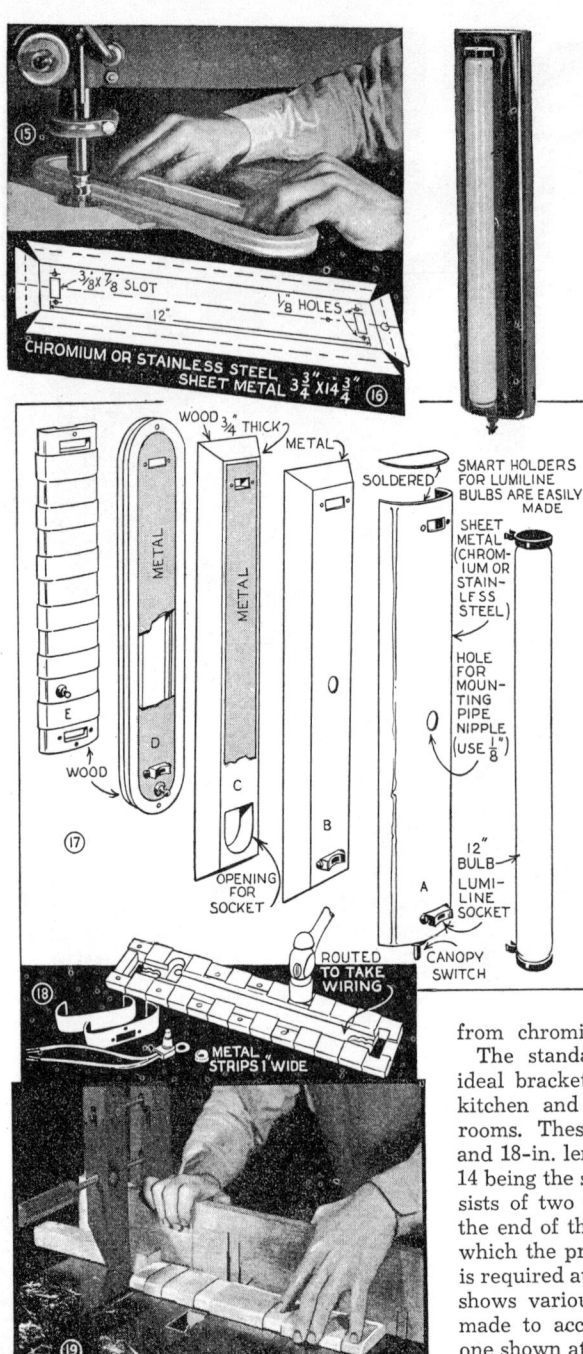

The standard "Lumiline" bulbs make ideal bracket lights for the bathroom or kitchen and can be used also in other rooms. These bulbs are obtainable in 12 and 18-in. lengths, the ones shown in Fig. 14 being the shorter type. The socket consists of two parts—a cap which fits over the end of the bulb, and a receptacle into which the prongs of the cap fit. A socket is required at each end of the bulb. Fig. 17 shows various wall plates which can be made to accommodate these bulbs. The one shown at A, Fig. 17, can be made from chromium or stainless-steel sheet metal, the end caps being soldered in place. Style B is also all-metal, and is cut according to the dimensions given in Fig. 16, which also show the correct dimensions for any other style of wall plate. Style C is a wood plate, cut out to accommodate the sockets and wiring, and covered on the face with sheet metal cemented in place. Stainless steel with fiber backing is ideal for this purpose. Style D is a wood frame, molded at the edges, Fig. 15, and cut out and routed in the center to take the sockets and a sheet-metal plate. Style E is a wood frame, shaped and grooved with molding head knives, and is alternately banded with strips of stainless steel cemented and tacked in place. All of the units use a rotary canopy switch, which can be mounted at any suitable point. The best finish for the wood brackets is baking enamel, although ordinary enamel or lacquer can be used. The units can be mounted with either an ornamental cap fitting a ⅛-in. pipe nipple, or by screws through the bracket. In all cases, the outlet box is immediately behind the fixture, although exposed lampcord wiring can be used in places permitting it. A simple ceiling fixture employing three bulbs is shown in Fig. 20. The ceiling box is cut to take the six receptacles wired in parallel, Fig. 21.

Lighting styles of current popularity are flush, cove, and louver lighting, Figs. 22 to 27 inclusive. The flush type of light can be fitted to either the ceiling or to the wall. Frosted or opal glass can be used although ⅛-in. translucent plastic is better adapted for intricate shapes. The glass is held in place with a suitable molding in either plastic or chromium. Cove or built-in lighting can be used above doors or windows, or in a vertical bank flanking mantels or bookshelves. The window cornice is perhaps the most popular and least expensive style to install. Louver lights are similar to flush lights except that a louver plate is used instead of glass, making a more shielded light. Louver plates in different sizes and in chromium or enamel finish can be purchased or made up from sections of enameled metal-window ventilators, as in Fig. 27. Other styles resembling small venetian blinds can be made up readily, and can be made adjustable.

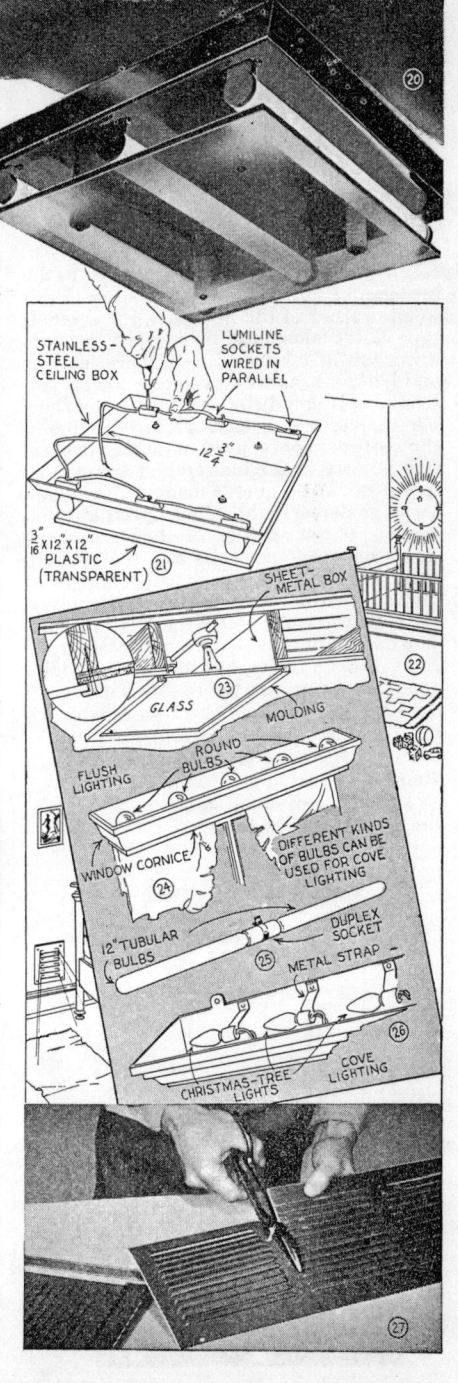

LINEN CLOSET

HERE IS ONE homeowner's answer to the problem of not enough closet space. Built in one corner of the bedroom, the cabinet-type closet takes little floor space, yet does a big job in taking care of shoes, clothes and hatboxes, as well as linens, towels and blankets. If you dislike using the open corner shelves for shoes as pictured, some of the shelves may be used to display decorative hosiery and handkerchief boxes, pictures, etc. The frame is made of 1 x 2s and 1 x 4s, as shown in the detail. Quarter round screwed to the walls and ceiling covers the joints between the walls and the cabinet. The remainder of the construction is of plywood, $3/8$-in. being used for the shelves, $1/4$-in. on the walls and $1/2$-in. for the doors. The doors are rabbeted, as indicated in the inset, to form a lip, and the door handles are made from quarter round. This is done for economy and also because the handles thus produced lend themselves nicely to the lines of the cabinet. The dimensions shown can be varied easily to suit the individual room in which the cabinet is to be built.

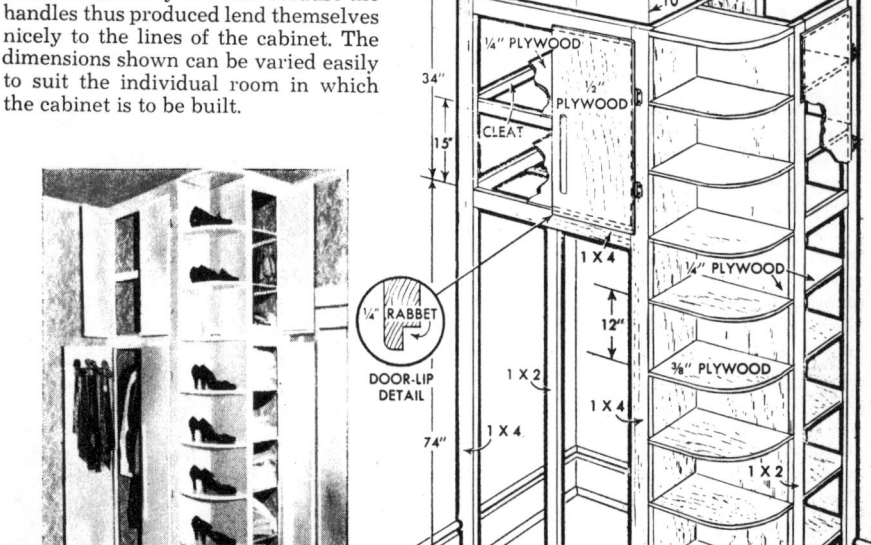

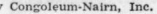

LINOLEUM and FLOOR TILE

LAYING LINOLEUM

LAYING your own linoleum is one of the most satisfying of all homecraft jobs because definite results can be accomplished in a comparatively short time. The money saved makes it well worth-while to do the work yourself and the approved procedures are so easy to apply that, by following a few simple directions, you can expect a good floor, even on the first job.

Types of linoleum: First of all, study the sketches on the opposite page which cover the fundamental procedure. Then decide what kind of linoleum you want. The cheapest material has a plain felt base and the pattern is usually a printed enamel. This type of floor covering must always be applied "loose lay," that is, not cemented or tacked to the floor, because it must be free to expand. A better grade comes with a duplex felt back and can be applied either loose-lay or cemented directly to the floor. The best grade of linoleum is built up on a burlap base and is always cemented over a lining felt. All linoleum is stiff and brittle when cold and should not be unrolled until warmed to a temperature of 70 deg. for at least 24 hours. When it is practical to do so, it also is a good idea to allow the linoleum to lie flat for another 24 hrs. after unrolling. In this way, all danger of breaks in the surface of the linoleum will be avoided.

Laying wall-to-wall: This is the most common application and is illustrated by the details on the following page, which also explain the basic operation of scribing. Scribing simply means marking the linoleum to a shape and size to fit the room in which it is to be laid. It is done with a pair of dividers of the type used in woodworking, or with special dividers made for linoleum work. The latter tool is called an overscriber. In laying linoleum floor covering wall-to-wall, begin by removing the shoe mold at the bottom of the baseboard. If the floor is rough, with offsets at the joints, it should either be sanded or planed smooth. Then, cut the first sheet of linoleum about 4 in. larger than needed and fit it in place against one wall, letting the ends ride up on the baseboard. Follow through with the procedure pictured in Figs. 1 to 4 inclusive, which show the methods of scribing quite clearly. A close fit against the baseboards is not advisable in ordinary wall-to-wall installation. It is best to allow at least ⅛ in., or better, ¼ in. all around. The gap will be covered when the shoe mold is replaced. If the doorway is of the arched type or if it is fitted with a swinging door, it is common practice to extend the linoleum halfway through. Otherwise, the linoleum is ended at the

TOOLS

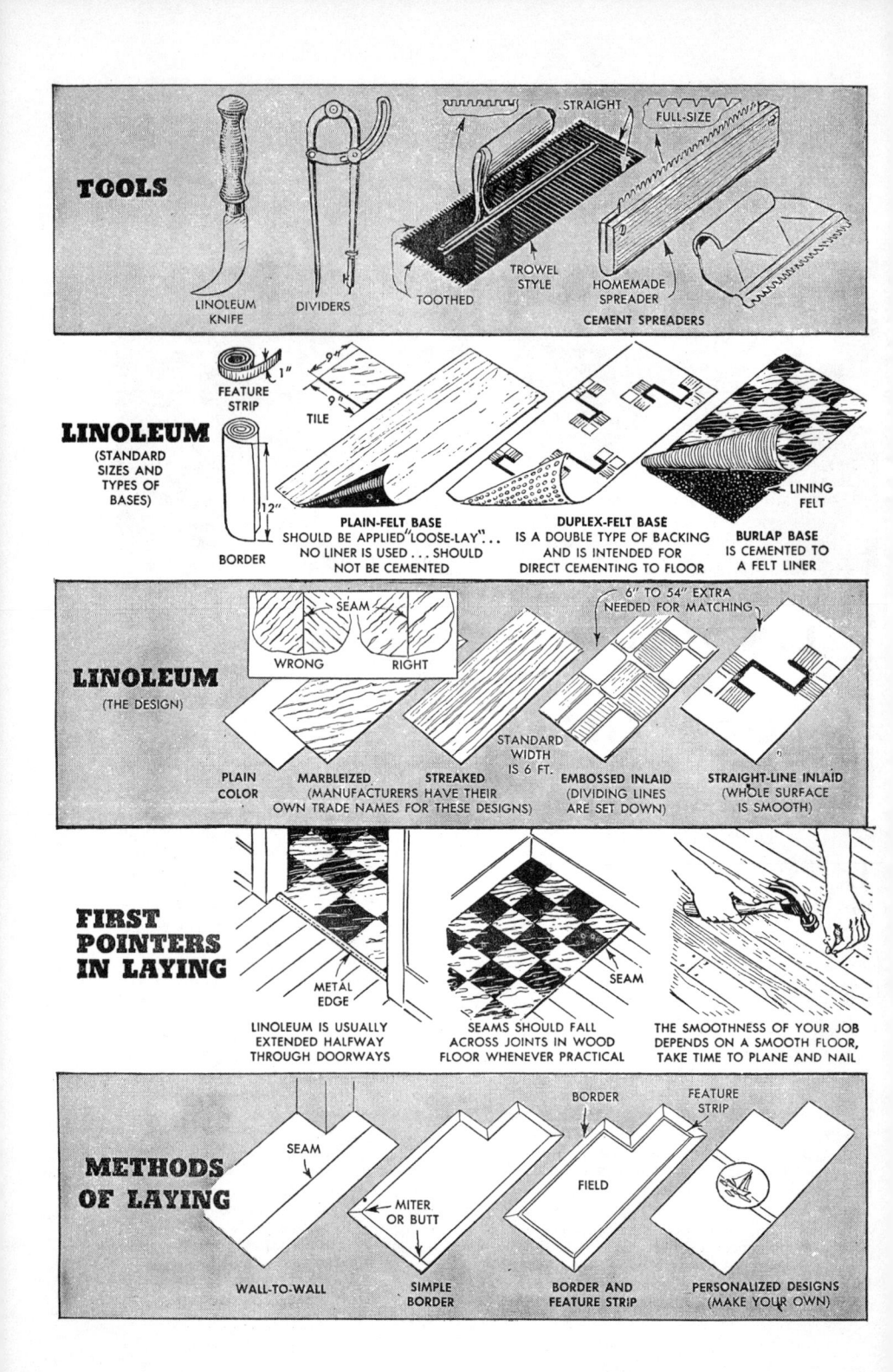

- LINOLEUM KNIFE
- DIVIDERS
- TOOTHED / TROWEL STYLE (STRAIGHT, FULL-SIZE)
- HOMEMADE SPREADER
- CEMENT SPREADERS

LINOLEUM
(STANDARD SIZES AND TYPES OF BASES)

- FEATURE STRIP
- TILE (9" x 9")
- BORDER (12")
- **PLAIN-FELT BASE** SHOULD BE APPLIED "LOOSE-LAY"... NO LINER IS USED... SHOULD NOT BE CEMENTED
- **DUPLEX-FELT BASE** IS A DOUBLE TYPE OF BACKING AND IS INTENDED FOR DIRECT CEMENTING TO FLOOR
- **BURLAP BASE** IS CEMENTED TO A FELT LINER
- LINING FELT

LINOLEUM
(THE DESIGN)

- SEAM — WRONG / RIGHT
- STANDARD WIDTH IS 6 FT.
- 6" TO 54" EXTRA NEEDED FOR MATCHING
- **PLAIN COLOR**
- **MARBLEIZED** (MANUFACTURERS HAVE THEIR OWN TRADE NAMES FOR THESE DESIGNS)
- **STREAKED**
- **EMBOSSED INLAID** (DIVIDING LINES ARE SET DOWN)
- **STRAIGHT-LINE INLAID** (WHOLE SURFACE IS SMOOTH)

FIRST POINTERS IN LAYING

- METAL EDGE — LINOLEUM IS USUALLY EXTENDED HALFWAY THROUGH DOORWAYS
- SEAM — SEAMS SHOULD FALL ACROSS JOINTS IN WOOD FLOOR WHENEVER PRACTICAL
- THE SMOOTHNESS OF YOUR JOB DEPENDS ON A SMOOTH FLOOR, TAKE TIME TO PLANE AND NAIL

METHODS OF LAYING

- SEAM — **WALL-TO-WALL**
- MITER OR BUTT — **SIMPLE BORDER**
- BORDER / FIELD — **BORDER AND FEATURE STRIP**
- FEATURE STRIP — **PERSONALIZED DESIGNS** (MAKE YOUR OWN)

SCRIBING IS BASIC OPERATION ON ALL JOBS

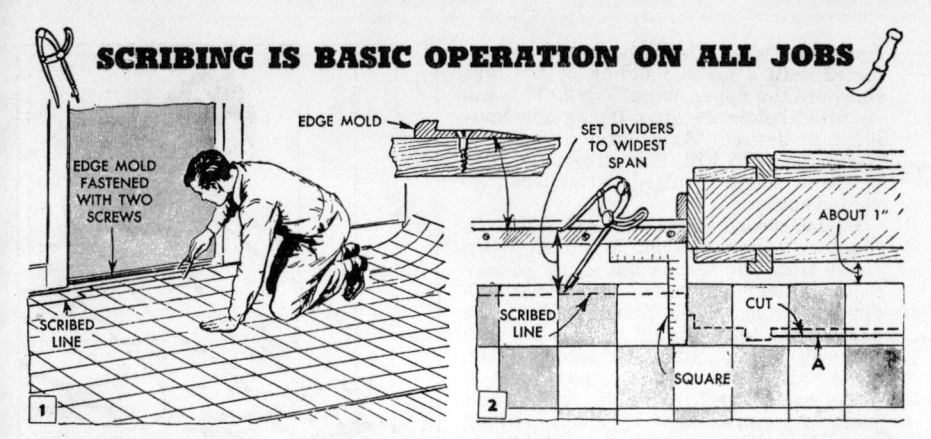

1. Cut linoleum 4 in. over required length. Lay on the floor with one side about 1 in. from wall and let both ends ride over baseboard. Set dividers at the required opening (Fig. 2) and, with one point of the dividers in contact with baseboard, scribe the line

2. When scribing to fit a doorway, metal edge molding can be temporarily fastened in place with two screws to serve as a guide. Shoe mold will cover the cut at A, so linoleum can be cut from 1/8 to 1/4 in. beyond the scribed-to-wall mark, making piece easy to fit

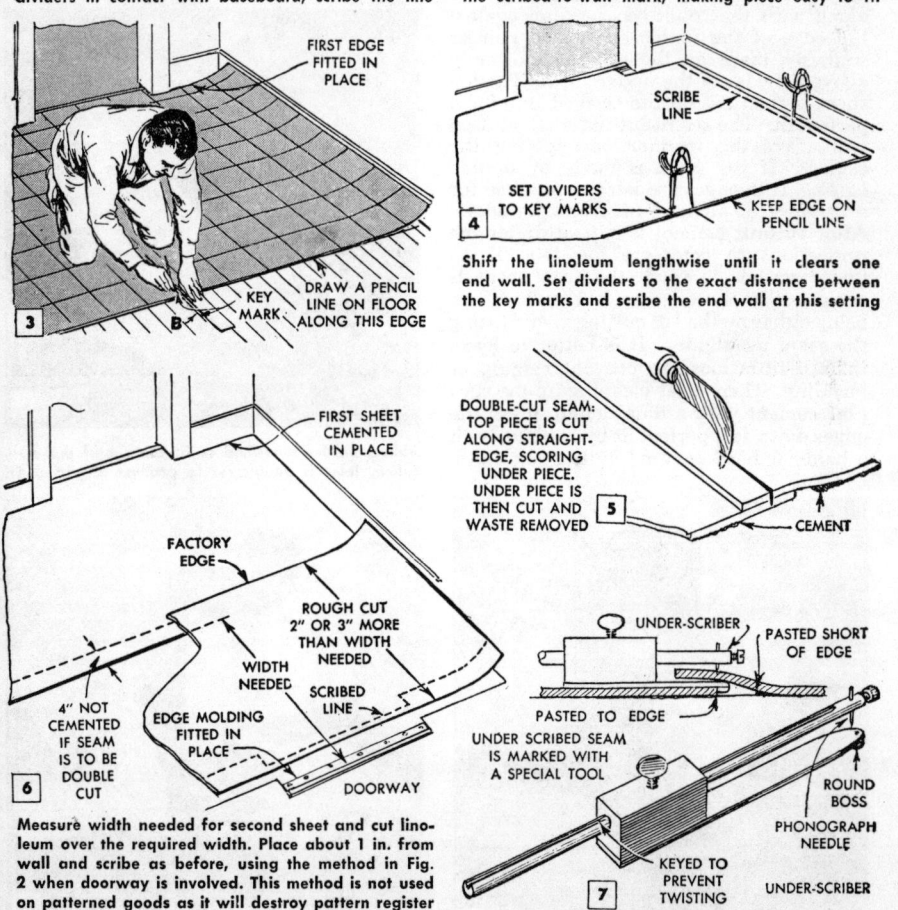

4. Shift the linoleum lengthwise until it clears one end wall. Set dividers to the exact distance between the key marks and scribe the end wall at this setting

6. Measure width needed for second sheet and cut linoleum over the required width. Place about 1 in. from wall and scribe as before, using the method in Fig. 2 when doorway is involved. This method is not used on patterned goods as it will destroy pattern register

doorstop, Fig. 2. The edge should be protected with a metal molding of the type shown in the upper detail, Fig. 2. This can be fitted before or after laying the linoleum, as desired. As suggested in Fig. 1, fasten the mold with two screws and then scribe to it. If required, it then can be shifted easily.

Laying second sheet: The second sheet is fitted as shown in Figs. 5 and 6. When rough trimming the second sheet, be sure to scribe and make the cut parallel with the opposite edge. When fitted in place, the edge at the seam should be parallel with the wall. After fitting the second sheet, there will be an overlap where the two pieces come together, Fig. 6. A neat butt joint, or seam, is made by double cutting as in Fig. 5. A recommended alternate method involves the use of a special tool called an under-scriber, Fig. 7. This tool speeds the scribing job and assures accuracy. When the under-scriber is pulled along with the round boss riding against the edge of the under sheet, the pointer scribes a mark on the top sheet directly above the edge of the under sheet. The top sheet is then cut on the scribed line for a perfect fit. The seam-cutting method used will affect the manner of applying the cement. If the seam is made by double cutting, both edges are left uncemented for a distance of about 4 in. from the edges. After cutting, cement is spread under the projections to complete the job. If an under-scriber is used, the first sheet is cemented to the edge as in Fig. 7. When using either method of cutting, avoid fitting the seam too tightly. It is better to leave them a little loose to prevent ridging or buckling. Then, after cleaning up the surplus cement with a damp cloth, press the edges down to a perfect fit by rubbing with a hammer head covered with a soft cloth.

Paper pattern is used for transferring odd shapes

Above, border is scribed to fit contour of the wall. Below, joint is double-cut to produce a perfect fit

LAYING BORDER AND FEATURE STRIP

Fitting with paper pattern: Sometimes you will run into a job that is difficult or impractical to scribe by ordinary methods. In this case, use of a paper pattern probably will be the solution. To build up the pattern, cut and paste pieces of heavy paper or cardboard together to form the exact shape you want the linoleum to be. Then apply rubber cement or linoleum cement to the face of the pattern and roll the linoleum onto it so that the two stick together firmly, Fig. 8. Then turn the linoleum over and you have the required cutting pattern cemented to the back. This method also works out well when fitting linoleum having a figure or design imprinted or inlaid. In this case, you cannot cut to form a seam in the usual way as this may destroy the design register. Using the paper pattern, the linoleum is cut a little over the required width, the second sheet is butted against the first sheet and then rolled to the wall where it picks up the paper pattern.

Fitting simple border: In nearly all border jobs, you have the choice of laying either the field or the border first. Figs. 9, 10 and 11 show the border laid first. Chalk a line all around the room an equal distance from the walls. Use either the scribing or paper-pattern method for marking the border pieces at doorways. After the pieces have been cut out, place them in position and check for fit. Then remove the pieces, one at a time, and spread cement under each one out to the chalk line. Replace all pieces and double-cut the corner seams, using either a miter or plain butt

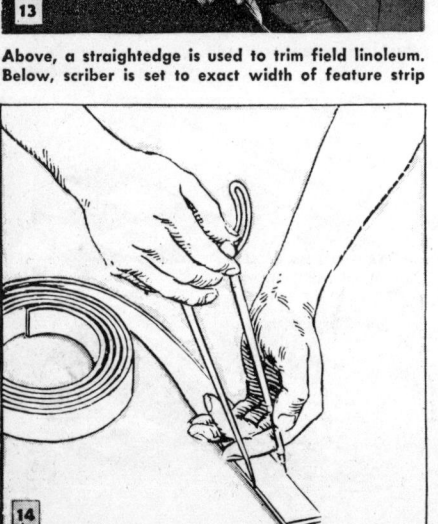

Cement for linoleum field is spread to the chalk line

Above, a straightedge is used to trim field linoleum. Below, scriber is set to exact width of feature strip

Place the border against field and scribe to wall, making sure strip does not buckle while being scribed

THIS IS THE MOST ATTRACTIVE OF THE FLAT-LAY STYLES AND IS EASY TO DO WITH READY-CUT STRIPS AND BORDERS

joint at the corners. Although the miter joint wastes more material, it is by far the neatest and most attractive. After the border has been cemented down, clean up all excess cement and then rough-cut the field material 1 to 2 in. oversize, except, if desired, one factory edge, which can be butted to the border. On the remaining edges use the under-scriber, Fig. 10, to mark a line for the cut above the border. Cut on the scribed lines and finish cementing at the edges of the field. Clean up with a damp cloth and then rub down all joints with a hammer head.

Border and feature strip: This attractive style, Fig. 17, requires only a little more time to lay than does the simple border. The method pictured in Figs. 15 to 20 inclusive calls for laying the field first, a procedure which many workers prefer for this installation. Chalk a line around the room a distance from the walls equal to the combined width of the border and strip. Butt one factory edge of the field to the line and trim the other edges about 1 in. oversize. Roll the material back and spread the cement, first on one half of the floor, then on the other, Fig. 12. Be sure to keep the cement inside the line. Bed the linoleum in the cement, pick up the chalk marks with a straightedge and cut the field to the required net size, Fig. 13. Then sandbag the field and clamp the edges with wooden strips nailed to the floor, Fig. 19. Allow to stand for 24 hrs. Border pieces are now trimmed to a width of about ½ in. less than the distance between the edge of the field and the wall, Fig. 18. Set the scriber (dividers) to the width of the feature strip, Fig. 14, and, with the border pressed tightly against the field, scribe as in Fig. 20. If all measurements have been made correctly, the border and feature strip will fit neatly into the opening as in Fig. 16. It should be noted that in border work, the border itself is scribed net to the wall without a gap. The slight pressure fit thus obtained helps to get a good, tight joint between the border and feature strip and between the strip and the field. In some jobs involving a border of the same pattern and color as the field, the linoleum is laid wall-to-wall and the feature strip is cut in. This procedure is sometimes advantageous, depending on the size of the room and the distance from the wall to the strip. Make suitable allowances at the baseboard for a prominent feature strip.

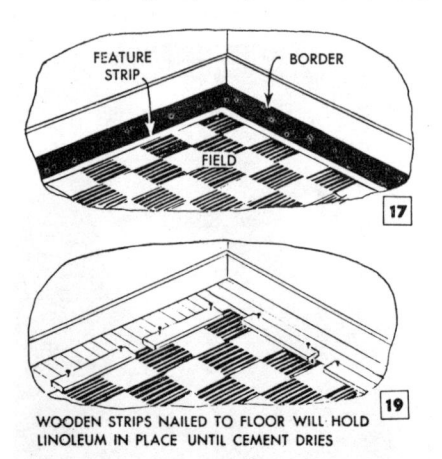

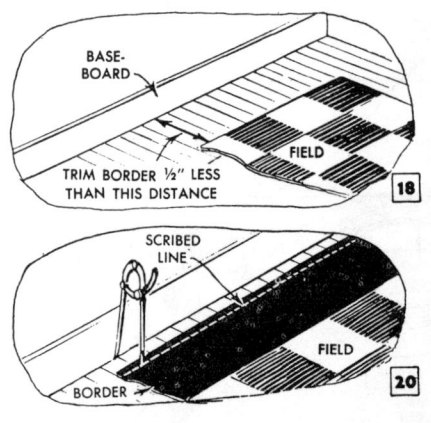

If essential measurements have been properly made, border and feature strip will fit opening accurately

WOODEN STRIPS NAILED TO FLOOR WILL HOLD LINOLEUM IN PLACE UNTIL CEMENT DRIES

Courtesy Douglas Fir Plywood Association

FLOOR TILES

Laying linoleum on floors from wall to wall, and the procedures for laying the border and the border-and-feature-strip designs have been explained previously, but that's only the beginning. Linoleum is supplied as either floor or wall tile to be applied in various sizes and pattern combinations and it's also used for wainscoting, all-over wall coverings and counter tops. Such applications require somewhat different procedures.

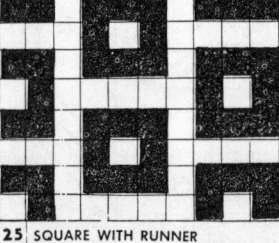

Floor tiles: If you design your rooms, considering the best floor covering that is fitted for the activities of your family, your efforts will be well repaid. The illustrations of some of the suggested floor designs will show just how successfully the ideas of some homeowners can be carried out. Linoleum tiles are easy to install over wood floors and, in some cases, over concrete floors that are dry at all times. The most common tile sizes are 9 in. square and 9 x 18 in. rectangular. The squares can be laid in one color only or in various patterns such as shown in Figs. 23 to 26 inclusive. Tiles are obtainable in a variety of plain and marbleized colors in standard linoleum

photos courtesy Congoleum-Nairn, Inc.

and also in a special asphalt-base product. The latter floor covering is intended for use over concrete floors below grade, Fig. 21. The ordinary linoleum tile should not be laid on damp basement floors. On wooden floors, the tile is regularly applied over a felt liner which has been previously cemented to the floor. On concrete floors below grade no liner is used. Instead, the tiles are applied directly to the concrete, using a special cement which is proof against dampness. It is important that the floor over which the tiles are to be laid be made smooth and level. Sagging floors should be leveled and reinforced. Some squeaky floors can be corrected by a few well-placed nails, without tearing up the floor.

Application of tile: Tiles usually are applied in a field-and-border design and the border width for any room is determined by using the simple calculation given in Fig. 22. After determining the border width required, the first border piece is laid out and cut, but is not cemented down. Then the second piece is left overwidth and is fitted to the second wall. The width of this piece is then marked at the point A, Fig. 28, and it is then squared by using the 3-4-5-ft. method, A, B, C, Fig. 28, or any other means which will assure a square corner. Never depend on the room being square. The second border piece is then cut and both the first and second pieces are cemented down. Now, start putting the tiles in place. Apply the cement with a toothed trowel over an area about equal to that covered by 8 to 10 tiles. Position each tile individually and press firmly in place. If

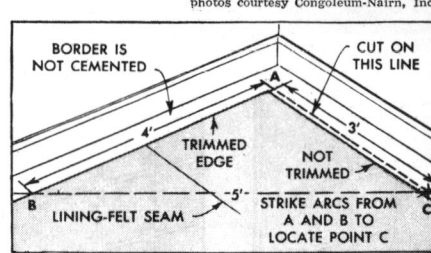

28 | SQUARING THE BORDER IS FIRST OPERATION

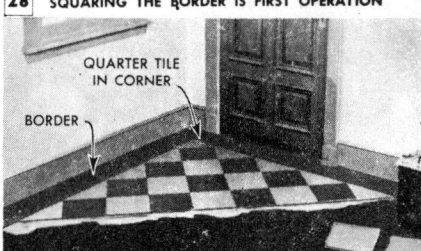

29 | DIAGONAL CHECKERBOARD STARTS WITH QUARTER TILE

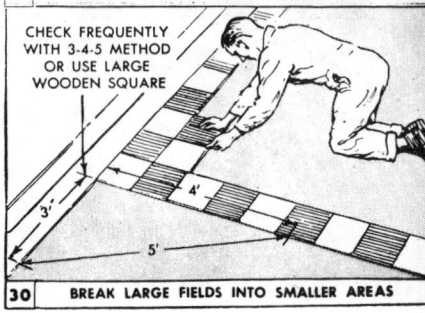

30 | BREAK LARGE FIELDS INTO SMALLER AREAS

LEVELED GUIDE LINE

LINOLEUM

METAL TRIM IS FITTED AFTER ONE WALL IS COVERED

31

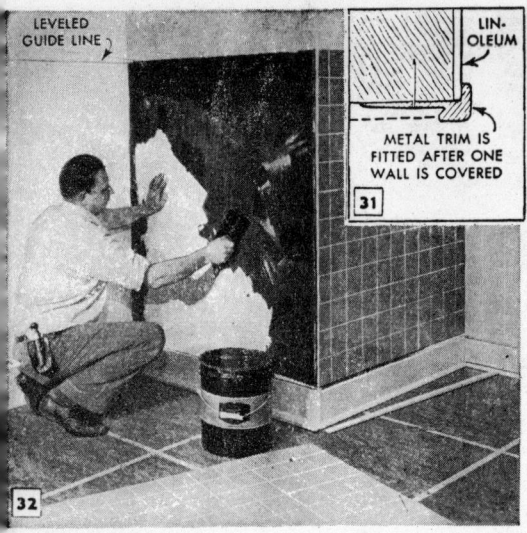

32

Standard wall linoleum comes 4 ft. 6 in. wide and is cemented directly to the wall, keeping it running level with a guide line marked on the wall. Detail shows how corner molding fits

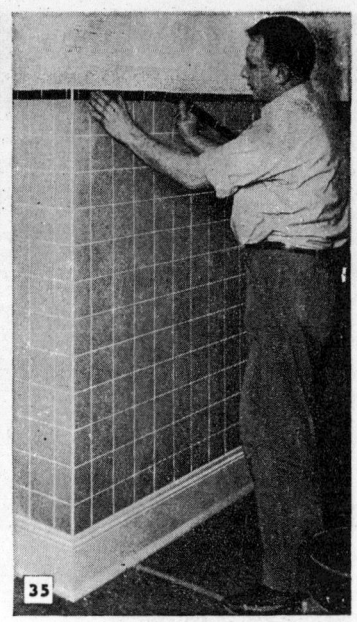

MORTAR "JOINT" IS LEFT ON FIRST PIECE

MORTAR "JOINT" CUT OFF

33

MOLDING

34

35

Top of linoleum is finished with cap mold

the field (room) is large, take a little extra time to square up smaller areas and then fill in these areas by working from outside to center as in Figs. 29 and 30. Wipe off all surplus cement with a damp cloth, and roll the tiles lightly as they are laid, using an ordinary rolling pin. Check carefully as each row of tiles is laid so that straight lines are maintained in both directions, Fig. 27. This precaution will assure that you finish with a straight line of tiles at the opposite walls. When the field is completed, scribe and fit the remaining border pieces. If you use a diagonal checkerboard pattern, the start in the corner is made with a quarter tile, Fig. 29. Of course, the edges will be finished with half tiles. Asphalt tiles will break cleanly along a scored line, but linoleum tiles must be cut through with a sharp knife. It will save time to make a simple wooden frame to serve as a template, or cutting jig, for this special purpose.

Wall installations: The standard material for wall application is linoleum sheets ruled into 4½-in. squares. Usually the sheets are 4½ ft., or 54 in. wide. If the baseboard is level, the job is simple. Always check the room and the baseboards with the level first. If the baseboard is not level all around then it will be necessary to scribe a leveled line around the room a little less than 4½ ft., or the width of the linoleum sheet, above the top of the baseboard. The top edge of the linoleum is fitted to the line, Fig. 32, and the bottom edge is

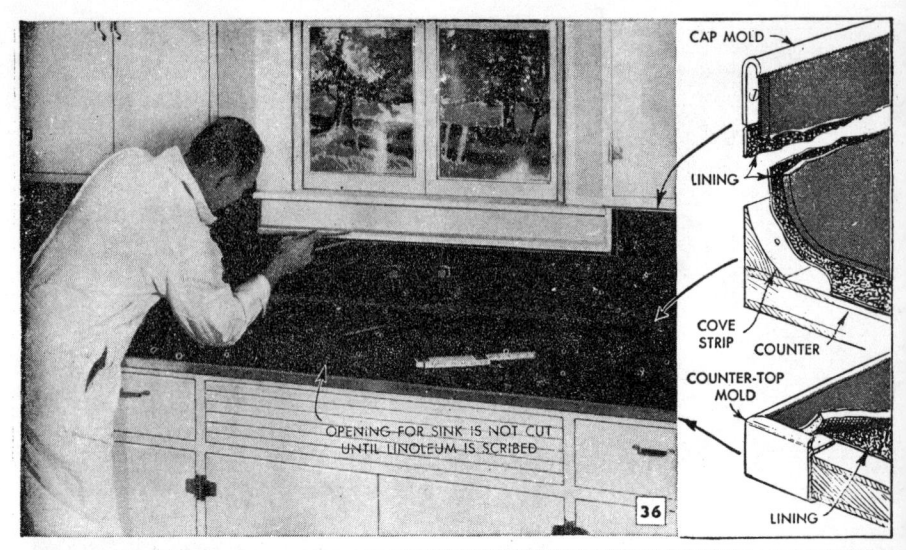

OPENING FOR SINK IS NOT CUT UNTIL LINOLEUM IS SCRIBED

CAP MOLD
LINING
COVE STRIP
COUNTER
COUNTER-TOP MOLD
LINING

Above, counter installation is made over a duplex felt lining. After the lining is fitted, it is used as a pattern for marking and cutting linoleum as at right

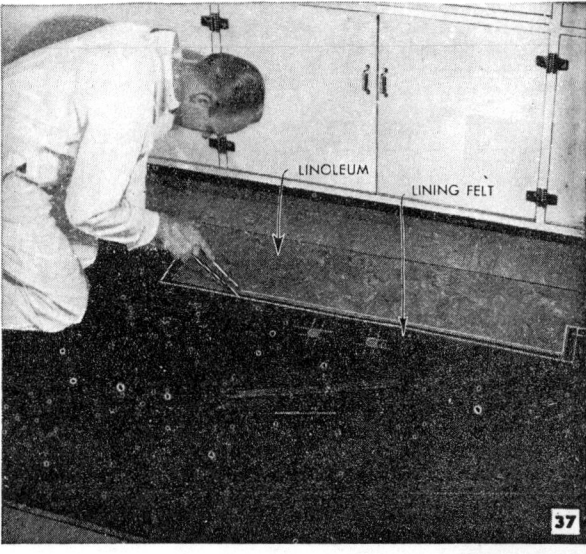

LINOLEUM
LINING FELT

scribed to fit the baseboard. Joints should be butted, Fig. 33, and if there is an outside corner, Fig. 31, it is fitted with a metal or plastic molding after one of the walls has been finished. The adjacent wall is then fitted and the edge is underscribed to fit the molding, Fig. 34. Inside corners can be butted, although special metal moldings available for this purpose make a neater job. After the linoleum has been applied to all the walls, the job generally is finished with a metal cap strip as in Fig. 35. Wooden strips and matching linoleum strips sometimes are used. If the baseboard is fitted with a top molding, this should always be removed before fitting the linoleum. In this case, rough fitting at the bottom edge is satisfactory as the edge will be covered when the molding is replaced. If the original baseboard is badly scratched it may be replaced with a new matching baseboard of rubber or plastic. These baseboards usually are applied with special cement.

Counter tops: Counter-top jobs call for the best-quality linoleum and careful workmanship. It's common practice to lay the linoleum top over a felt liner, which provides not only a good cementing base but also a template for cutting the linoleum accurately. Start the job by installing all metal trim, Fig. 36. Then, if the linoleum is to extend up the wall to provide a splash back, apply a flexible plastic cove strip in the corner between the counter top and the wall. Next, fit a length of liner felt ½ in. short at the corners, edges and metal trim. Do not cut out for the sink well; this will be taken care of later. After rough-cutting the felt, fasten it securely with thumbtacks so that it will not move during the scribing operation which is to follow. Be sure that

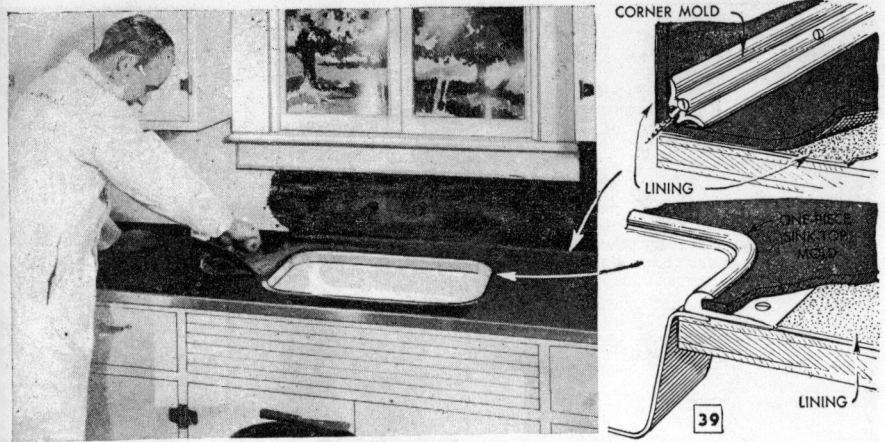

|38| DAMPPROOF CEMENT IS USED FOR BOTH LINING AND LINOLEUM

|39| CORNER MOLD — LINING — ONE-PIECE SINK-TOP MOLD — LINING

|40| THE LINOLEUM IS UNDERSCRIBED TO FIT THE SINK MOLDING

the felt is pushed tightly against the cove strip. Set the scriber (dividers) to approximately a 1-in. opening and scribe up to all edges except the sink well and the front edge, Fig. 36. Now, remove the felt and spread it over the linoleum. Tape it in place to prevent movement. Then, with the same divider setting, retrace the scribed lines to transfer the outline to the linoleum as in Fig. 37. Next, place the felt back on the counter top and fit it carefully to the flanged metal trim. Then cement in place, using a moistureproof cement. Cut the linoleum to the scribed lines, spread moistureproof cement over the felt, Fig. 38, and roll the linoleum in place. Make certain of a good contact over the cove strip in the corner. Go over the entire linoleum surface with a rolling pin to assure perfect contact. Underscribe the overhanging front edge and roll it down last. Cut out the sink well with a sharp knife about 1 in. oversize. Then underscribe to a neat fit in the sinktop molding as in the lower detail, Fig. 39, and also in Fig. 40. For this particular work the underscriber saves a lot of time where neatness is so essential. The upper detail in Fig. 39 shows an alternate back-corner treatment which is somewhat easier to work out than using the continuous top and splash back over a corner cove. In this, the top and back are applied separately, the back being butted to the top in a right-angle butt joint. A corner mold of plastic or metal finishes the joint. Finish all counter tops with paste wax rubbed to a high polish.

Patching: Very often existing linoleum installations on floors, walls and counter tops can be made to give much longer service by patching the worn areas. Greatest wear comes in narrow traffic lanes through doors, in front of the kitchen stove and sink, and also on the counter top at both sides of the sink. Since most linoleum patterns are stock items, it generally is possible to purchase the amount necessary to make the repair. If the linoleum is figured, be sure to allow material for matching the pattern. Cut out the worn area, following the pattern lines wherever possible. Then match the new piece over the opening. Fasten with thumbtacks or tape and rough-cut to about 1 in. oversize. Then use the underscriber to obtain a perfect cutting line. Spread cement in the open area and press the patch firmly in place, rolling it down until it makes full contact. Clean up all excess cement and, finally, rub lightly with fine steel wool to dull the surface so that it will match the old as closely as possible to avoid casual detection.

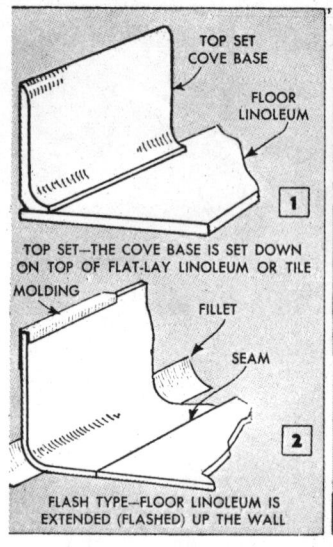

TOP SET—THE COVE BASE IS SET DOWN ON TOP OF FLAT-LAY LINOLEUM OR TILE

FLASH TYPE—FLOOR LINOLEUM IS EXTENDED (FLASHED) UP THE WALL

Lay Your Own LINOLEUM COVE BASE

EXTREMELY NEAT and colorful, the cove-base linoleum floor has the added advantage of forming a smooth, easily cleaned joint between floor and walls, an especially desirable feature in kitchens, laundry and utility rooms. Although the simplest installation is top set, Fig. 1, the true cove base is the style in which the floor material is extended (flashed) up the wall. Flashed cove base can be installed in a variety of decorative schemes, the plainest design being the same color and pattern as the floor material, Fig 3. More often, the cove base is a solid color repeating the predominant shade of the floor pattern. A distinctive separation line between the cove base and the floor (or field) linoleum often is provided by using a feature strip, Fig. 14, of a contrasting color.

Fittings: Two essential items for flashed cove base are the cap molding and the fillet. The cap molding is a metal or plastic shape which provides a trim and stop on the top edge of the cove, and is nailed 4 to 6 in. above the floor. The fillet is a cove molding, Fig. 2, which is available in plastic

Above, simplest installation of cove base uses same color and pattern of linoleum as the floor, or field

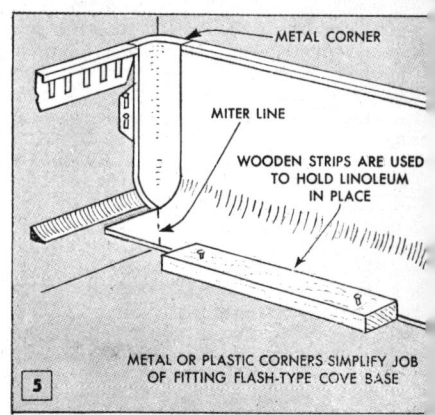

Figs. 4 and 5 show method of using metal or plastic inside and outside corners to fit flash-type cove base

6 PAPER PATTERN · CONTRASTING FEATURE STRIP

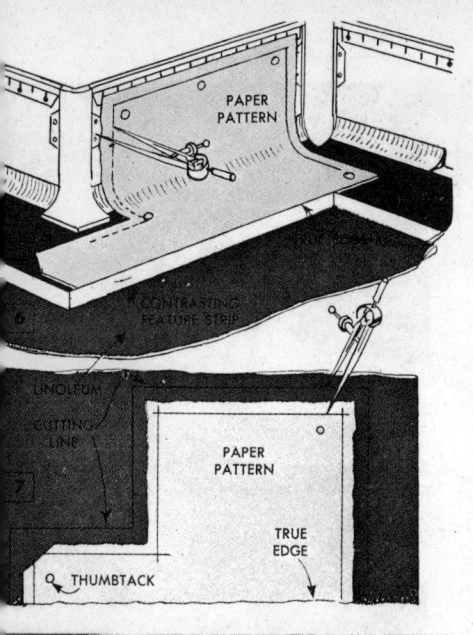

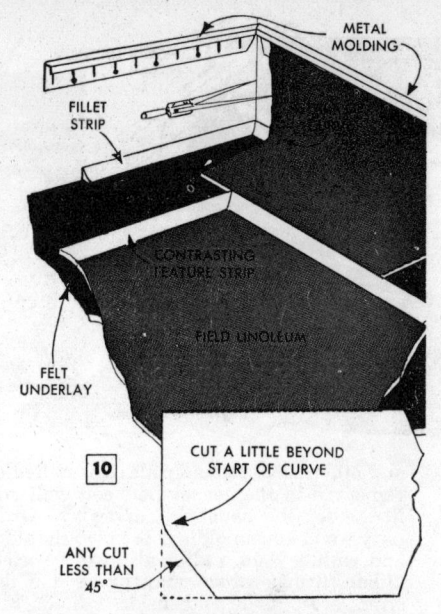

7 LINOLEUM · CUTTING LINE · PAPER PATTERN · TRUE EDGE · THUMBTACK

10 METAL MOLDING · FILLET STRIP · CONTRASTING FEATURE STRIP · FIELD LINOLEUM · FELT UNDERLAY · CUT A LITTLE BEYOND START OF CURVE · ANY CUT LESS THAN 45°

8

Figs. 11 and 12 show how an inside corner is cut and fitted when metal or plastic corners are not used

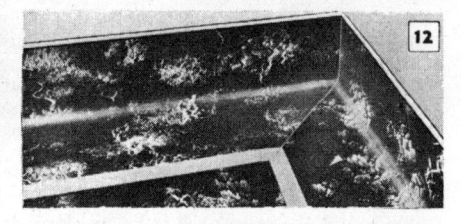

11

12

Above, when patterns are used to cut cove, allowance must be made for fitting under cap molding and corners

Below are shown tools that can be purchased or made which aid fitting and shaping cove base in corners

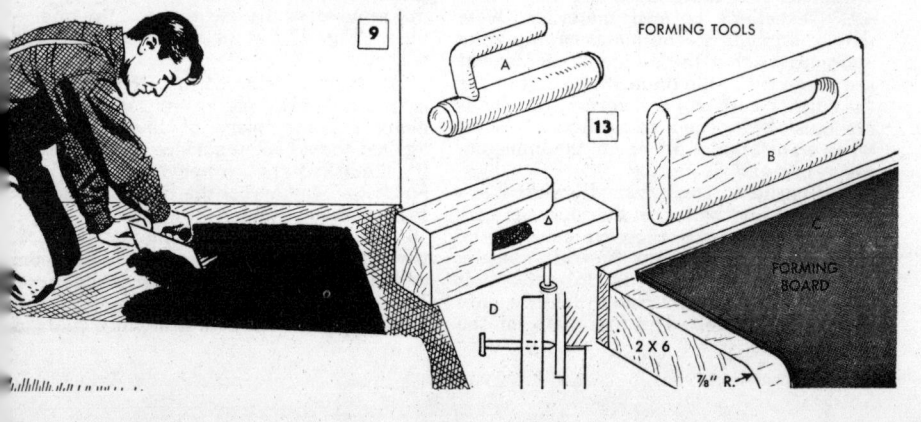

9

13 FORMING TOOLS · A · B · C · FORMING BOARD · D · 2 X 6 · 7/8" R.

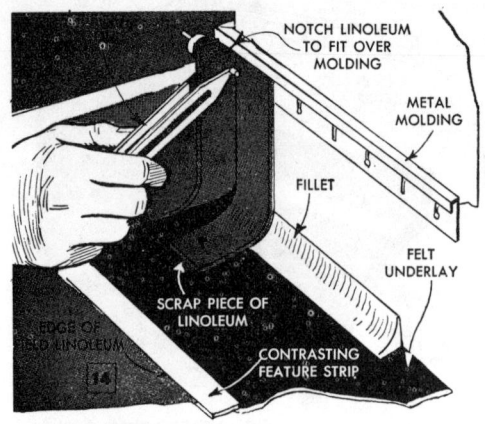

Above, linoleum on self corner is cut at 45-deg. angle down to middle of fillet on outside corners

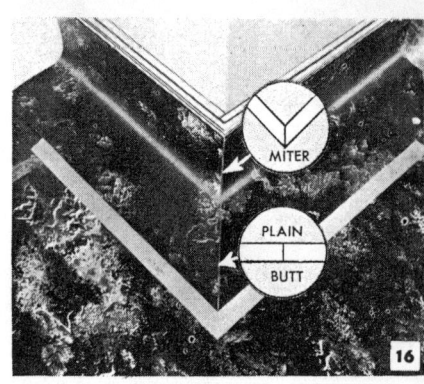

Above, from center of fillet to corner of field linoleum edge is cut at right angle to form plain butt

and other flexible materials. It is nailed or cemented in the corner between wall and floor over the usual felt underlay. Ordinary wood-cove molding, 7/8 in. on the sides and with a 7/8-in. radius also can be used. Other fittings which may be used if desired, are outside and inside corners, Fig. 4, and right and left-hand end stops. They are available in either metal or plastic, 4½ or 6 in. high. In all cases, when installing cove-base floor, the conventional wood baseboard and base shoe, if present, are removed.

Installation with corner pieces: As with a flat-lay floor, either border or field can be laid first. Fig. 9 shows the field laid first, while Fig. 5 shows the cove border laid first. Wood strips nailed to the floor are used to crowd the cove strip against the fillet while the cement dries. The cove base itself is simply a strip of border-strip linoleum trimmed to the desired width and shaped to fit into the corner. The miter line on the floor in the corner can be cut singly as each piece is applied, Fig. 5, or the two corner pieces can be overlapped and cut together.

Either direct or pattern scribing can be used to determine the shape of the linoleum. If a paper pattern is used, Fig. 6, it should be fitted snugly in place without wrinkles. Felt underlay and lining paper make excellent pattern materials. Note that in scribing to the metal cap molding and corners that the dividers ride against the edges. Since the linoleum is to fit under the edge of both cap and corner, the dividers must be opened an additional 1/8 in. when transferring this line to the linoleum, Figs. 6, 7 and 8.

Self corners: This style of corner takes more skill and patience, but since no corner pieces are used, the cove base can be flashed any distance up the wall, and the cove molding located accordingly. Self corners are usually scribed directly onto the linoleum itself. The first strip for the corner is pressed into place and scribed from the top down to about halfway across the fillet, Fig. 10. The second strip then is pushed into position, Fig. 11, and scribed and cut in the same manner. Both strips are cemented in place, then the cut is continued down across the fillet and out to the corner of the floor linoleum. After double-cutting in this manner, the top piece of linoleum is lifted slightly to permit removal of the waste end of the underpiece. Fig. 12 shows the finished inside corner.

To fit an outside corner, the first strip is pressed into place and underscribed along a scrap piece of linoleum fitted against the adjacent surface, Fig. 14. After it is marked the corner is cut down to about halfway across the fillet, the cut being a 45-deg. miter, as shown in Figs. 15 and 16. After the linoleum is cemented in place, the cut is continued across the fillet to the corner of the floor linoleum. Note here, Fig. 16, that the cut edge is a 45-deg. miter down to the fillet, then gradually changes

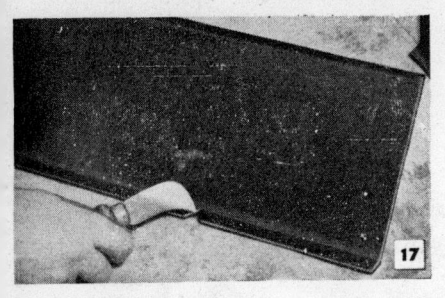

Above, asphalt tile used for top-set cove is notched, heated and bent to form inside corner as shown below

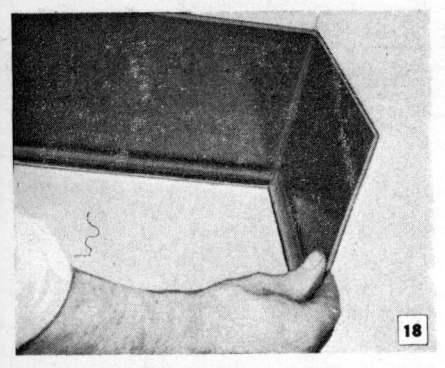

Below, score back of asphalt tile and heat to form outside corner. Also cut slit in the surface on floor

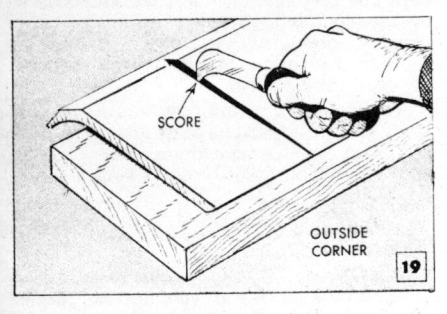

Below, close-coupled inside and outside corners present neat appearance when shaped from one piece

across the fillet to become a right-angle cut to form a plain butt the rest of the way across the cove.

Whether you use metal or self corners, the most difficult part of laying a base cove is shaping the linoleum to the required bend. The linoleum must be warm. If the room is cool, use a heat lamp to warm the linoleum before bending it. Form the bend as well as possible by hand before fitting. A better method is to use a forming board, shown in detail C, Fig. 13. Be sure the linoleum fits snugly against the fillet before scribing it. Also, double-check your cementing job; a few sandbags will help keep the linoleum in place while the cement is drying. Detail A in Fig. 13 shows a manufactured metal tool that is used to press the linoleum firmly against the fillet strip. Detail B is an easily made wooden tool used for the same job. Detail D shows a simple scribing tool that can be made if you do not have a regular one.

Top-set cove base: This style is easy to lay, being the same as the flashed type, except that the cove base is merely set on top of the floor covering, Fig. 1. If an asphalt material is used, corners are made by heating the spot with a blowtorch or heat lamp. To make an inside corner, notch the front of the molding, Fig. 17, apply heat and bend, Fig. 18. For an outside corner, score the back, heat and bend, Figs. 19 and 20. If the corner breaks slightly when bent, apply heat and mold shut with a hot knife, Fig. 21. Dust scraped from a scrap piece of asphalt can be used to fill in and "weld" a wider gap.

Rubber cove base must be cut and fitted the same as linoleum. Rubber does not cut as easily as linoleum and is usually fitted by using a special deep miter box and a fine-toothed saw. Some types of rubber tile are obtainable in preformed inside and outside corners, thus eliminating the problems of mitering. Preformed corners are cemented in place first and base strips are installed and fitted between them. ★ ★ ★

Scrap piece of asphalt tile is welded into opening left when outside corner is bent. Hot knife is used

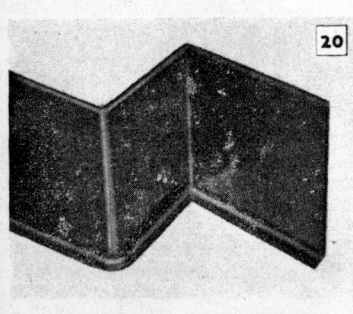

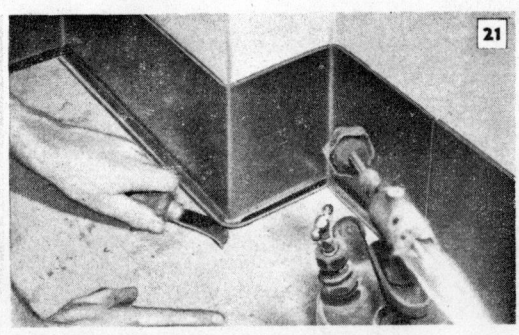

HOW TO USE LIVE BAIT

USUALLY GAME fish can be induced to strike at almost any properly selected artificial lure you can toss at them. But when you've given every plug or spinner in the tackle box a try without getting a single strike—and every fisherman has had such days—it's your cue to dust off a few tricks with live baits. Record catches have been made with natural lures hooked and fished by approved methods and every angler knows of at least one wise old lunker that always ignored every artificial lure thrown at him but finally fell for a properly presented live bait.

Few panfish, and even trout, can resist a lively worm or nymph and if a run of bad luck in good bass water forces you to turn to a frog or crayfish for bait you may be in for a surprise. One of the most popular of the live baits is the common earthworm, or angleworm. It will be taken by nearly all fresh-water fish and is easily obtained by digging in rich, damp soil at any time from early spring to late fall. It also can be kept and propagated if desired. The largest of the common earthworms is the dew worm, or nightcrawler. This worm is one of the best live baits for black bass, panfish, big trout and catfish.

As a rule you'll find dew worms on the lawn or in the garden only after nightfall and most anglers seek them after a heavy rain. By walking softly and using a small flashlight to locate them, you can obtain all the dew worms you'll need for a full day's fishing. Common earthworms can be

Spotted salamanders are highly prized by live-bait fishermen but are rather difficult to find. Look for them under stones along banks of streams and lakes
L. W. Brownell photos

The hellgrammite, an insect in larval form, is rated as tops in live baits for black bass, trout and panfish. It is hooked under the band for the best results

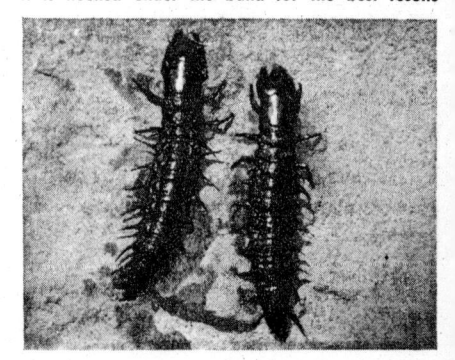

HOOKING WORMS

GRASSHOPPER

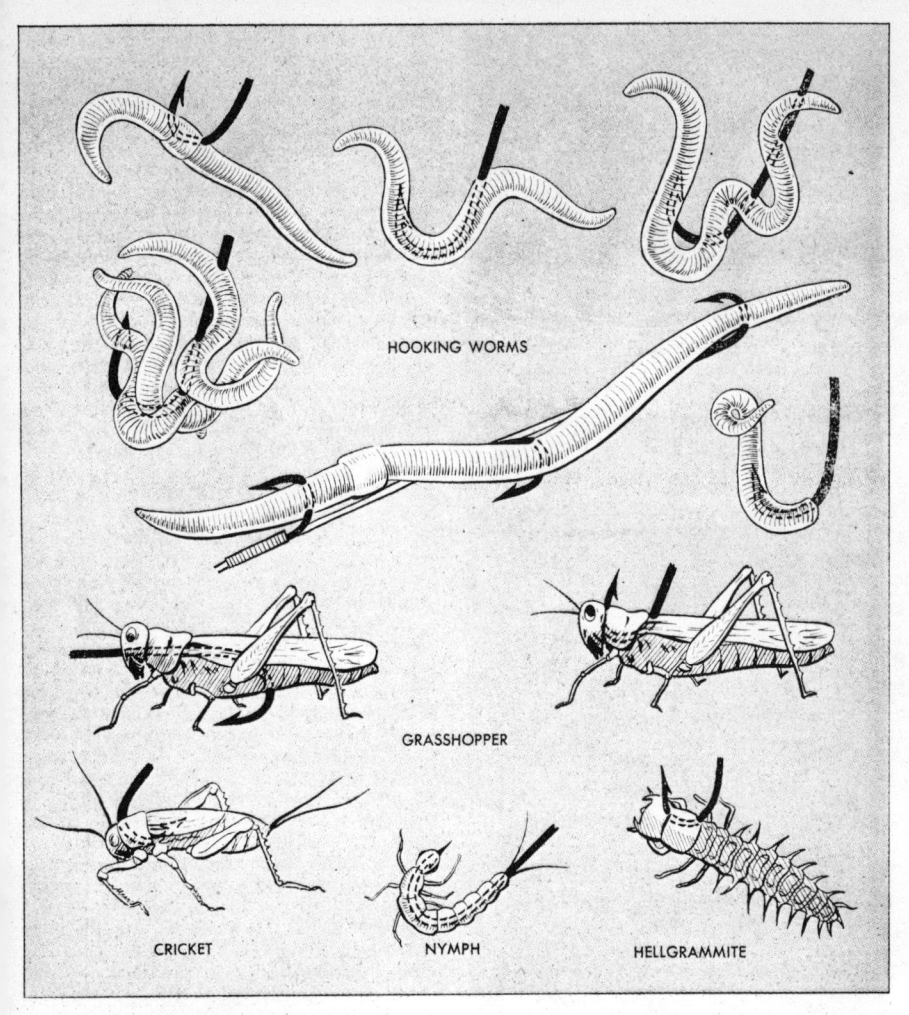

CRICKET　　**NYMPH**　　**HELLGRAMMITE**

Leopard frogs, also other kinds, are favorite live baits for bass fishermen. They are easily caught along stream banks by using a landing net or a dip net

A grasshopper dropped lightly onto the surface of a quiet pool may take fish when all other baits fail. The smaller ones usually are more acceptable to game fish

American Museum of Natural History photos

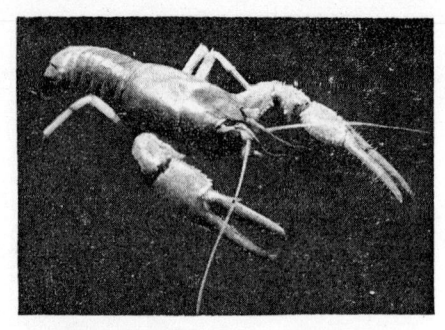

A sure-fire bait for smallmouth bass and big trout is the soft-shelled crayfish. For best results it should be lashed securely to the hook with a fine thread

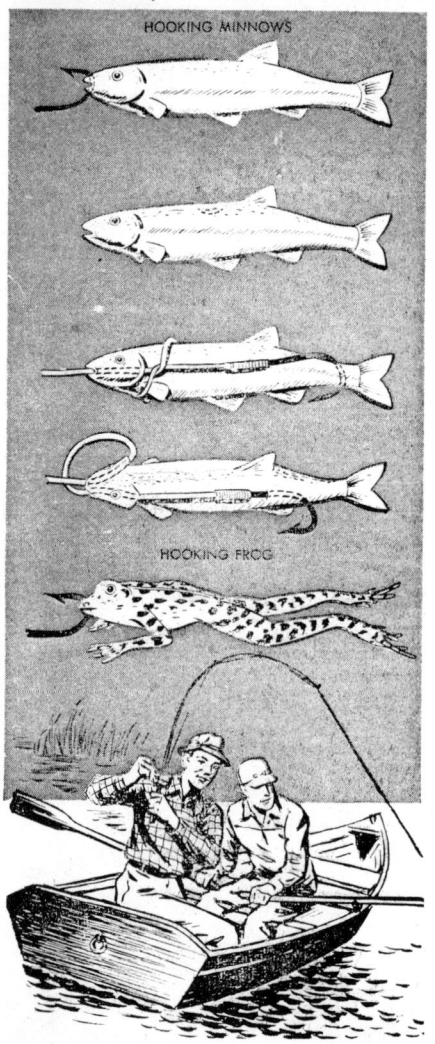

HOOKING MINNOWS

HOOKING FROG

kept for long periods of time in large containers filled with rich soil if they are supplied with foods such as corn meal or poultry mash. The soil should be kept damp, but not wet, and the food should be buried 2 to 3 in. beneath the surface.

Methods of hooking and fishing worms will depend to some extent on the fish you are after, the condition of the water and the techniques used. For trout in moving water, hook the worm under the band as in the upper left-hand detail on the preceding page. But when fishing in still water, thread the worm along the bend and shank of the hook, leaving both ends of the bait to wriggle in a lifelike manner. For panfish, loop a small worm along the bend and shank of the hook in such a way as to conceal the point and barb. For bass, catfish and bullheads, loop several worms on one hook. If you use gang hooks, hook a large worm as in the center detail on the preceding page.

Next in popularity and effectiveness among the live baits are minnows varying from 2 to 10 in. in length. They can be taken with seines or drop nets in nearly all small fresh-water streams, lakes and rivers. Various types of baited funnel traps also are used. Before trapping or netting minnows in public waters be sure to check the state laws, as most states have strict regulations governing the seining or netting of minnows. Minnows can be kept alive for a day or more in an ordinary minnow bucket, but for longer periods they must be kept in a "live" box, which is simply a wooden frame covered with galvanized wire mesh and sunk either partially or wholly below the surface in shallow water. The minnows should be fed regularly with a prepared fish food.

Details at the left show various methods of hooking minnows. For still fishing hook them through the lips or the back near the dorsal fin. For casting, drifting or trolling, minnows are lashed or "sewed" on the hook as in the two lower details. The size of the minnow used for bait depends on the fish you are after. For panfish, use minnows 1½ to 2 in. in length. For small and medium game fish use minnows 3 to 5 in. in length. For fresh-water giants such as pike, muskellunge and the big-river catfish, minnows 5 to 10 in. long are preferred. There also are many other small fish which are not true minnows but which make good live baits. These include the suckers, stonecats, lampreys and darters, and in some cases, yellow perch and bullheads.

Many fishermen prefer water and land insects as live baits. One favorite for bass is the hellgrammite which is found under rocks and logs along the stream banks. This insect, which is a larval form, is hooked under the collar, or band, for best results.

Every angler is familiar with the bass and perch bugs of early spring. These are nymphal stages of dragonflies and damsel flies and are good bait for many of the smaller game fish. Most water insects are caught by using a wire or cloth screen placed in the riffles of small streams. Lifting or overturning rocks in the stream bed above the screen dislodges the insects and the swift current will sweep them into the screen. The land insects such as grasshoppers, crickets, grubs and caterpillars are good live baits for most game fish. Many fishermen find it easy to catch the livelier ones, such as grasshoppers and crickets, with a landing net into which has been fitted a lining of coarse cloth or netting. The landing net also is quite effective for catching frogs along the stream banks and in weedy inlets of small lakes and ponds.

The leopard frog is one of the best live baits for bass, pickerel and pike. One approved method of hooking frogs is shown on the opposite page, the hook passing through both lips with the barb on top. There also are special frog harnesses that hold the frog without injury, yet permit it free movement while in the water. Ranking with the various frogs in effectiveness as a live bait is the crayfish, particularly in the soft-shelled stage. Look for crayfish in shallow water along the stream bank, and under stones and among weeds at the water's edge of small lakes and ponds. Soft-shelled crayfish must be tied securely to the hook with fine thread. Hard-shelled crayfish are simply hooked through the tail with the point of the hook up.

In addition to those already pictured and described, many other live baits are used by fresh-water fishermen. These include the salamanders and newts, leeches, fresh-water clams, snails, slugs and even small snakes and mice. Fishermen also use salmon eggs when angling for steelhead trout in Western lakes and streams. However, some states have outlawed the use of salmon eggs, so check the state laws carefully.

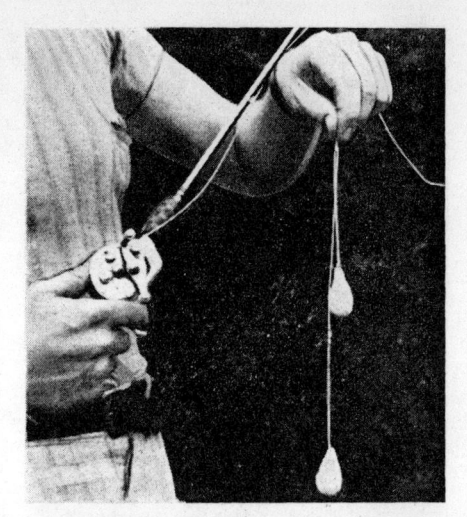

Doughball baits are familiar to big-river fishermen who go after carp, catfish and buffalo. Dough should be stiff enough to mold smoothly over the hook

The best-known of the so-called prepared lures are the doughball baits used mainly for carp, catfish and buffalo. These consist of mixtures of flour, corn meal or bran, and a sweetening such as honey or corn syrup. A common recipe calls for equal parts of plain flour and corn meal and sufficient syrup to make a rubbery mixture when boiling water is added. It is important that the mixture be sufficiently tough to be molded onto the hook as pictured above. For catfish, fishermen sometimes add cheese, ground meat, oil of rhodium, oil of anise and asafetida to the dough when mixing. Prepared lures of this type usually are fished deep in still water, also in rivers where sluggish current conditions permit fishing the bait on or along the stream bed. When fishing, the bait must be renewed frequently. ★ ★ ★

Keep crawfish alive in this improvised "icebox." Pack the bottom of a pail with ice, then cover it with screen wire and a layer of grass. The crawfish are placed on top of the grass and, in turn, are covered with several thicknesses of newspaper and a wet burlap sack

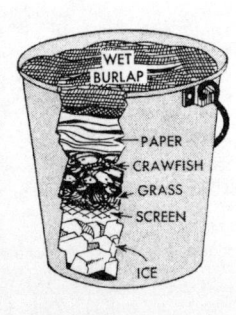

Keep live grasshoppers or crickets in a jar with a piece of inner tube stretched tightly over its top. Cut a slit in the inner tube wide enough to permit inserting your thumb and forefinger for grasping

Tempt 'Em With Live Bait

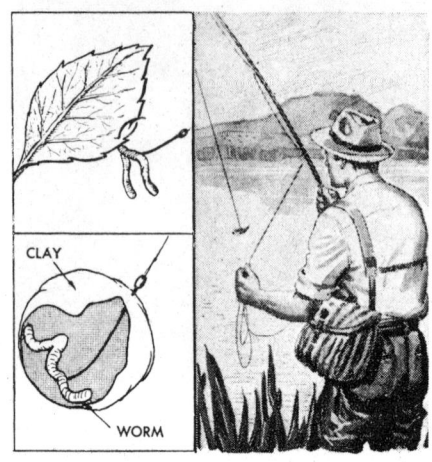

Even in the most hard-fished streams, trout find it difficult to resist a worm offered in either of these two ways. One successful method utilizes a twig with a leaf or two attached. After the worm-baited hook is pushed through the edge of the leaf, as in the upper detail, the twig is allowed to float downstream. When the leaf drifts over a likely looking pool, the hook is jerked free by a twitch of the rod tip. The bait then sinks slowly to the bottom, coaxing the trout into striking. The other method is to mold a ball of clay or mud over the worm and hook, as in the lower detail. The ball is lowered slowly into the water and washes away to expose the worm

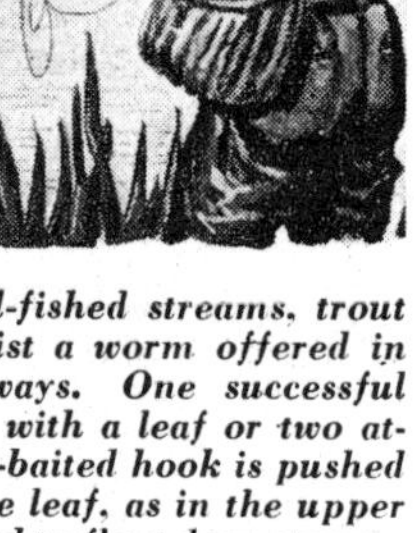

The difficulty of seining crawfish from a pond or stream that contains vegetation, where they usually are plentiful, is eliminated with this seine, which consists of a wooden scoop having a burlap sack at the rear end. If the seine has a tendency to float or not rest heavily enough on the bottom of the pond to flatten the growth as it is pulled along, weight it with a stone or heavy piece of metal. Drilling a number of small holes in the bottom of the scoop also will help it sink to the bottom

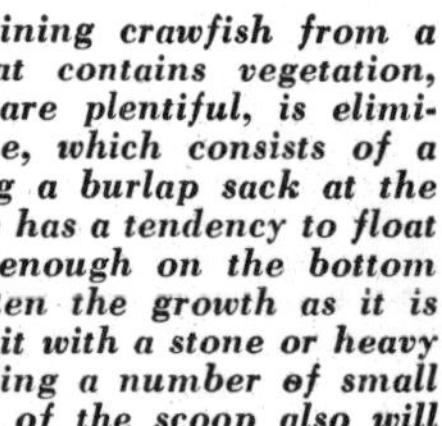

The problem of carrying a jar of salmon egg fish bait so that it is easy to get at is solved with this simple holder, which slips over your trouser belt. Take an ordinary clothespin and screw a couple of strips of heavy sheet metal or spring steel to it as indicated, making the top one so that it pivots easily. Then cut a notch in the pin just above the circular metal strip to take the edge of the jar lid. When you want to remove some of the eggs, just swing the pivoted metal piece to one side and take off the lid

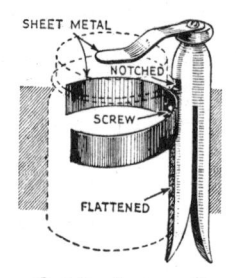

A handy minnow net consists of a wire coat hanger, which is straightened and bent to form a hoop at one end, and a circular piece of plastic screen mesh sewed to the wire hoop

The next time you have a hard time finding worms for fish bait, force the tines of a pitchfork into the ground and twang the handle. This sets up vibrations that cause the worms to come up to the surface

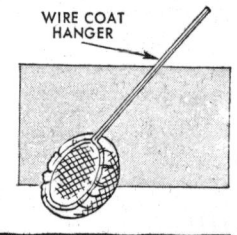

Minnows used as bait are often washed from the hook when fishing in swift water. By wrapping and tying a piece of cellophane around the line and sliding it down so that it covers the minnow on the hook, the bait will be preserved without affecting its efficiency as a lure. This idea may be used to protect worms and grasshoppers, also

LIVE BAIT BOXES

Designed to hook over the transom of a rowboat, this convenient live box rides partially submerged to assure the fish you catch a constant supply of fresh water. Being located at the rear of the boat, the box will not impose any appreciable drag when rowing. Construction details show how it is assembled. Redwood is the most durable wood to use, although pine will do. The rounded side of the box is covered with hardware cloth and trimmed with molding. Flat-iron brackets, which hook over the transom at the top, fit in sockets at the bottom and are anchored with retaining pins. Note that the cover is made large enough to overlap the opening on all three sides.

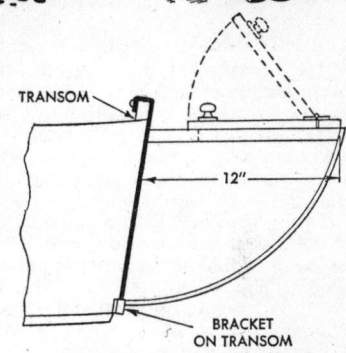

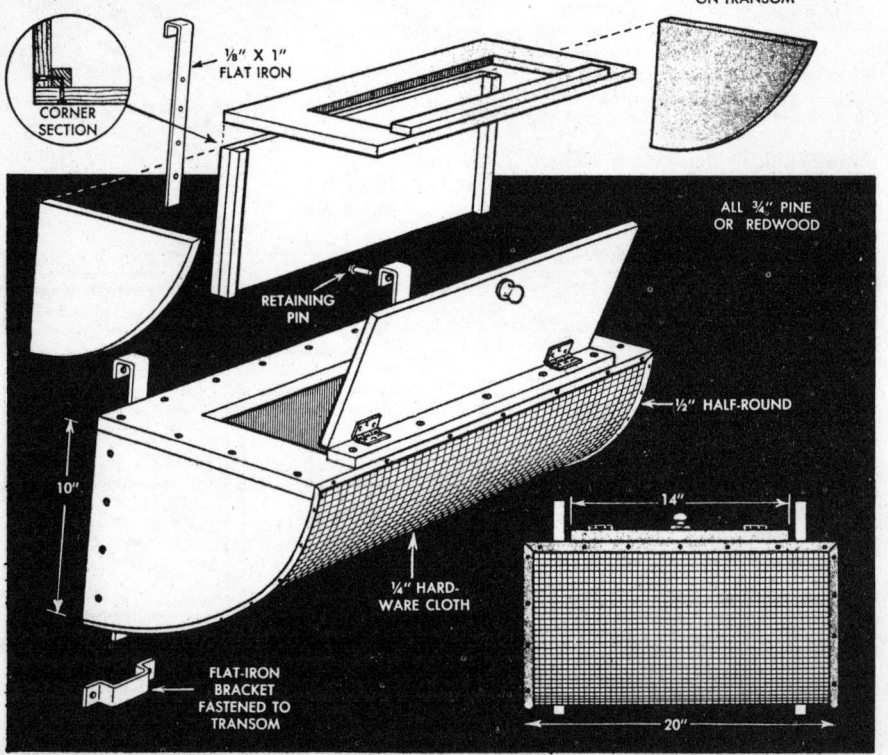

Living-Room FURNITURE

Part One

MODERN as tomorrow, here's a smartly styled ensemble that will transform your living room into a room of distinction. It's exclusive, colorful and, to top it off, this furniture is easy to build. Designed expressly with the home craftsman in mind, all the pieces of the group can be duplicated with just a jointer and a saw. Combining plywood with solid stock further simplifies construction, and the use of separate upholstered units enables the inexperienced worker to do a first-class job of upholstering the sectional units and easy chair.

The suite, most of which is shown in full color on preceding pages, includes a sectional sofa, easy chair, hostess chair, step table, commode, cocktail table, corner table, desk, table lamp and floor lamp. The desk chair shown was an extra piece and is not included in the group. However, a dining chair of the style presented in the dining-room furniture suite illustrated in another part of this book will provide a

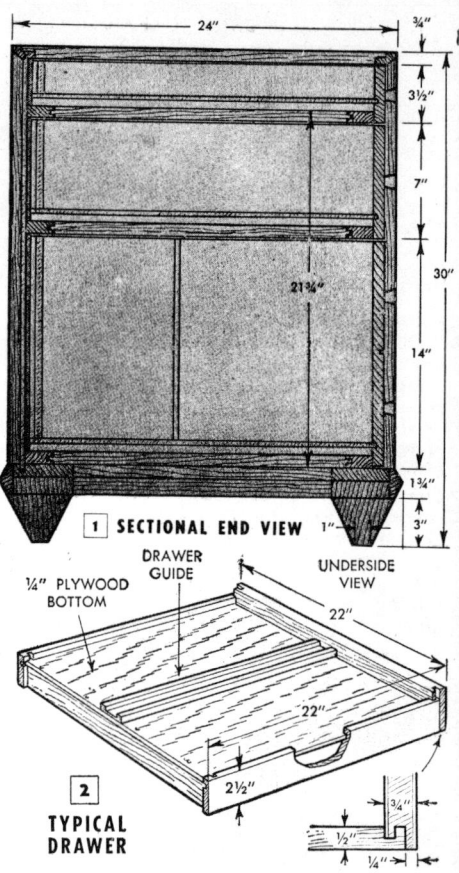

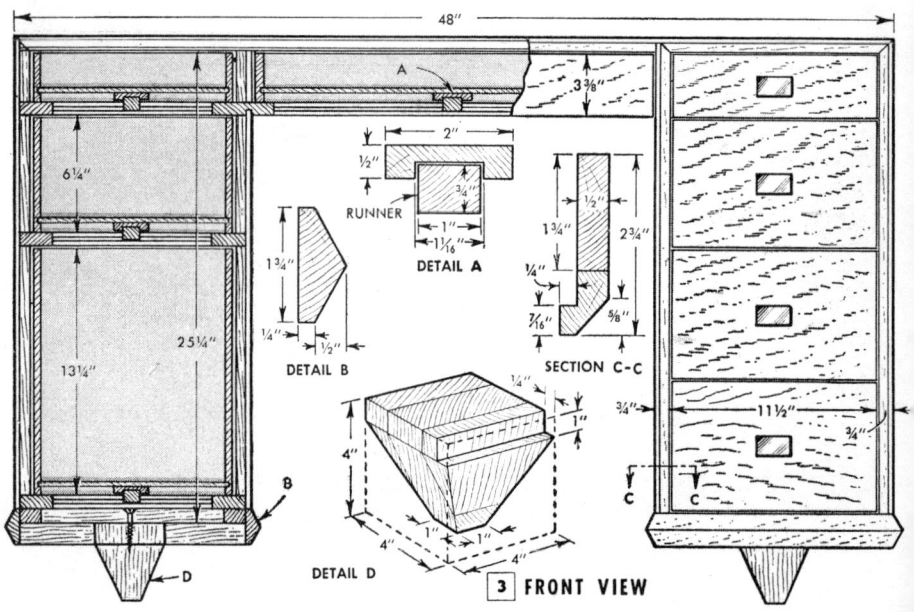

4 CUTAWAY ASSEMBLY OF DESK

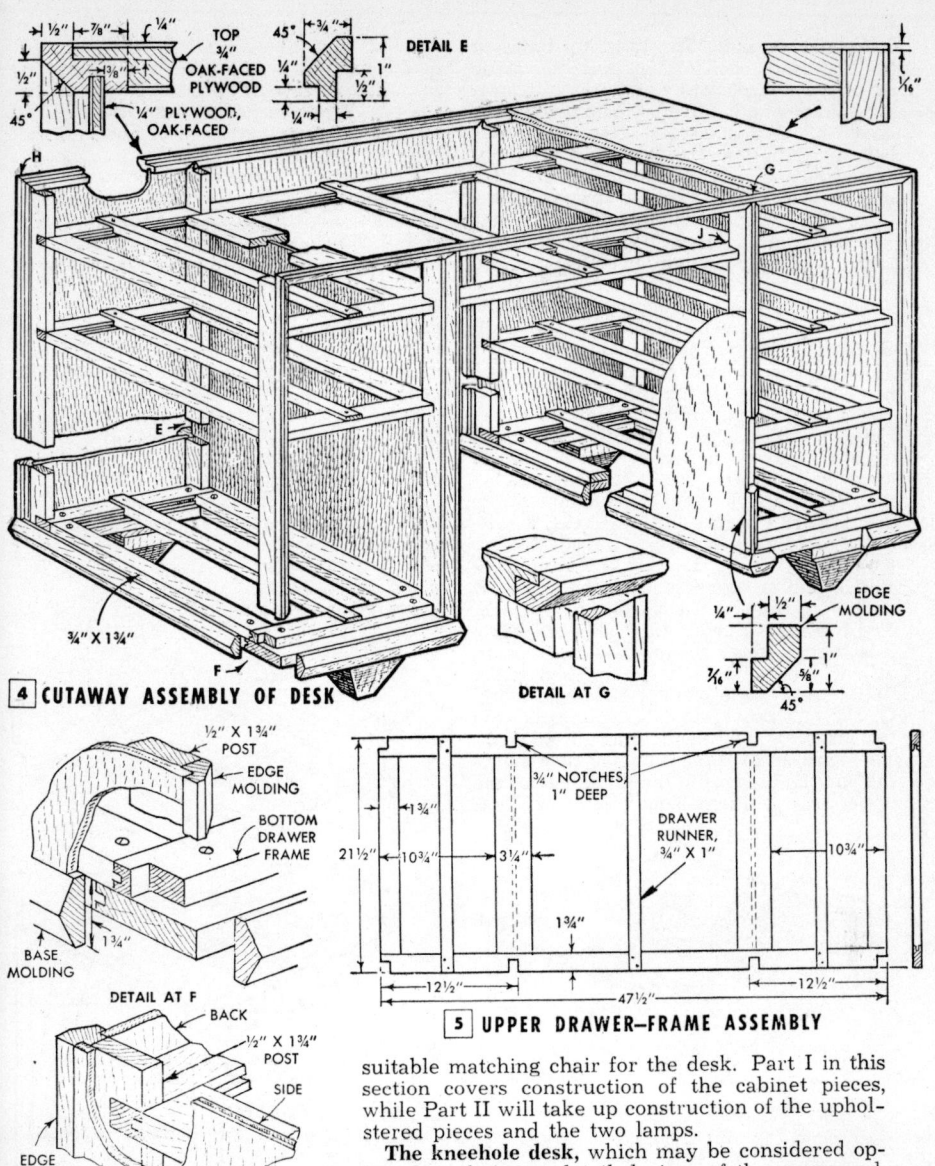

5 UPPER DRAWER—FRAME ASSEMBLY

suitable matching chair for the desk. Part I in this section covers construction of the cabinet pieces, while Part II will take up construction of the upholstered pieces and the two lamps.

The kneehole desk, which may be considered optional, is the most detailed piece of the group and, therefore, requires the most explanation. Figs. 1 and 3 show sectional views, while the cutaway view in Fig. 4 gives a general idea of the assembly. The original furniture was made of oak-faced plywood and solid stock and finished in limed oak, but birch, maple or other hardwood can be used. The upper drawer frame of the desk is made first, Fig. 5, and although the drawer runners are shown in place, these can be added later. The dotted lines represent ¼-in. grooves which are made on the underside. As no part of this frame is exposed, it can be made of inexpensive wood. The built-up bases at the bottom of each drawer unit are made exactly alike, consisting of a drawer frame, screwed to a bottom frame. The two frames must fit flush at the sides. The front member of the bottom frame should be about 4 in. wide and cut from finished stock. Note in Fig. 6, detail F, that the drawer frame is notched at the corners for ½-in. posts which also are notched. Next, the posts are joined to the upper and lower frames, and intermediate drawer frames are installed.

Now, a molding is applied around the base of each drawer unit, as in Fig. 3, detail B. This is mitered at the corners, and is glued and screwed to the bottom frame to bring it flush with the top. When this is done, the four tapered legs can be fitted. Fig. 3, detail D, shows how they are built up and rabbeted across the front edge to come flush with the face of the base molding. Glue and a long, flat-headed screw are used to attach each leg. Next, a rabbeted molding is applied to the edge of the front posts. This is detailed in section C-C, Fig. 3, and is edge-glued to the posts. A similar molding cut according to detail E, Fig. 4, is edge-glued to the back posts. Note that the molding strips on the inner posts, at both front and back, are notched at the upper ends in the manner shown in detail G, Fig. 4. The ¼-in. plywood panels on the inside of the kneehole are fitted next. These are cut to fit the rabbeted edge molding and the grooves in the underside of the upper drawer frame. You'll notice in Fig. 6,

detail J, that the plywood is cut L-shaped at the top to continue it to the top of the front posts. In gluing and clamping in place, the panels are pushed up into the grooves so that the lower edge rests flush on the edge of the base molding. The panels covering the outside of the drawer units are installed similarly, except that they run to a point about ¹⁄₁₆ in. down from the mitered end of the edge molding. This permits the desk top to overlap the end plywood and conceal the plies in a rabbeted joint. Plywood panels which enclose the rear ends of the drawer units are slid into grooves in the edge molding, and then strips of rabbeted molding are run across the top of the desk at the front and back. Note that the rear strip differs slightly from the front one. The top, which is ¾ in. thick, is rabbeted all around, leaving about the thickness of one ply at the ends and ¼ in. at the sides.

Fig. 2 details drawer construction. The bottom left-hand one is a deep single drawer which is grooved across the face to represent two smaller drawers. Guides are detailed in Fig. 3, A.

The commode, detailed in Figs. 7 and 8,

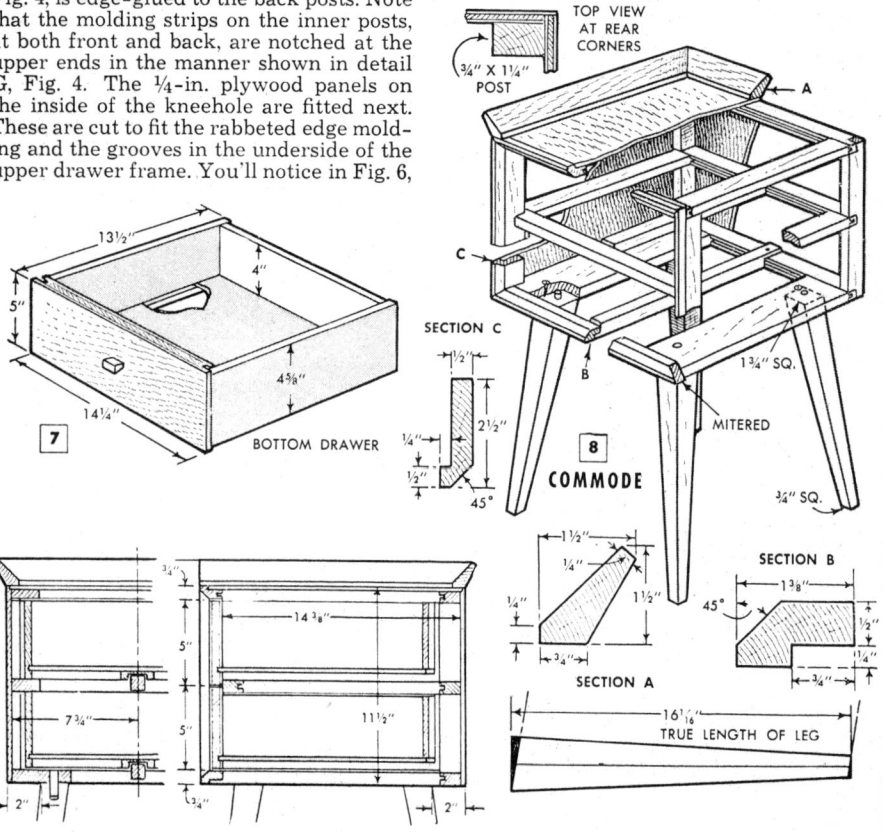

9 LEG

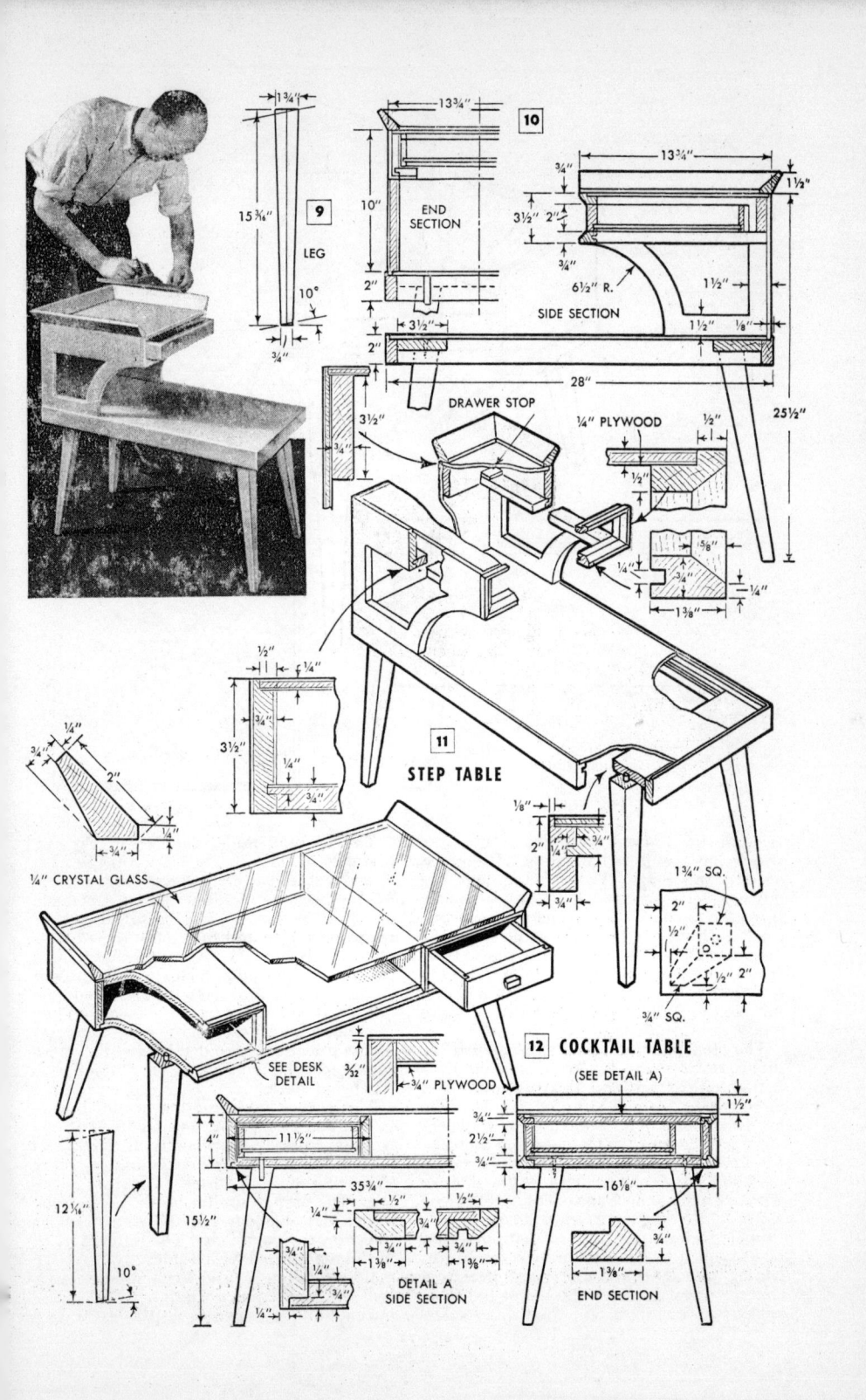

- 1¾"
- 15 3/16"
- 10°
- ¾"

10 END SECTION / SIDE SECTION

- 13¾"
- 10"
- 2"
- 3½"
- ¾"
- 1½"
- 2"
- 3½"
- ¾"
- 6½" R.
- 1½"
- 1½"
- ⅛"
- 28"
- 25½"

DRAWER STOP

¼" PLYWOOD

11 STEP TABLE

- ½"
- ¼"
- ¾"
- 3½"
- ¼"
- ¾"
- ⅛"
- 2"
- ¾"
- ¼"
- 1¾" SQ.
- 2"
- ½"
- 2"
- ¾" SQ.

12 COCKTAIL TABLE (SEE DETAIL A)

¼" CRYSTAL GLASS

- ¼"
- ¾"
- 2"
- ¾"
- ¼"

SEE DESK DETAIL

- 3/32"
- ¾" PLYWOOD
- 4"
- 11½"
- 2½"
- ¾"
- 1½"
- 15½"
- 35¾"
- ½"
- ½"
- 16⅛"
- ¾"
- ¾"
- 1⅜"
- 12 5/16"
- 10°
- ¼"
- ¾"
- ¼"
- 1⅜"
- ¼"
- ¾"
- 1⅜"

DETAIL A SIDE SECTION

END SECTION

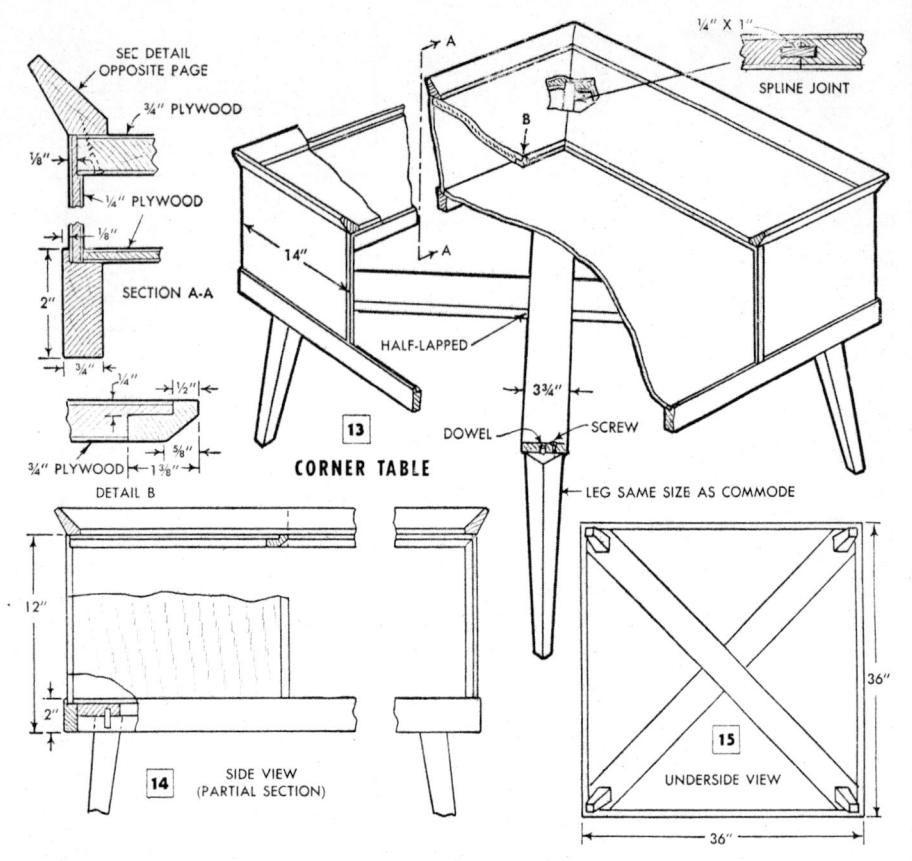

SEE DETAIL OPPOSITE PAGE
¾" PLYWOOD
⅛"
¼" PLYWOOD
⅛"
2"
SECTION A-A
¾" ¼" ½"
⅝"
¾" PLYWOOD 1⅜"
DETAIL B

¼" X 1"
SPLINE JOINT

14"
HALF-LAPPED
3¾"
DOWEL SCREW
13
CORNER TABLE
LEG SAME SIZE AS COMMODE

12"
2"
14 SIDE VIEW (PARTIAL SECTION)

36"
15 UNDERSIDE VIEW
36"

is built up around a rough framework which has a mitered molding edge-glued around the front. The tapered legs are doweled and screwed 2 in. in each way from the corners. The framework is covered with ¼-in. plywood which is rabbeted, as shown, to conceal the plies at the rear corners and around the top. The top is finished off on three sides with a slanting molding and, finally, two drawers are made and installed.

The step table, detailed in Figs. 9 to 11 inclusive, is made by grooving and rabbeting solid stock and then assembling it on edge to form a mitered frame. The grooves house tenons on two crosspieces to which the legs are fastened and the rabbet around the top edge receives a ¼-in. plywood panel. The latter is glued flush with the top of the frame. The scrollsawed sides, which are cut from solid stock, are rabbeted along the top and bottom edges and grooved to take the drawer frame. An edging glued around the front frames a shallow drawer. After the top of the table is covered, edging is added on three sides.

The cocktail table, detailed in Fig. 12, features a glass top, a center pocket for magazines and two small drawers. Except for the legs and molding, the table is made of ¾-in. plywood. The bottom board is rabbeted all around the edges, the rabbets at the ends being a reverse of those along the sides. The outer sides of the drawer compartments are likewise rabbeted and are glued and nailed from the bottom. Then the endpieces for the backs of the compartments are tenoned and rabbeted to take a rabbeted edge molding which frames both front and back.

The corner table for the sectional sofa is detailed in Figs. 13, 14 and 15. The top is made of ¾-in. plywood while sides and bottom of the table are ¼-in. plywood. Construction is apparent from the drawings. The base is similar to that of the step table, and the legs are the same size as those on the commode. The mitered joint in the top is fitted with a spline and an edge molding, detail B, Fig. 13, is used to conceal the laminations. Here again, a slanting molding is added to finish off the top.

Living-Room Furniture

Part Two

WITH THE cabinet pieces of the living-room group completed as described earlier, you can begin the sectional sofa and chairs. Normally, conventional upholstering that requires tying springs and sewing welt seams is a job for the experienced worker, but here, through the use of sagless-type springs and "box" cushions, the work is simplified to the point where an amateur can do a first-class job. The upholstered pieces of the group include a three-piece sectional sofa, an easy chair and a hostess chair, the latter being made with either wooden or padded arms. The two photos on the opposite page show how the hostess chair will look in each case. Basically, the easy chair, shown at the top of the same page, is a duplicate of the sectional units, except that it is fitted with two arms. Seat and back cushions of each unit are built up on separate frames which are then screwed to the chair frame. In the case of the hostess chair, the upholstering method differs in that the seat and back are sprung and padded right on the chair.

The sectional sofa is highly functional and can be arranged in a number of ways. The units may be grouped around a corner table as pictured on the title page, or they may be used side by side and flanked with a pair of end tables as shown on the preceding page. Still another suggested arrangement is given in the photo above. Here, the two end units having the arms are placed together and the center unit is used as an individual chair. In addition to the upholstered pieces.

Frames for the sectional sofa, easy chair and the hostess chair are detailed in Figs. 18 to 22 inclusive. The original pieces were made of solid oak to match the rest of the furniture. Parts of the frames not exposed can be made of any scrap hardwood. Both end units of the sofa are made as pictured in Fig. 19, one being built with a right and the other with a left-hand arm. The center unit of the sofa, and the easy chair, are exact duplicates of the others, the center unit, of course having no arms. Construction of the hostess chair, Fig. 21, is basically the same, differing primarily in size and the addition of a tacking rail across the back. As previously mentioned, this chair can be fitted with either wooden or padded arms. In both cases, the arms are made up separately and doweled through the holes in the fabric and into the frame after the upholstering has been completed. Then screws are driven into each arm from the inside. Fig. 18 details the framework for the padded arms and Fig. 17 shows them being trial-fitted, registering holes being made for the dowels. Note that the seat corner joints of all the frames are glued and doweled and then drawn up rigidly with lag screws. The front legs of the frames are half-lapped into the side rails, glued and screwed. The arms on the easy chair and the end units of the sectional sofa are screwed permanently to the frames before upholstering. Holes for the screws are counterbored through the width of the arms. When the frames are completed, all exposed surfaces of the chairs are sanded and finished as desired. The rear

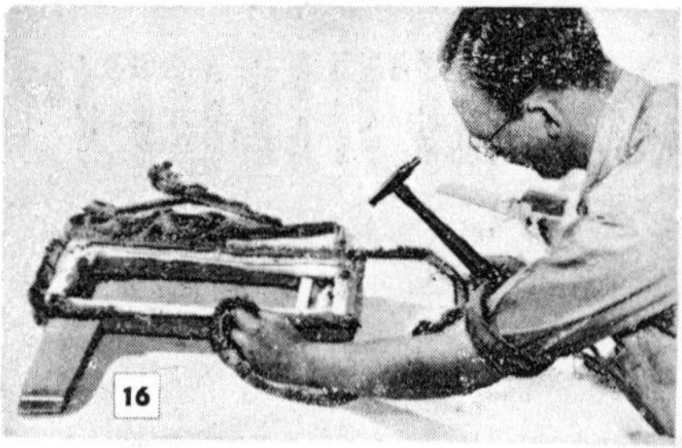

16

17

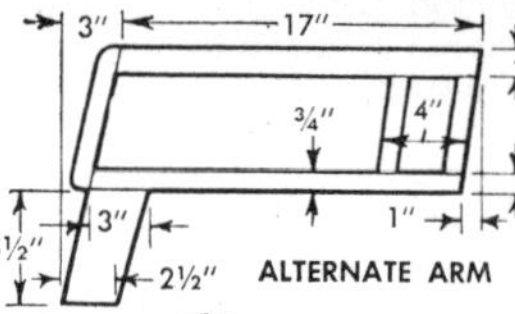
3"
17"
1¼"
6½"
¾"
4"
5½"
3"
1"
¾"
2½"

18 ALTERNATE ARM

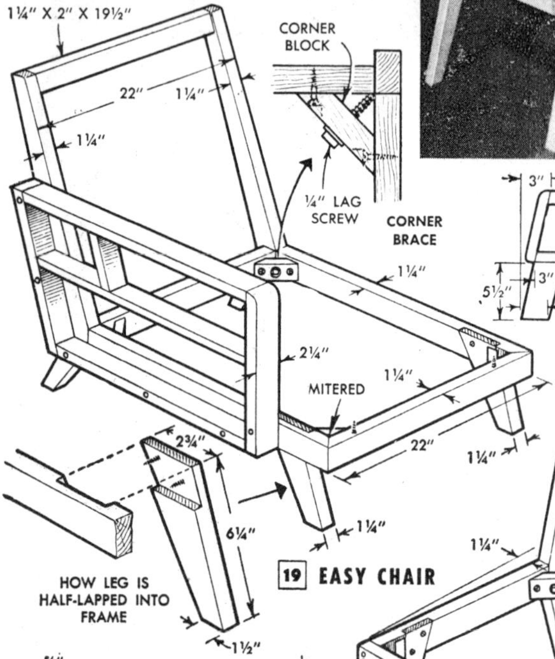
1¼" X 2" X 19½"
22"
1¼"
1¼"
CORNER BLOCK
¼" LAG SCREW
CORNER BRACE
1¼"
2¼"
1¼"
MITERED
22"
1¼"
2¾"
6¼"
1¼"
1½"
HOW LEG IS HALF-LAPPED INTO FRAME

19 EASY CHAIR

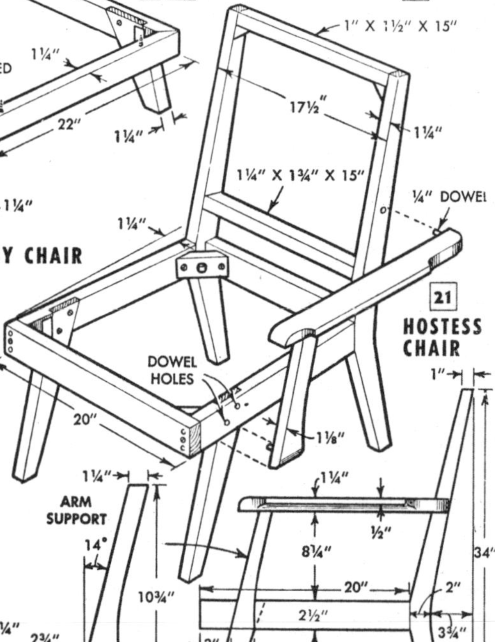
1" X 1½" X 15"
17½"
1¼"
1¼" X 1¾" X 15"
¼" DOWEL
1¼"
DOWEL HOLES
20"
1⅛"

21 HOSTESS CHAIR

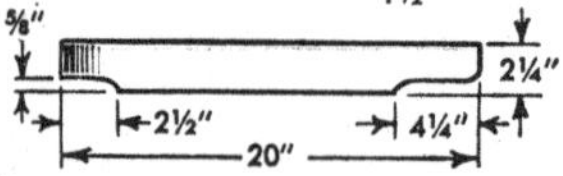
⅝"
2¼"
2½"
20"
4¼"

HOSTESS-CHAIR ARM

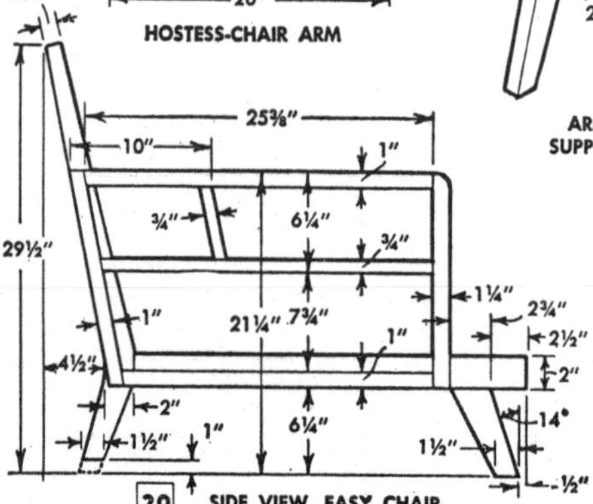
1¼"
25⅝"
10"
1"
29½"
¾"
6¼"
¾"
1¼"
2¾"
1"
21¼"
7¾"
2½"
4½"
1"
2"
2"
6¼"
14°
1½"
1"
1½"
½"

20 SIDE VIEW, EASY CHAIR

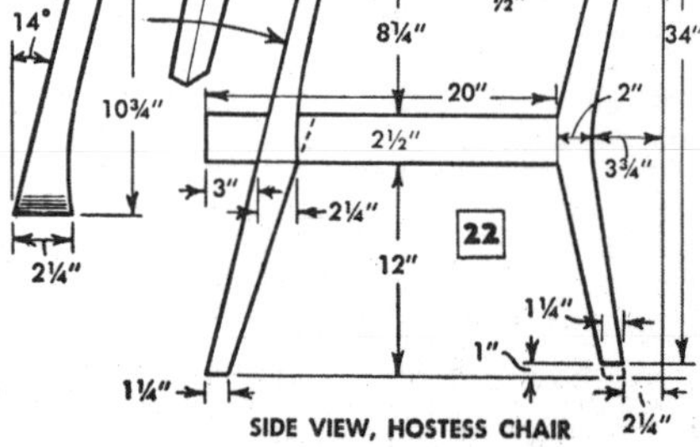
1¼"
ARM SUPPORT
14°
10¾"
2¼"
1"
1¼"
½"
8¼"
34"
20"
2"
2½"
3¾"
3"
2¼"
12"
1¼"
1"
1¼"
2¼"

22 SIDE VIEW, HOSTESS CHAIR

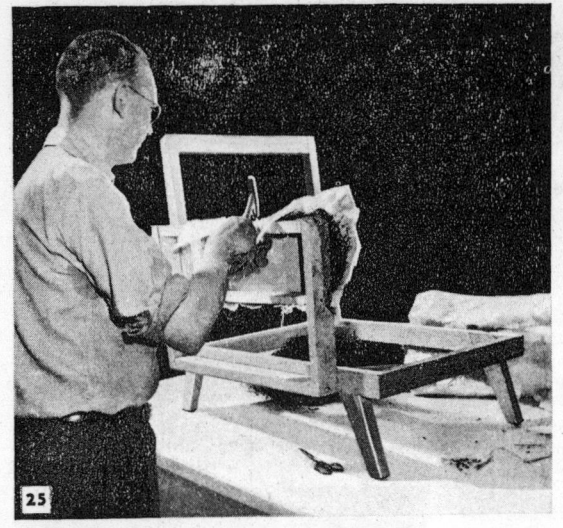

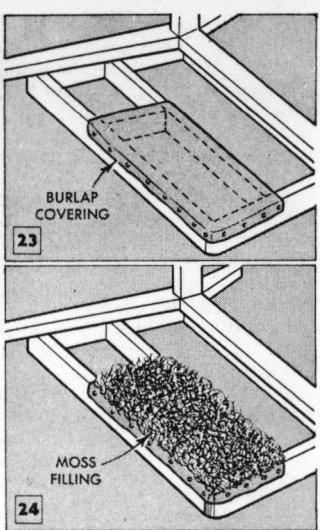

Above, this operation shows the padded arm of one of the sofa units being covered with muslin. Below, pocketed coil springs sold in strip form are sewed together to make up the required-size marshall unit

Below, the marshall unit is placed on top of the spring unit to which it is sewed around the four sides. Below, right, moss or tow filling is placed over marshall unit and then two layers of cotton

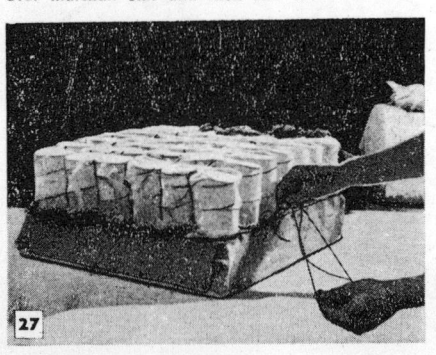

legs of the chairs can be cut off now to give the correct slant or after the units have been upholstered and tested.

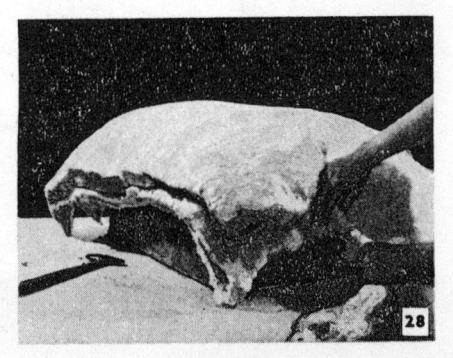

Upholstering of the sofa units and the easy chair is done exactly alike. First, the arms are covered on the inside with burlap as in Fig. 23. Next, the burlap is padded with moss or tow filling which is held in place with long stitches through the burlap, Fig. 24, and then a layer of cotton placed over the filling is covered with muslin which is pulled firmly and tacked to the arm, Fig. 25. Edging material called "brush" is applied to the inside and outside edges of the arms. The details in Fig. 31 show how this is done. On the inside, the brush is tacked to the face of the arm on three sides so that it overhangs the padding. On the outside, the material is tacked to the edge of the arm on all four sides so it stands upright. The space around the edge of the arm between the brush trim is covered as follows. A piece of stiff cardboard long enough to reach around three sides of the arm is padded with cotton and covered with fabric. Gimp nails are used to tack the strip in place, driving them along each edge and then concealing the heads by pulling the fabric up over them. This leaves the outside of the arm which is covered similarly with a cardboard panel, tacking it in the same manner as just described.

Cushions for the seat and back of the sofa and easy chair are built up separately on flat hardwood frames, half-lapped at the corners, Fig. 32. The frames are made the same size as the over-all size of the chair frames. The seat frames require five 8-ga. sagless-type springs, 26 in. long, while the back frames take five, 12-ga. springs 16½ in. long. The springs are anchored to the front and rear edges of the frame with special clips made for the purpose. These are first nailed to the wood and then crimped over the spring wire before renailing. Note that the ends of the springs are bent back to prevent them from slipping out of the clips. In all installations, it is

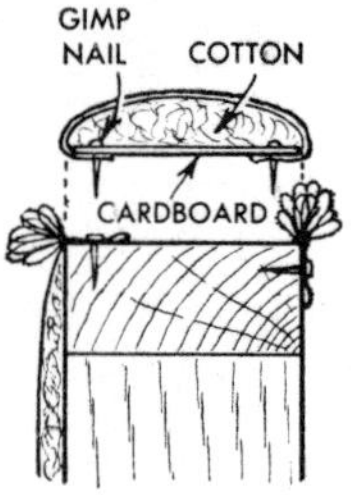

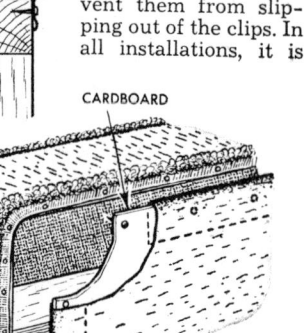

Above, here the seat and back cushions of the easy chair are being tested for satisfactory fit prior to covering with fabric. Screws hold cushions in place

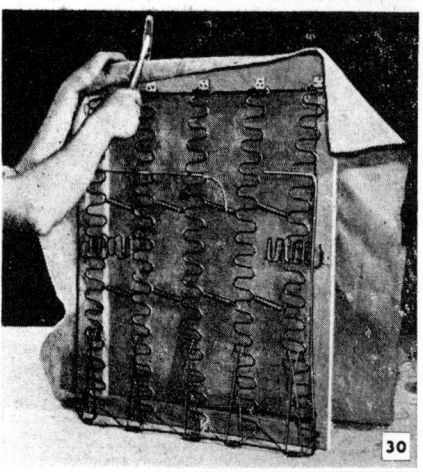

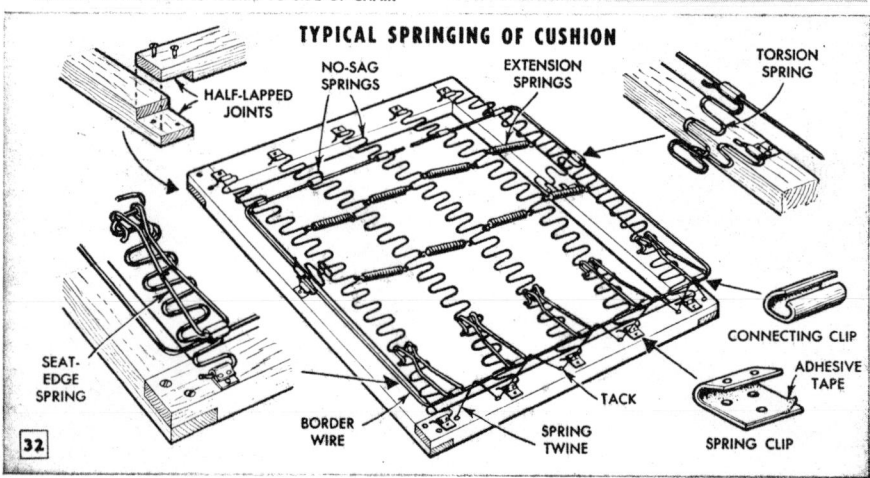

METHOD OF ATTACHING FABRIC TO SIDE OF CHAIR

TYPICAL SPRINGING OF CUSHION

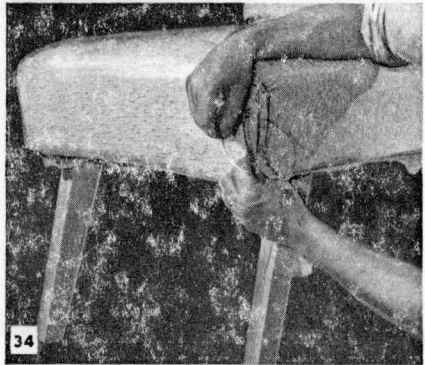

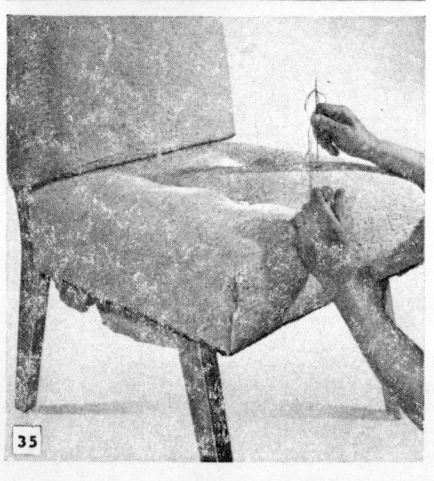

recommended that the direction of the bent ends be alternated. If the bent end of the first spring points to the right, the bent end of the next spring should point to the left. This permits the connecting extension springs to be applied in a straight line. The springs are applied bowed and are cross-tied with eight extension springs. Edge springs are hooked into the sagless springs across the front and these serve to support an 8-ga. border wire which forms the box shape of the spring unit. The border wire is attached to the edge springs with special clips which are crimped over the two parts. This type of clip also is used to fasten the rear edge of the border wire to the sagless springs. At each side, a torsion spring is used to support the border wire, clips being applied as before to anchor it to the wire and the wooden frame. You'll note that the border wire does not extend to the rear of the frame but stops about 8 in. or so from it. This is done to allow room for the back cushion to fit neatly into the seat cushion. Finally, the border wire across the front edge is pulled down under tension with regular spring twine and tacked securely to the wooden frame. The back cushions are sprung in exactly the same manner, the front edge of the seat becoming the top edge of the back.

The next step is to cover the spring unit with burlap as shown in Fig. 30. This is pulled firmly over the border wire and tacked to the edges of the frame. Now, the spring unit is covered with a marshall unit which is made up of a number of individual coil springs encased in cloth pockets. These are purchased in strips and you merely build up the unit to the required size by sewing them together as in Fig. 26. The seat requires a unit seven coils wide and nine coils deep, while the back takes seven wide and five deep. The completed unit is sewed to the spring unit, using a curved upholsterer's needle and sewing around the border wire and the base of each individual coil spring, Fig. 27. Next, moss or tow filling is placed on top of the marshall unit and then two layers of cotton, Fig. 28. This is followed by covering the cotton with muslin, pulling it down smoothly and tacking it to the underside of the wooden frame. Fig. 29 shows checking the cushions for fit after the muslin covering has been finished. Box corners are formed at the front corners of the seat and at the top corners of the back cushion by hand-stitching the fabric covering as shown in Fig. 34. The cushions are held in place in the chair with flat-headed screws. In the case of the seat, the screws are driven up from below in counterbored holes in the rails. The back cushion is held with four screws which are driven through holes in the back of the chair frame and capped with regular wooden screw plugs or short lengths of dowel rod. The back cushion differs from the seat in one respect. With the underside of the frame exposed, the back must be fitted with a fabric-covered cardboard panel before fastening the cushion to the chair frame.

The hostess chair is upholstered in the same general manner, except that the springing is done right on the chair frame as indicated in Fig. 37. Five 9-ga. sagless springs, 20½ in. long, are required for the seat and five 12-ga. springs, 14½ in. long, are needed for the back cushion. Also, 10 additional seat-edge springs are required as these are installed at both front and back, Fig. 37. The fabric covering is brought down over the rails of the seat and tacked to the underside. The back is covered similarly, Fig. 33. The rear of the back is enclosed with a fabric-covered panel in the same manner as previously described, nailing it with gimp nails and then pulling the material up over the heads. If padded arms are to be used on the chair, brush edging is applied around the outside edge of the arms, Fig. 16, and a cardboard panel is used on the outside.

Tufting the seat and back cushions gives them a professional touch. Four buttons are placed in the seat cushions of all the pieces and two in the back cushions. Plain wooden buttons made for this purpose are covered with scraps of fabric and then stitched through to the springs, Fig. 35. The tufts are made by pulling down on the buttons and tying the sewing twine securely to the sagless springs, Fig. 36. Finally, cambric dust covers are tacked to the underside of each seat cushion.

The lamps are detailed in Figs. 38 and 39. The table lamp features four 6-in.-sq. ceramic tiles framed in a box-shaped base. These can be purchased in any large department store and, as the exact size of the tile varies somewhat, buy them first and build the base around them. The sectional view shows how the wooden members are grooved for the tiles which are installed as the assembly progresses. Corners of the base are mitered and holes are provided for a 30-in. length of ⅛-in. threaded pipe. The exposed portion of the pipe is covered with ½-in. brass tubing as indicated. Then, with a standard cap and canopy attached to the top, a locknut concealed in the base of the lamp ties the assembly together.

The wooden column of the floor lamp, Fig. 38, is made up of two separate pieces to provide a center hole for the lamp-cord pipe. A ¼ x ½-in. groove is run lengthwise in each piece and then the two are glued and clamped together before tapering. Assembly of the top portion of the standard is shown in the detail above Fig.

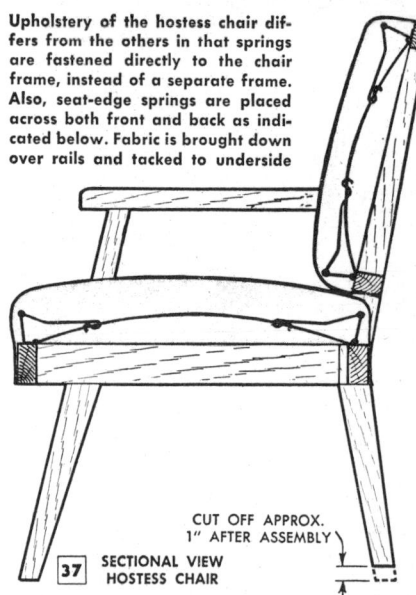

Upholstery of the hostess chair differs from the others in that springs are fastened directly to the chair frame, instead of a separate frame. Also, seat-edge springs are placed across both front and back as indicated below. Fabric is brought down over rails and tacked to underside

37 SECTIONAL VIEW HOSTESS CHAIR

CUT OFF APPROX. 1" AFTER ASSEMBLY

38. The brass fixture at the top of the wooden column is not a stock item but is made up specially from lengths of brass tubing soldered together.

The fixture is designed to take a standard mogul-type socket which is screwed to the upper end of the ⅛-in. pipe. A hole is provided in the side of the fixture to accommodate the socket turn button and the top ring is fitted with three thumbscrews to support a standard 10-in. glass bowl. The base of the lamp consists of four 1¾-in.-thick pieces planed to ¾ in. at the edge and mitered at each corner. These are glued together and then a flat surface is planed at the apex of the four pieces to receive a ¾ x 25/16-in.-sq. block. Then a ½-in. hole is bored through the center and the hole is counterbored on the underside of the base to take a washer and nut. A small tapered leg is glued and screwed to each corner of the base and then the lamp column is tightened securely to the base by means of the nut in the bottom.

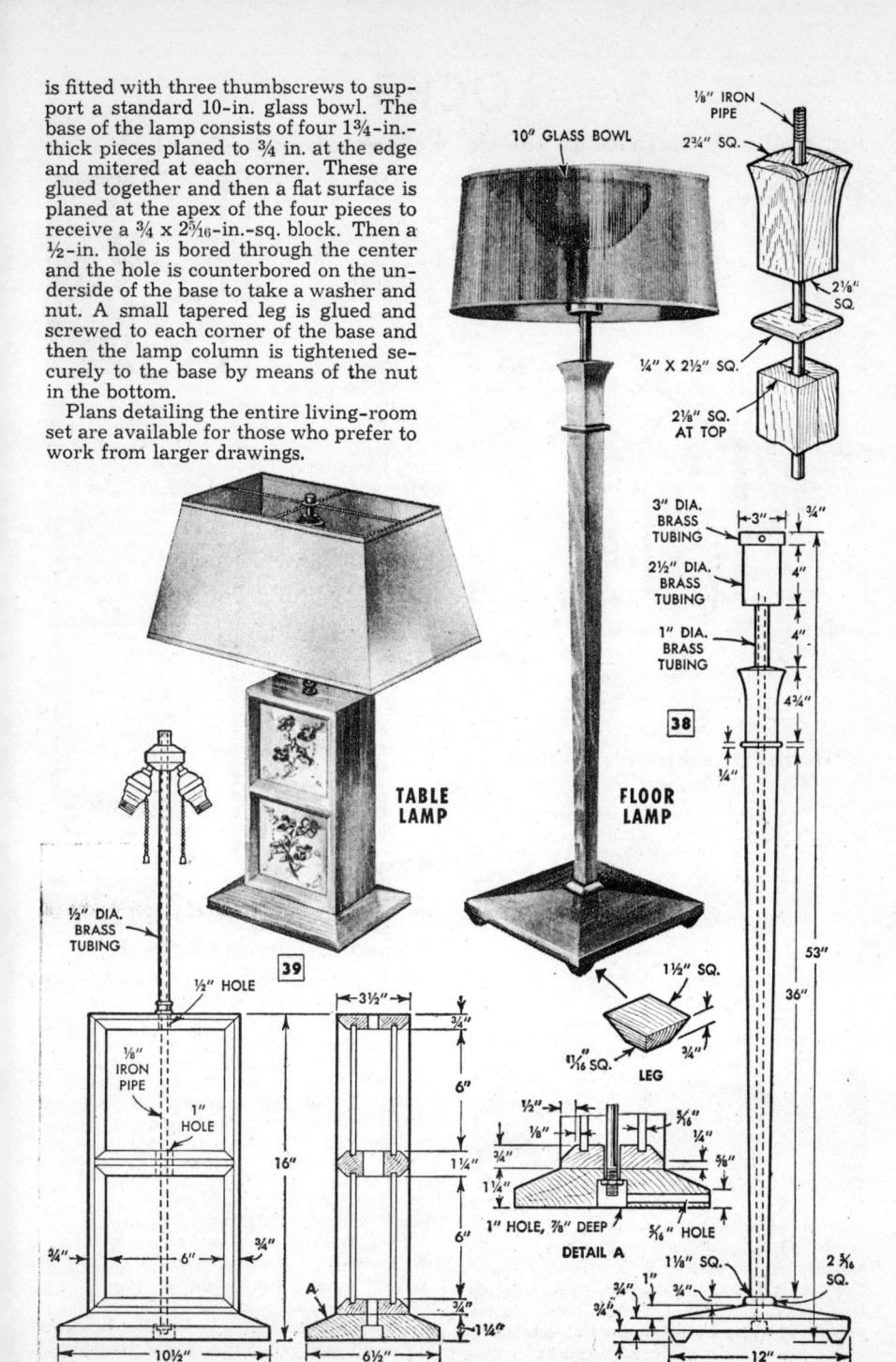

Plans detailing the entire living-room set are available for those who prefer to work from larger drawings.

LOCKS

Improving the Hasp for a Padlock

When a hasp of the type shown is used with a padlock, any sagging of the door or settling of the frame will throw the parts out of alignment so that the lock cannot be inserted. If one part of the hasp is removed from the door, and the bolt hole made into a slot, as indicated in the drawing, the part can then be raised or lowered to align with the other one.

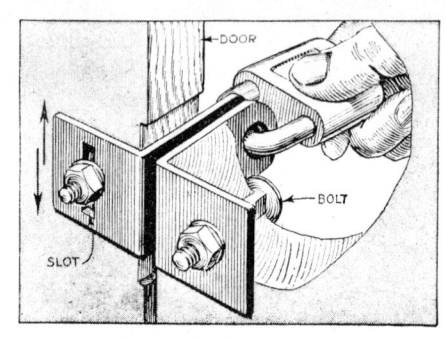

Altering Shackles on Padlocks To Suit Special Purposes

It's not difficult to alter the shape of a padlock shackle to suit some unusual locking job, such as locking oars to the side of a boat, a telephone receiver to its hook, etc. All you have to do is cut off the shackle close to the lock, and either reshape it to suit or else make a new shackle from small-diameter rod, and weld it to the shackle stubs at the lock. Before doing this, cut V-grooves in the stubs to receive the V-shaped ends of the new shackle, thus providing a good joint when welded.

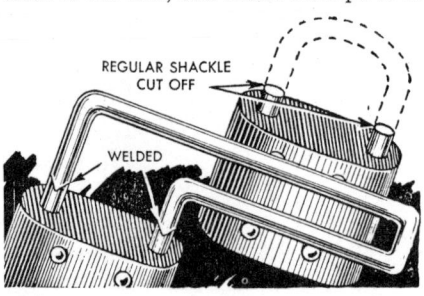

Keyhole Guard for Door Lock

If you have a lock of the type shown on an outside door, you can make a guard that will prevent a skeleton key from being used to gain entrance. Cut a slot in one side of the keyhole plate so a strip of metal can be inserted to cover the keyhole. This is on the inside of the lock. Then file off the end of the key so that when it is flush against the metal strip—inserted from the outside—the key will engage the lock to open the door.

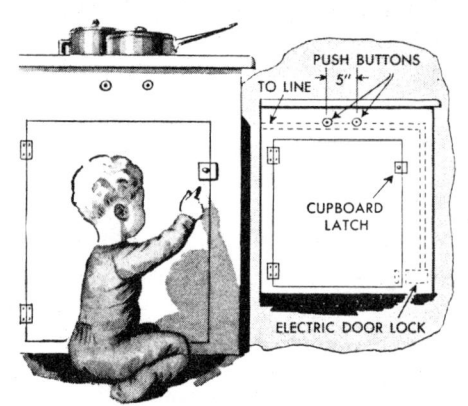

"Baby Proof" Lock for Cabinet Safeguards Inquisitive Child

One of the most distressing problems of parents who have a curious toddler investigating every nook and cranny in the house is that he might sample bleach, lye or some other poisonous cleaning solution. To set his mind at ease, one parent installed a "baby proof" lock on the door of the storage cabinet in which cleaning materials are kept. In addition to the regular cupboard latch, he fitted an electric lock on the cabinet door, using a standard electric door lock. Two push buttons, wired in series, must be depressed simultaneously to open the lock, and these are spaced 5 in. apart—too far for the spread of the tiny fingers of a baby's hand. If the child uses both hands to depress the push buttons, the cupboard latch still keeps the door closed. An adult should be able to depress easily both buttons with one hand while opening the latch with the other.

Care and Repair of MORTISED DOOR LOCKS

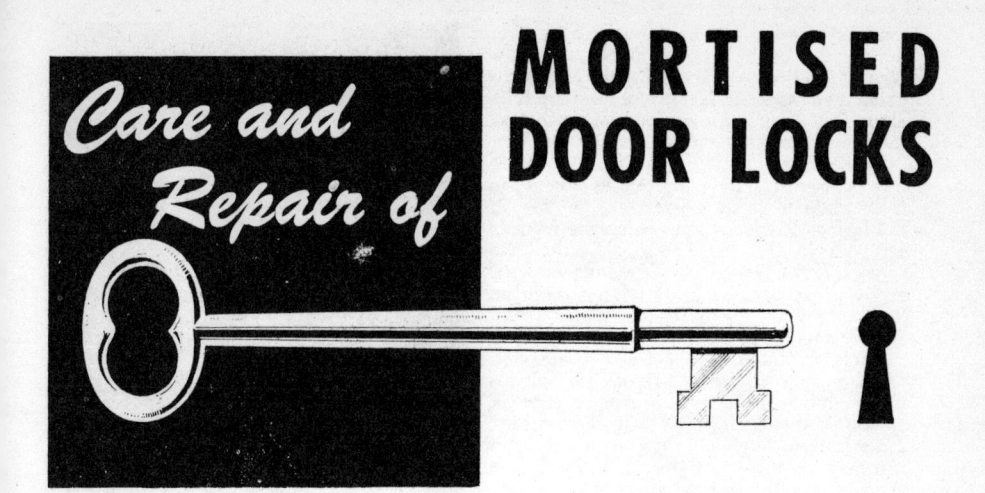

BECOMING acquainted with the workings of the door locks in your home is the first step in keeping them troublefree. Although commonly called locks, those faithful guardians mortised in the doors of your home actually consist of two distinct mechanisms—a latch and a lock. The drawings below picture four lever-type tumbler locks in general use, including the popular free-action latch. The various parts of a typical lock are labeled in the lower right-hand detail on the opposite page.

How the latch mechanism works: When the doorknob is turned in either direction, movement is transmitted to the hub via the spindle. The hub, in turn, transmits the movement to the latch bolt by means of two cams which operate through a latch lever or a shoe. This retracts the latch bolt and puts tension on the latch-bolt spring. All the various latch-spring actions illustrated below are designed to place a strong tension on the hub, so that when the knob is released, it snaps back to an idle position.

How the lock mechanism works: In some locks, the tumbler is on the case side of the lock, while others are made with the tumbler on the cap, or cover, side. The key must raise the tumbler to exactly the right height to allow the tumbler stump on the key bolt to pass between the upper and lower shoulders of the gate in the tumbler. If the tumbler is lifted too high or too low, one shoulder will prevent shooting the bolt, as shown in the detail on page 36.

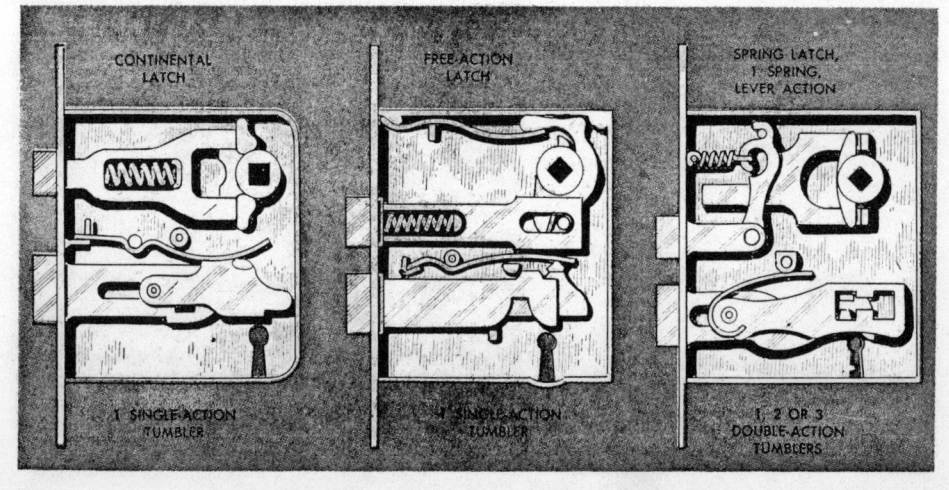

Where to obtain repair parts: As a rule, hardware stores do not stock extra parts for locks, other than knobs, spindles, trim and key blanks. Thus, in most cases, it is necessary to obtain replacement parts direct from the manufacturer of the particular lock. If the lock has not become obsolete, the part needed can be ordered by giving the number of the lock and enclosing a pattern of the broken part. When buying a replacement lock, it is always best to take along the old lock for comparison as to size, thickness, backset and spacing.

Loose front plate: Vibration resulting from slamming a door sometimes causes the front plate to work loose from the case. If the lock is of the type in which the front is screwed to the case, simply tighten the screws. If the front is held by lugs, retighten by pressing the lugs down between the jaws of a vise. To tighten a riveted front, simply peen over the rivets.

Sluggish latch-bolt action: Generally, this condition is due to lack of either cleaning or lubrication, or both, or to a weak spring. If the latch bolt won't extend, look for a broken latch-bolt spring. If this isn't the case, try loosening the cap screw. When the screw is turned too tightly, it will sometimes buckle enough to bind the latch bolt.

Broken tumbler spring: If the mechanism is well lubricated, gravity may move the tumbler down even though the tumbler spring is broken. However, the remedy is to cut and form a new spring and slip it in place. If the lock is of the type having a double-action tumbler, the spring may allow the tumbler to work up and stick at a point where the lower shoulder in the gate will prevent passage of the tumbler stump on the key bolt, making it impossible to operate the lock. If this is the case, remove the broken spring from the notch in the tumbler. If the replacement spring is undersize, it can be enlarged slightly to fit the notch by peening it with a hammer. In the event it is oversize, file down the end for a tight fit and press it in place in a vise.

Above, replacing broken lever spring in free-action latch. Long-nose pliers are used to bend new spring

New spring for single-action tumbler is slipped in place, above, after it is cut to length and formed

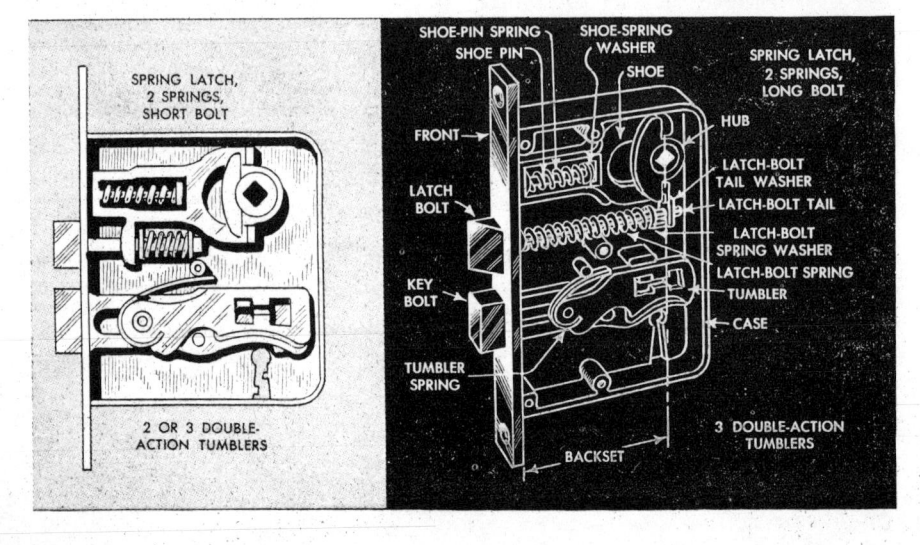

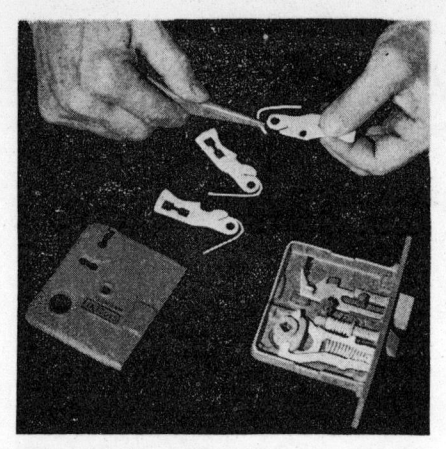

Spring in double-action tumbler, above, should fit tightly in notch. Burrs are filed from both sides

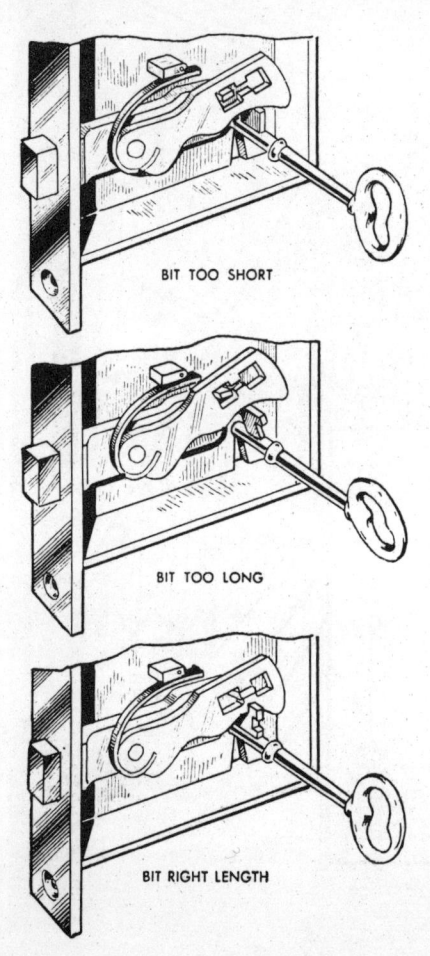

BIT TOO SHORT

BIT TOO LONG

BIT RIGHT LENGTH

Replacing springs: An important thing in replacing a broken spring is to see that it is of the same tension as the original one. Flat spring stock is available in lengths of 12 in. and longer and is easily cut and bent with pliers. Compression coil-spring stock is sold in lengths of 12 in. and in a variety of diameters. Extension coil springs with eyes formed at both ends are manufactured in assorted sizes and tensions, and some locksmiths have bending machines with which they make coil springs from wire.

Tumblers rusted together: In a seldom-used lock having two or three tumblers, the tumblers sometimes rust together, making it impossible to shoot the bolt. The remedy is to remove the tumblers, clean off the rust with sandpaper and use light machine oil to lubricate the sides of the tumblers which rub together. This is the one instance where oil may be used in mortised locks. Use the same method to restore the usefulness of a lock in which the tumblers have stuck together as a result of gummed oil or prolonged idleness.

Lubrication: Use fine, dry-flake graphite to lubricate a mortised lock. Never use oil or grease, as these catch dust which clogs the mechanism and causes the lock to work sluggishly. Lubricating with graphite can be done without taking the lock from the door. Turn the knob to retract the latch bolt, and blow a few whiffs of graphite into the hole in the front on the beveled side of the latch-bolt head. Release the knob and then inject a little of the graphite into the keyhole. Do not use too much graphite, as a light dose about once a year will afford better results than excessive lubrication at less frequent intervals.

Cleaning: Many lock troubles caused by gummed oil and dust in the mechanism can be cured by cleaning. After removing the lock from the door, soak the lock for several hours in mineral cleaning solvent or mineral paint thinner. Take the lock from the solvent occasionally and work it by inserting a screwdriver or spindle in the hub and moving the latch bolt. Also use the key to lock and unlock the key bolt. After draining the fluid from the lock, dry the mechanism with compressed air or by allowing it to stand overnight in a warm place. Finally, give the mechanism a light dusting with graphite. Cleaning may be accomplished in much less time by disassembling the lock, removing the springs first, and washing the individual parts in solvent and drying them with a cloth. But before lifting out any of the parts be sure to make a diagram of the mechanism, using the cap as a template and drawing as many outlines of the case as there are layers of parts in the particular lock. As you remove each part, trace it on the outline.

Build your own

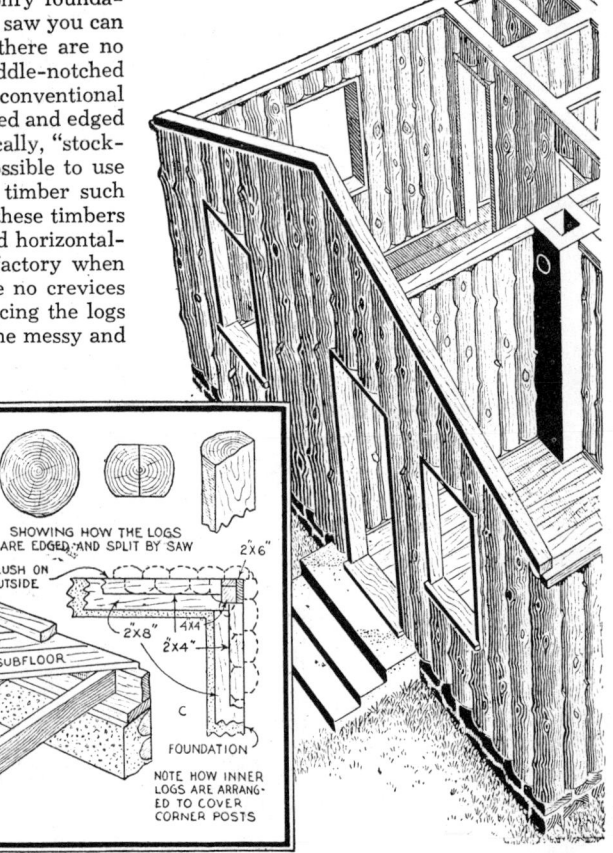

A PICTURESQUE year-round cabin of logs, a week-end house, or a permanent hunting and fishing camp far off beaten trails can be yours at very little cost if you follow this simple method of construction. If you're fairly handy at building a masonry foundation and handling a hammer and saw you can do the whole job yourself, for there are no heavy materials and no tricky saddle-notched corners such as are found in the conventional log cabin. Here the logs are halved and edged as in Fig. 1 and assembled vertically, "stockade" fashion, which makes it possible to use small logs from second-growth timber such as aspen and poplar. Although these timbers do not endure for long when laid horizontally, either will be entirely satisfactory when placed vertically since there are no crevices to catch and hold moisture. Placing the logs vertically also does away with the messy and difficult job of chinking.

Log Cabin

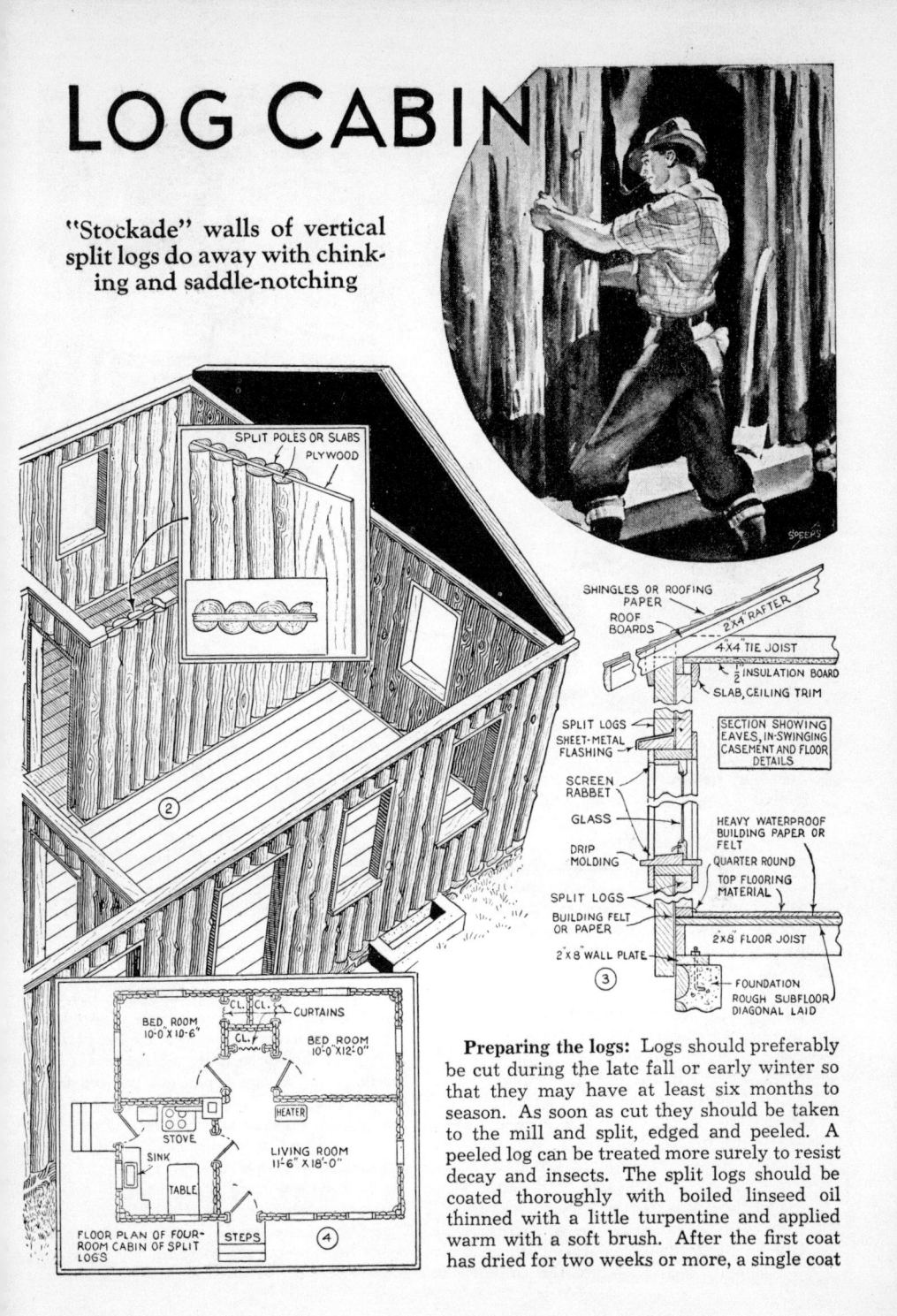

"Stockade" walls of vertical split logs do away with chinking and saddle-notching

Preparing the logs: Logs should preferably be cut during the late fall or early winter so that they may have at least six months to season. As soon as cut they should be taken to the mill and split, edged and peeled. A peeled log can be treated more surely to resist decay and insects. The split logs should be coated thoroughly with boiled linseed oil thinned with a little turpentine and applied warm with a soft brush. After the first coat has dried for two weeks or more, a single coat of pure linseed oil should be given. A third coat may be applied after the cabin is completed. Logs in contact with the earth should be painted with a creosote solution before being used. Creosoting the sills and floor joists is also advisable.

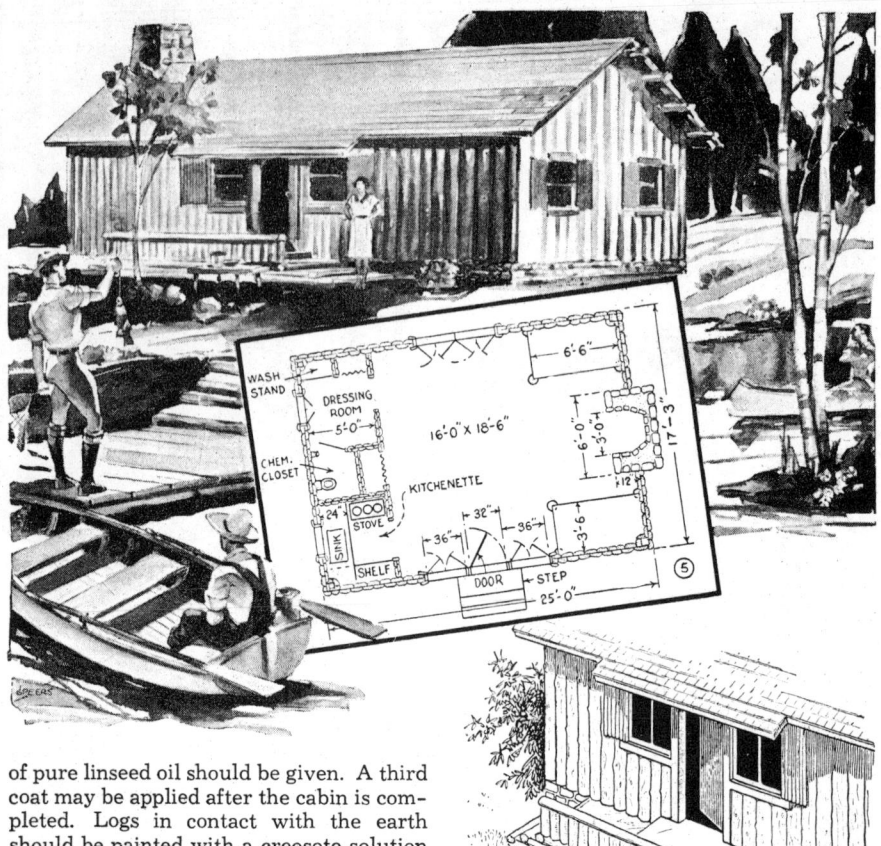

The foundation: Having chosen your site carefully in regard to drainage, water supply, view, etc., level the surface of the ground and mark out the locations of the walls and fireplace, Figs. 7 and 8. Then dig the foundation trench about 3 ft. deep and 16 or 18 in. wide. Excavate the space for the fireplace also. Have the sides of the trench neatly squared and you will not need a form up to the grade level. The footing should be about 6 to 8 in. deep and should cover the fireplace area as well. Use a mixture of cement, 1 part, to sand, 2 parts, and gravel, 3 parts. Once the footing has set, continue the wall about 1 ft. above the grade, using wood forms. If native rocks are handy, an attractive wall can be built by incorporating these with the cement, Fig. 7. While the concrete is soft, long bolts or rods threaded at the top are embedded in the mixture at regular intervals to provide anchors for the plates. Care must be taken to have the plates perfectly level. Large rocks are packed into the fireplace pit on top of the footing already laid, Figs. 10 and 11. This layer will bring the level up to about 18 in. from the surface. Top this with a layer of sand and then pour on 6 in. of concrete. Add a 1-ft. layer of small stones and cover with concrete up to the grade level. Use selected rocks laid in courses to bring the fireplace foundation up to the height of the walls.

Leave the work at this stage until later. Two or more vents should be left in the foundation wall to allow ventilation beneath the floor. These openings should be covered with a fine-mesh wire netting.

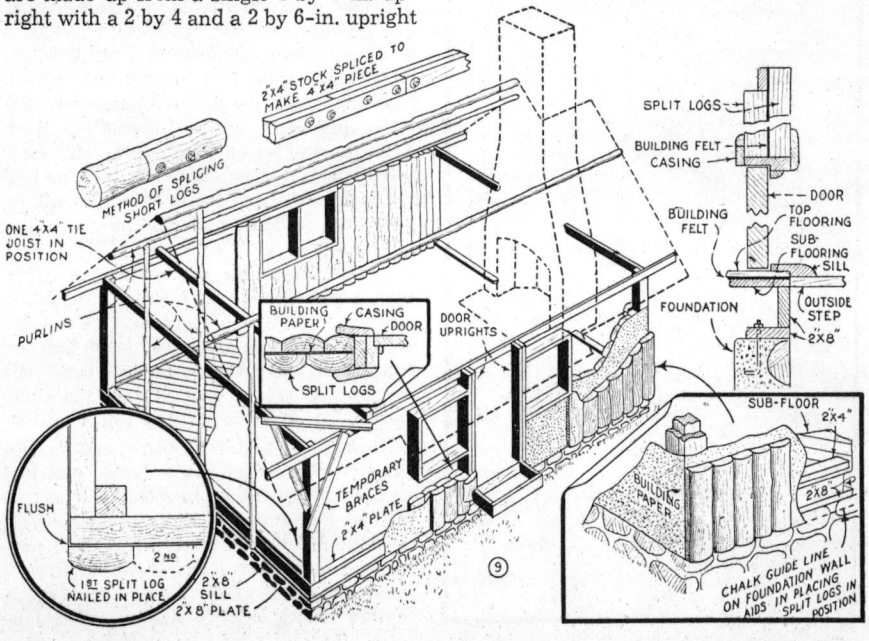

The framework: After the plates have been bolted to the foundation, the sills and joists are put on as shown in Figs. 1, 3, and 9. Note that the outside edge of the sills is flush with the outside surface of the foundation walls. The joists are spaced 16 in. on centers. Corner posts are next installed to hold the framework, Fig. 9. These are made up from a single 4 by 4-in. upright with a 2 by 4 and a 2 by 6-in. upright spiked on it, Fig. 1. Now the sub-floor, of ¾-in. unmatched boarding, is laid over the joists in the diagonal position shown. The framework pieces to which the outside vertical logs are spiked are next nailed in place. These horizontals are fastened to the corner posts at a height of 8 ft. above the floor and held solid by temporary braces on one side and on the other by the two uprights on either side of the door, Fig. 9. The 4 by 4-in. tie joists are now placed in position spaced about 2½ ft. apart. Use diagonal braces to keep this frame square and true until the outside logs are nailed on. Lastly, a 2 by 4-in. plate is nailed to the floor flush with the outside edge all around except across the door openings. Later the inside logs will be toenailed to this plate, Fig. 1, A.

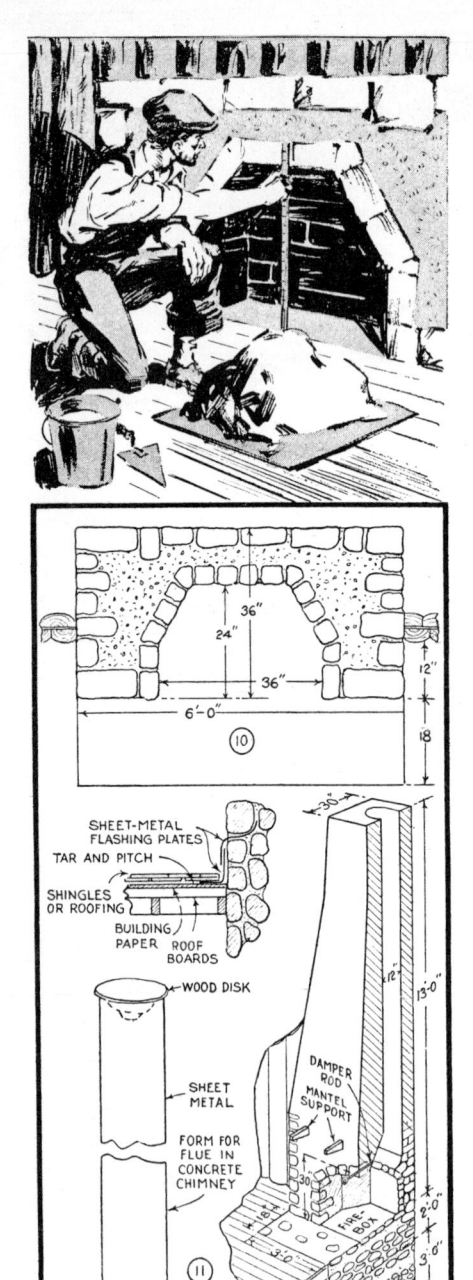

The split-log walls: The frame is now ready for the outside logs. Starting at one of the front corners, two 9-ft. lengths of heavy building paper or felt are stretched between the top frame and the sill. The first length is carried around the corner about 10 in. and tacked on. The second piece overlaps the first, making a weatherproof corner. Locate the first log with the vertical edge flush with the corner post. The lower end should extend down over the sill and cover a few inches of the stonework, Fig. 1, B. It is then spiked to the sill, wall plates and the top frame. Continue nailing the logs in place, working toward a doorway, hanging sheets of building paper ahead of the logs. As you approach the window opening, nail the frame in place against the door uprights, Fig. 9. Then using short logs, cut to length, place them in under the window and nail the frame to them. On the end opposite the fireplace, begin at the center with the longest split log. This supports the ridge pole and should be located in the exact center of the wall. Of course, a length of building paper is first stretched from the nailing joist to the sill. Finish from the center log to the corners, running the paper only up to the nailing joist. After the end wall is completed, rafters are spiked to the upper ends of the logs and the latter trimmed flush with the rafters. Then tack building paper to the logs above the nailing joist. The height of the center pole determines the pitch of the roof. The roof shown in Figs. 2, 3, and 9 rises 12 in. in 3 ft., a pitch which is about right for a cabin of this type and size. Note that the logs which support the purlins carry the weight of the ridge roof down to the foundation. These logs should be full length.

Now back to the fireplace again. The ridge pole and part of the frame will rest on the stonework as in Fig. 9. The side wall of logs must be cut to fit closely to the masonry to make a weatherproof seal. A form for the flue can be made from a sheet of metal bent around two wood disks, the lower disk slightly smaller to give the form a taper, Fig. 11. Wrap the form with paper so that it can be withdrawn from the concrete easily. After finishing the chimney to the ridge line allow to set thoroughly before the ridge pole and the two purlins are put in place. These purlins will have to be trimmed to fit onto the supporting masonry. Both purlins and ridge should extend 2 ft. over the walls to form wide eaves. With these parts in place the chimney can be finished to desired height.

Now for the interior logs. Begin with a narrow split log in a corner and make sure that the joint between the last log and the one succeeding will fall on the flat face of the outside log in the same fashion as the "broken" joints in masonry or brick work, Fig. 1, C.

The roof: Before setting up the rafters finish up the gable ends with short logs. Center rafters can extend a few feet in front to provide a short roof overhang for a porch, Figs. 5 and 6. Use ¾-in. stock for roof boards. A layer of building felt or paper goes on under the shingles. Flashing is built into the chimney in the manner shown in the upper detail, Fig. 11.

Finishing the interior: Figs. 3 and 9 show how the doors, windows and screens are installed and the trim put on.

After the top floor is laid, over building paper, the partitions are installed. Floor plans shown in Figs. 2, 4, and 5 are suggested interior arrangements of rooms. Partitions are built up of split, peeled saplings which are nailed to each side of plywood sheets, Fig. 2.

HAND LOOM TECHNIQUES

A FEW hours of fascinating work at a hand loom will give you from 10 to 15 yds. of beautiful fabric finished ready for use in your own home or to sell. Materials for the finest work are inexpensive and can be obtained most anywhere.

Simple Looms: Of the several types of looms suitable for home work, the simplest consists of a board or frame, Figs. 1 and 3, around which threads are wound evenly, the cross threads woven in with a darning needle or a shuttle, Fig. 4. Although the principle of weaving is the same on all looms, some have more elaborate devices than others to speed up the work. In brief, weaving is simply the process of drawing transverse threads, known as the "woof" or "weft," through alternate lengthwise threads known as the "warp."

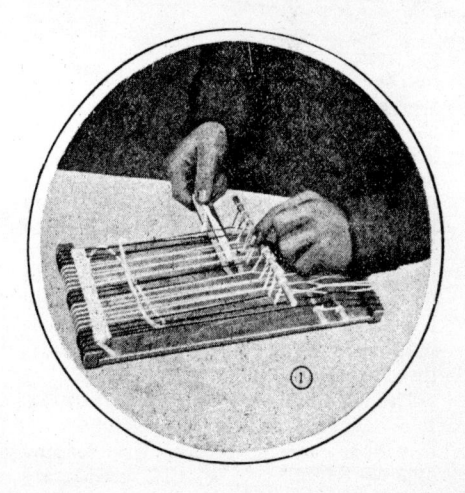

Top details in Fig. 3 show another type of loom made on a flat board. A row of pins near the ends held the warp, and a

shown spaced much farther apart than in actual practice. The warp is wound on, with two dowels, or lease sticks forming what is known as the "cross." The flat-pointed stick, or "sword" is used in tying strings to alternate warp threads. By tying tightly you assure that all these strings will be of the same length. The loose ends are then tied around a dowel, as in the lower right detail in Fig. 3, the whole thing serving as a heddle harness. Note how one set of warp threads is raised to form a space between through which the shuttle

the simple looms

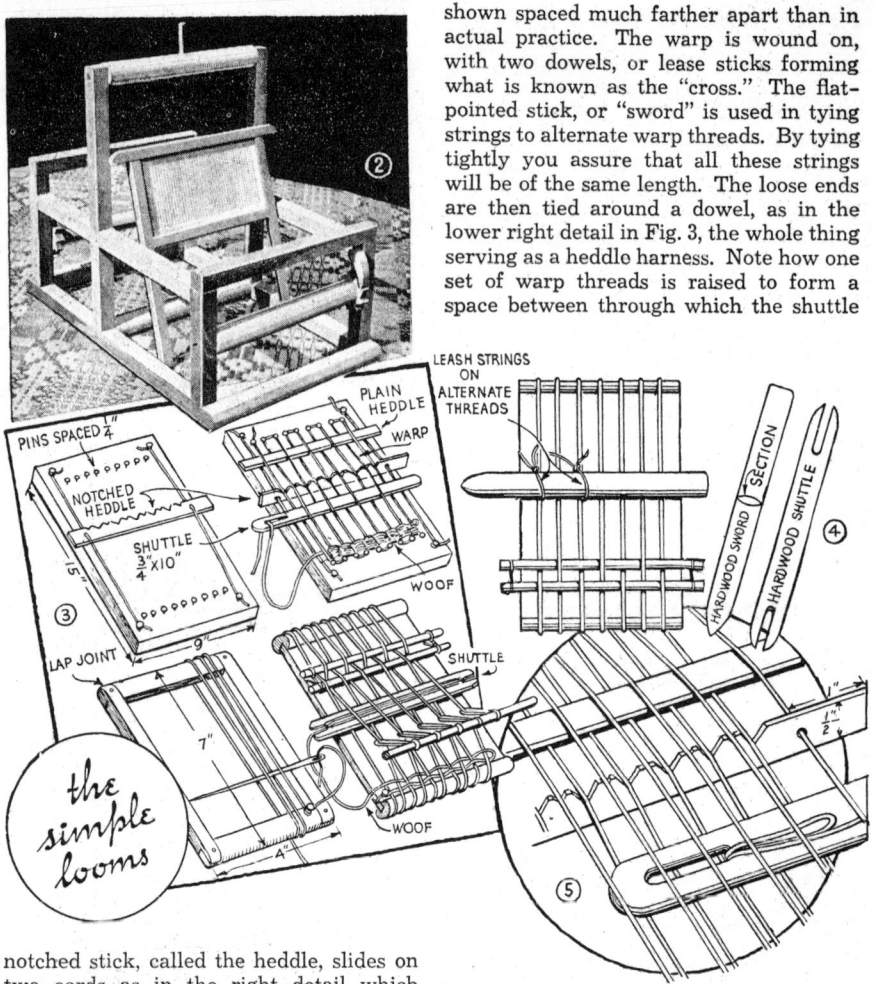

notched stick, called the heddle, slides on two cords as in the right detail which shows the loom complete with weaving in process and with the plain and notched heddles in position. The heddle, although it varies in form, serves the same function in every loom, that of separating alternate warp threads so that the shuttles can be passed between them. In Fig. 5 you'll see how the small notches of the heddle hold one set of warp threads higher than those in the deeper notches, permitting the shuttle to pass between.

The hand loom shown at the left in Fig. 4, is especially easy to operate, and once threaded, work can be done with reasonable speed. Of course, the threads are is passed. This space is known as the "shed." The shuttle passes from right to left through the first shed, then the heddle harness is lifted and the shuttle is passed from left to right through the second shed, the process repeating in the same order.

Table Looms: The table loom is a somewhat more elaborate affair as you will see from Figs. 2, 6 and 7. It consists of a framework of ¾ by ¾-in. hardwood with two rollers or beams carrying the heddles, which are moved with a lever as in Fig. 6. Then there is the slotted panel, called the reed, and a frame in which it is installed. The unit is known as the "beater" and is for the purpose of beating the woof threads tight after the shuttle has passed through the shed.

Structural details are shown in Figs. 6 and 7. Rollers with ratchets are provided at each end of the frame, one for winding up the cloth as it is woven, the other to unwind the warp threads. To show the construction more clearly only two wire heddles are shown on each harness, whereas in practice several dozen may be used, depending upon the width of the cloth to be woven and the size of threads. The right-hand detail of Fig. 6 illustrates the action of the heddles. After the shuttle has passed through the shed in one direction, the beater is pulled forward and returned, the handle on the top roller is moved back to reverse the position of the warp and the shuttle comes back through the shed again.

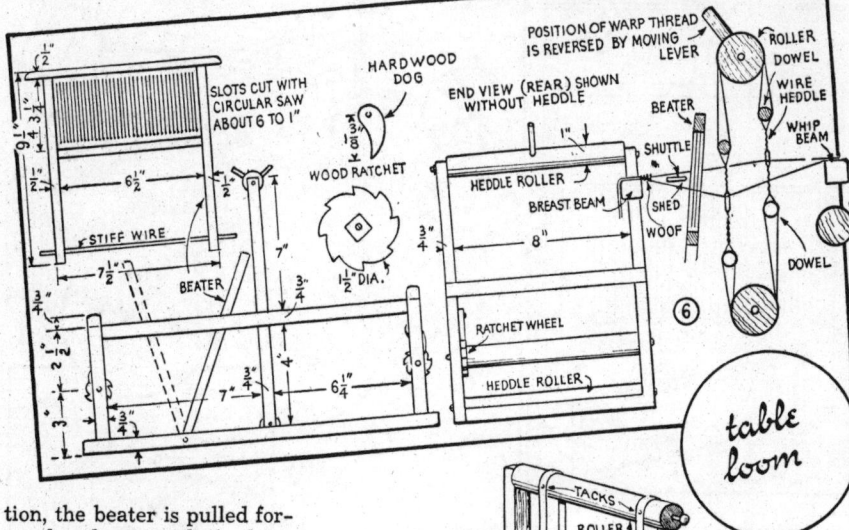

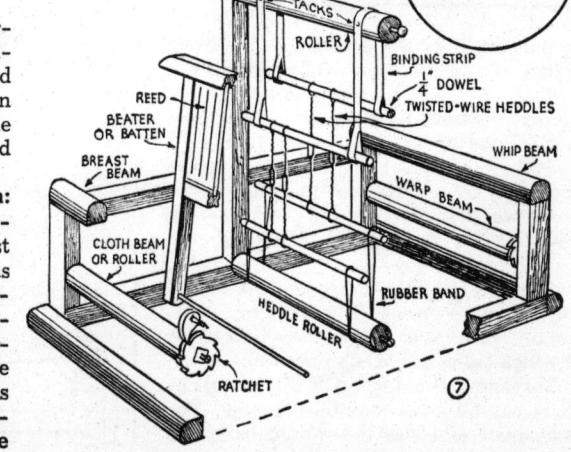

"Four-harness" Loom: Structurally this loom is essentially the same as that just described, except that it is fitted with four sets of heddles instead of two and is capable of weaving more intricate patterns. It should be kept in mind that hand looms are fitted with as high as sixteen heddles, but these can be handled only by an experienced operator. Figs. 8 to 11 show how this type is made and how it works. Note that the heddles slide in grooves in the uprights and are moved up and down by levers at the right side of the frame. The "15-dent" reed, Fig. 8, is satisfactory for ordinary work. The term "15-dent" usually refers to the number of strips or wires per inch. In ordinary pattern weaving two heddles are moved at a time, that is, they work in pairs, but not always the same pair. The combination to use is determined by the "pattern draft" which is purchased along with the material for weaving. A sheet-metal dog, Fig. 9, is installed in the top of the frame to hold up the heddle levers as required. When one or two levers are raised, the dog is released, which allows all other levers to drop. Heddle frames, or "harnesses" are generally made as in Figs. 8 and 17, each frame being connected to the lever bar by a wire link. One end of the frame, or side member, pulls off to permit stringing on the heddles. These are of light wire, the eye in the center being twisted very tightly to prevent the thread binding in the crotch. The crank for winding the rollers is usually made as in Fig. 10, the idea being to have it removable.

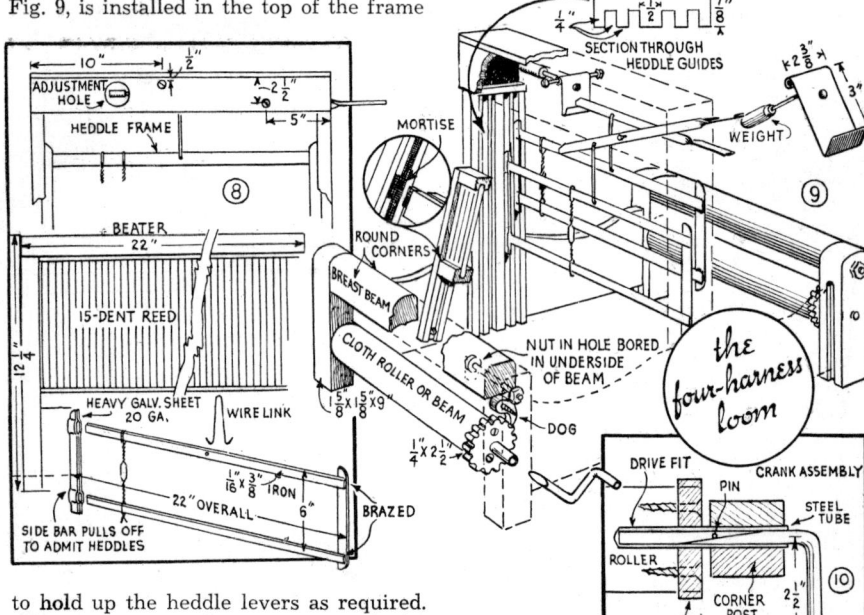

Threading the Loom: This involves more careful work than the weaving. The warp material comes in hanks, balls, spools and reels, but in any case it must be wound carefully on a warping board, Figs. 12 and 13, to prevent its becoming tangled when threading on the loom. To wind the warp, slip a large loop at the end of the thread over the peg A, Fig. 13, bring it under B and over C, and then around the other pegs in the manner shown until reaching E, then back over D, around the other pegs the same as the first thread until reaching C, which the thread goes under, then over B, around A as shown by the dotted lines, and return again until as many threads are wound on as the pattern draft calls for. Now tie cords through the crosses, Fig. 14, between C, B and A, as well as E and D, and carefully remove from the warping board. Next, you make what is called a "warp chain." The successive steps are illustrated in Fig. 16, A, B, C, and D. The purpose of this chain is to prevent the threads from becoming entangled before threading on the loom. These chains may be made up from a few to one or two dozen strands, depending upon the pattern, and there will be a number of them for each job.

This done, tie the tail end of the warp chain to the breast beam of the loom, Fig. 15, and having inserted two lease sticks in the crosses as shown, tie the sticks to the loom in the manner indicated, so that they will remain in place while proceeding with the work. Cut the ends of the threads and with a reed hook inserted in the back of the reed, draw the first thread through the first slot, Fig. 16, the next through the second, and so on, tying every few threads temporarily with a slip knot at the back of the reed as in the upper-left detail in Fig. 19.

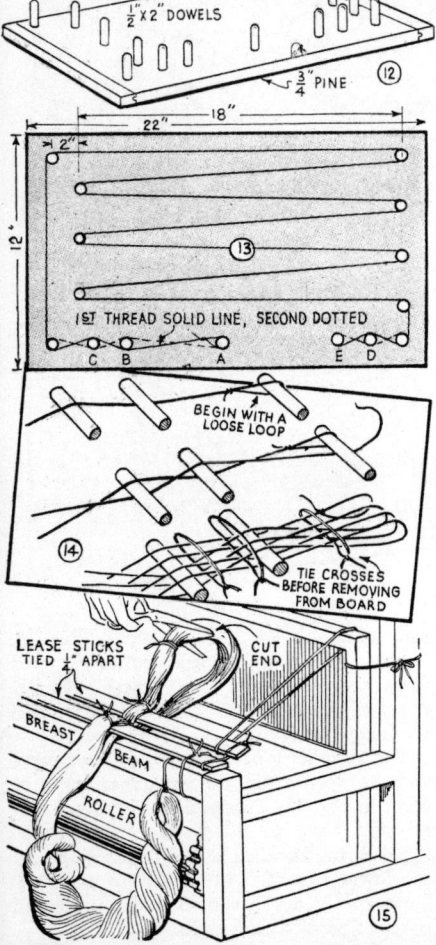

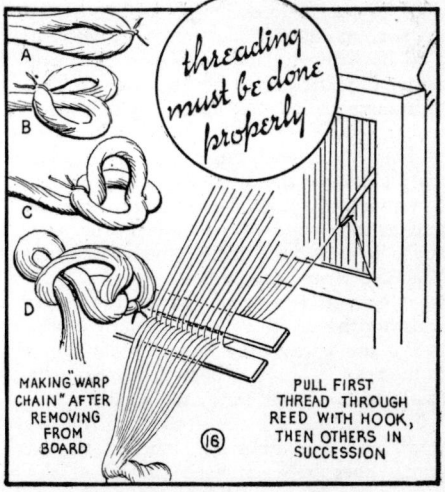

When all are threaded through the reed, start bringing them through the heddles. The pattern draft purchased with your material will indicate which heddle each

woof. At the start of weaving, thin strips of cotton rag can be worked in the same as woof to form a backing to work against, then several strands of thread, which should be sewn criss-cross later, for selvage. Plain or tabby weaving, Fig. 18, is produced by simply passing the shuttle back and forth between movements of the heddles. Of course, to form a pattern you have to change the color of the warp when threading the loom, and then weave in another color of woof to get variations like that shown at the upper right in Fig. 18. If the woof is pulled too tightly, it will form what is known as a "waist," Fig. 18. A bit of practice will easily overcome this tendency, however.

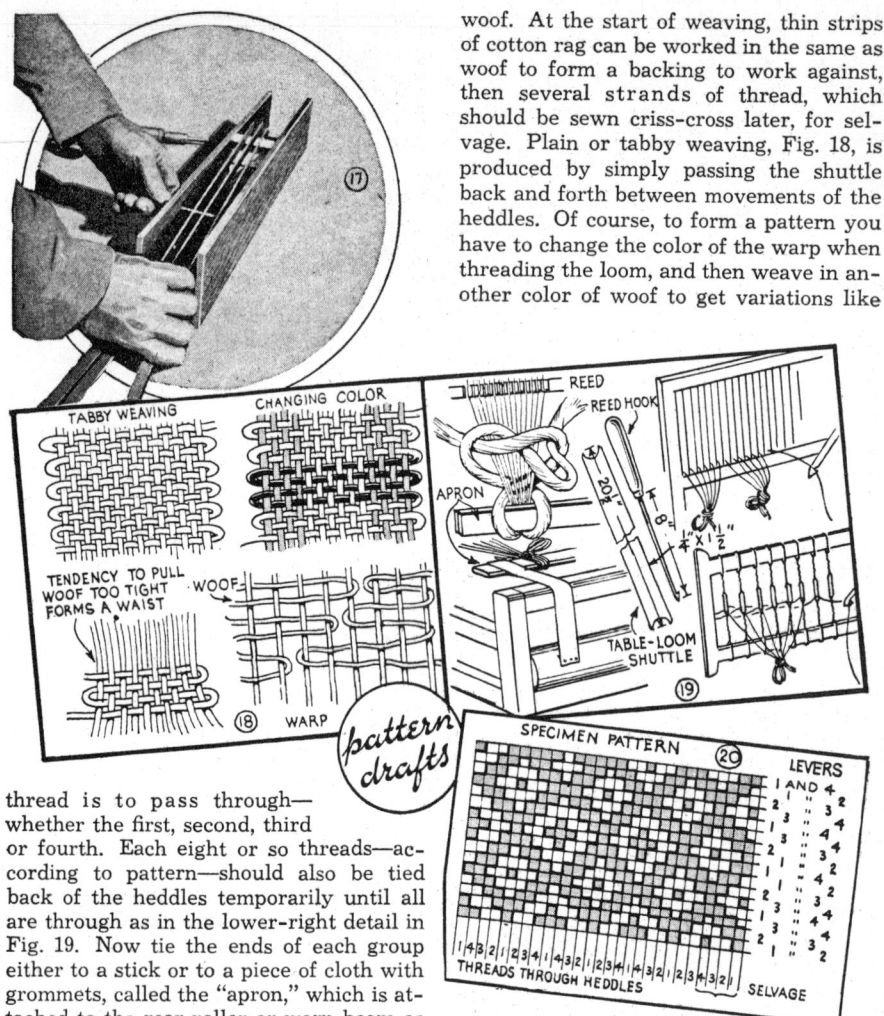

thread is to pass through—whether the first, second, third or fourth. Each eight or so threads—according to pattern—should also be tied back of the heddles temporarily until all are through as in the lower-right detail in Fig. 19. Now tie the ends of each group either to a stick or to a piece of cloth with grommets, called the "apron," which is attached to the rear roller or warp beam as in Fig. 19. When you start winding, the warp chains at the other end will gradually unwind, through the reed and heddles, and, as they wind on the roller at the rear, sticks are inserted from time to time to keep the strands separated. When all the warp thread is wound, attach the other ends to the front roller, or cloth beam, in the same way they were first attached to the rear. If this has been done properly, the loom is warped ready for weaving.

Two shuttles are required—one for the woof, and the other for binding thread which goes between each thread of the

The simplicity of threading and weaving will be seen from the specimen pattern draft in Fig. 20. This appears involved but all you have to do is run the threads through the heddles in the order given by the lower row of figures, and, in weaving, move the levers in the sequence indicated by the figures at the right.

LUMBER STORAGE RACK

A STURDY lumber-storage rack placed where there is an adequate circulation of air will aid greatly in keeping your cabinet lumber in perfect condition. Old pipe which will fit standard elbows and tees is ideal for construction. The size of the racks will be determined by the space available and the average length of lumber stock ordinarily kept on hand. Supports should be placed not more than 3 ft. apart to prevent warping of stock. If you have a basement workshop, use expansion sleeves for fastening the floor flanges to the concrete. In storing the wood, put spacer blocks over each frame crosspiece and between layers to air stock.

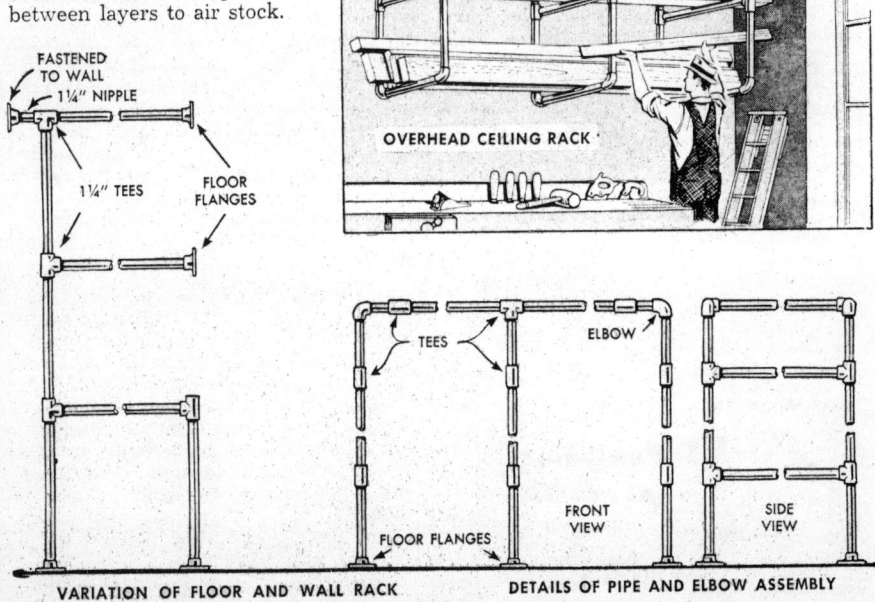

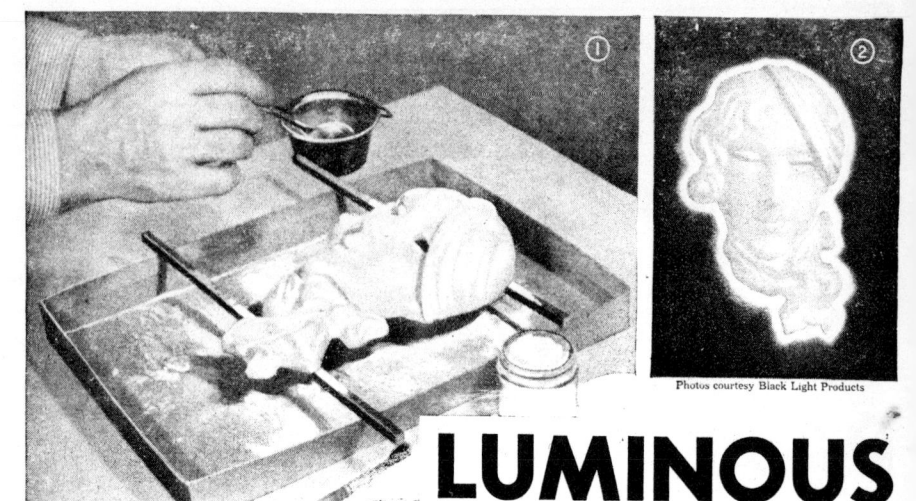

LUMINOUS

Photos courtesy Black Light Products

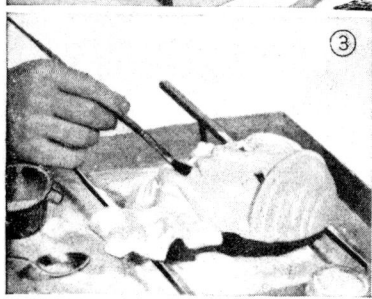

Application by dusting is done by sifting luminous powder over wet varnish coat to which the powder adheres

(4)

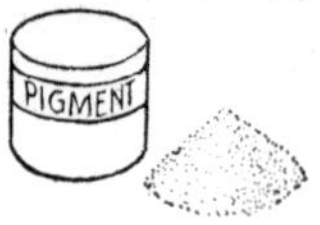

PIGMENT ... IN POWDER FORM VARIOUS COLORS

GUMMED TAPE ... WIDTHS TO 3"

COATED PAPER AND PLASTIC SHEETS

PREPARED PAINT

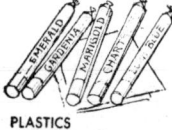

Luminous products ... ARE VARIED AND INCLUDE A WIDE ASSORTMENT OF MATERIALS

PLASTICS ... FLEXIBLE SHEETS AND PIGMENT-FILLED TUBES

POPULARIZED by varied uses, luminous paints and allied products find a steady and increasing use in industry, the home and "just for fun." Luminous paint is available in several in-the-dark colors, as is luminous pigment, and has a usable glow life of 15 minutes to 10 hours, depending on the kind used. A few of many products available are shown in Fig. 4.

Application by dusting: One of the most direct and best methods of applying luminous pigment is by dusting it on paper, novelty figures or other items. Applied to a mask, this process is shown in Fig. 1. The mask is coated first with a white-varnish enamel. After this has set for about 5 min. and while still tacky, the luminous powder is dusted on by sifting through a tea strainer or other sieve of about 40 mesh. After standing another 20 min., the excess powder is brushed off, as in Fig. 3, and the surface gently burnished. The object, when thoroughly dry, is finished with a clear top coat of shellac. This method gives a uniform coating which glows when the object is placed in the dark, Fig. 2.

Shadows are fun: In a darkened room, take a sheet of luminous paper and hold it behind your head, as shown in Fig. 6. Have someone turn on a light (a 100-watt lamp) so that your head casts a shadow. Hold the pose for about 5 sec. Turn out the light. In the dark the protected part of the sheet will show your shadow, Fig. 5, while the rest of the sheet, after exposure to light, will glow. This makes an interesting "guess what" party stunt by lumi-photographing hands or small objects. The luminous paper should be treated just like photoprinting paper—keep it in the dark or face down until you are ready, then expose and, finally, view it only in the dark. The best paper is one with a brilliant initial glow,

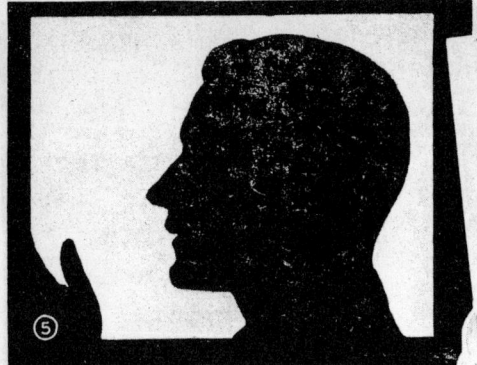

PAINT

but having a short life of about 15 min. This will permit repetition of the stunt after a short interval so that three or four sheets will keep the performance going indefinitely.

Making paint: Paint should be mixed at the rate of 4½ lbs. of pigment for each gallon of vehicle. Use the chart on page 124 for smaller quantities. Coverage is about 100 sq. ft. per gallon. The vehicle can be any clear neutral lacquer, any clear spirit varnish or most kinds of clear synthetic. Vehicles containing linseed oil, lead, manganese or cobalt should not be used. Obtain twice the amount of vehicle required to make the paint. Keep one half of this amount for an undercoat and overcoat. Whenever a white base is indicated, such as over metal, wood or glass, lithopone should be added to the undercoat until the required opacity is reached. The overcoat serves to protect the paint from moisture and cleaning operations.

Purchased paint: The vehicle commonly used is a synthetic—methyl methacrylate, vinyl acetate, alkyd resin and others. When you buy luminous paint, be sure to order thinner to match.

Activation: The luminous coating must receive light before it will glow in the dark. The period of activation need not be more than from several seconds to a minute and prolonged activation does not increase the glow. Any light source such as sunlight, electric light, flashlight or even the light of a match will activate luminous material. It is well to remember, however, that the weaker the

Shadows are fun — Here's one time you can walk away from your shadow if you cast it on luminous paper

⑦ SIGN LOOKS LIKE THIS BY DAY

⑧ THE LETTERS GLOW IN THE DARK

Cutout letters and figures of luminous paper provide a simple and inexpensive way of decorating rooms, making signs, etc. You can purchase plain or gummed-back paper for this purpose

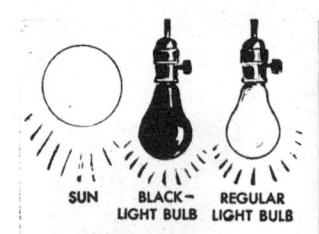

ACTIVATION
All luminous pigments require activation before they will glow in dark. Source can be sun, black light or regular bulb

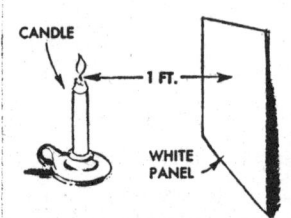

SUN — BLACK-LIGHT BULB — REGULAR LIGHT BULB

INTENSITY OF GLOW
Brightest luminous pigments have illumination value of about ½ foot-candle. A foot-candle is illumination on white panel from candle 1 foot away

CANDLE — 1 FT. — WHITE PANEL

DARK-ADAPTED EYES
Walk outdoors on a dark night. Everything will at first look black, later become visible. Most luminous paints can be seen clearly only after eyes have become adapted to dark

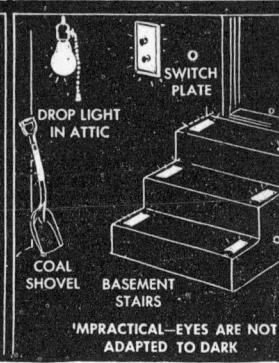

GOOD EXAMPLES OF ACTIVATION
- ILLUMINATED SIGN
- PAINTED BULB

AN IMPRACTICAL EXAMPLE OF ACTIVATION
- BASEMENT STAIRS

BRIGHT GLOW SHORT LIFE PIGMENT NEEDED
- FISHING LURES
- HAND SHADOWS
- WIG-WAG BATS

LONG LIFE MORE IMPORTANT THAN BRIGHT GLOW
- LAMP SHADES
- SWITCH PLATE
- DOORBELL
- STREET NUMBER

PRACTICAL—EYES ARE DARK-ADAPTED
- CLOCK
- INSTRUMENT PANEL
- DARTS AND OTHER GAMES

IMPRACTICAL—EYES ARE NOT ADAPTED TO DARK
- DROP LIGHT IN ATTIC
- SWITCH PLATE
- COAL SHOVEL
- BASEMENT STAIRS

WORK SURFACE
Surfaces to be luminous painted should be white or receive a coat of white paint. Luminous paint will not cover dark colors or dirty marks on a white panel

MIXING PROPORTIONS
All pigments take 4½ lbs. pigment per gallon of vehicle

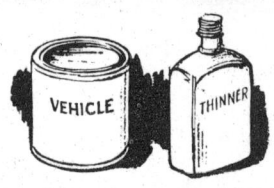

VEHICLE (Liquid Measure)	PIGMENT (By Weight)
1 gallon	4½ lbs.
1 quart	1 lb., 2 oz. (18 oz.)
1 pint	9 oz.
½ pint (8 oz.)	4½ oz.
¼ pint (4 oz.)	2¼ oz.
1 ounce	½ + oz.

VEHICLE and THINNER
Vehicle for homemade paints can be clear lacquer, spirit varnish or synthetic varnish. Factory mixtures use special vehicles—be sure that you use a compatible thinner

light, the longer is the required period of activation. Dip coating of light bulbs is practical and has little effect on regular lighting intensity. An example of poor activation is shown on page 124; basement stairs seldom get enough light of any kind to activate them strongly.

Intensity of glow: Some applications require the brightest possible glow with duration of glow being secondary. On the other hand, all outdoor and many indoor applications must glow all night, making duration more important than intensity.

Dark-adapted eyes: Dark-adaptation of the eyes is perhaps the most important factor to consider in luminous applications. The glow of the pigment is relatively weak when compared with white light, and in most instances is valueless unless the eyes are dark-adapted. You are aware of how black the interior of a theater looks when you first enter, yet after your eyes become accustomed to the dark the illumination seems quite strong. Exactly the same thing applies to all luminous paints: a degree of glow which would be almost invisible to eyes not quite dark-adapted becomes quite strong when the eyesight is better adjusted to darkness. Perfect applications for dark-adapted eyes include instrument panels, clock hands, etc. Some impractical applications are shown on page 124. In no instance can you walk from a well-lighted room into a dark room and expect to see the glow of a luminous object immediately —it just can't be seen.

Carrying power: Under normal conditions of brilliance, luminous objects are nearly invisible at 100 feet. Even when visible, the glow has no character. Signs, house numbers, etc., all appear only as a blur of light when viewed from a distance, and the object takes on character and outline only when viewed at a comparatively close range. If you plan any luminous display to be observed at a distance, it is advisable to coat a sheet of cardboard with paint and then observe the effect under actual working conditions.

Making paint is simply a job of stirring pigment into vehicle, such as lacquer or clear synthetic varnish.

CHARACTERISTICS OF PHOSPHORESCENT PIGMENTS

Pigment	Color 1 in Dark	Particle Size 2	Intensity of Glow	Period of Glow	Stability
Strontium	Blue	Coarse	High	8-10 Hrs.	4 Very Sensitive to Moisture
Zinc	Emerald	Medium	High	1 Hour	Stable
Calcium	Violet	Coarse	Medium	10-12 Hrs.	4 Very Stable Indoors
Zinc and Cadmium 3	Orange-Yellow	Coarse	High	15-30 Min.	Stable
Calcium and Strontium 3	Blue-Violet	Medium	Medium	10 Hrs.	4 Moisture Sensitive
Strontium and Zinc 3	Green	Coarse	High	8-10 Hrs	4 Moisture Sensitive
Strontium 3 and Calcium	Green-Blue	Coarse	Medium-High	10 Hrs.	4 Very Sensitive to Moisture

1. Fades gradually to white in all cases. Day color of all pigments is off-white.
2. Runs from 10 to 40 microns. Very much coarser than ordinary paint pigments.
3. In all mixtures, pigment named first predominates.
4. Must always be protected with clear top coat . . . loses glow when exposed to moisture.

Brightness of various pigments represented by squares, shows intensity and life of luminous pigments (17) (18) (19)

Tools Are Kept Handy on the Job In This Canvas Carrier

This multi-pocket tool carrier will be found handy about the shop or house as the tools are easily accessible, and the carrier can be hung where they are within convenient reach, such as on the rung of a ladder, on top of a stepladder, or over a fence, sawhorse or the back of a chair. Made of canvas, the carrier can be any size desired with as many pockets as needed. A piece of flat iron sewed in the center prevents sagging and provides a rigid place for riveting a leather handle.

Toolbox Capacity Increased

The space at the bottom of machinist's toolbox for storing the removable front cover is an excellent place for keeping small tools such as taps, drills and punches. These are placed in shallow trays of light sheet metal of the proper dimensions, and each tray is fitted with a knob on the front edge so that it can be pulled out easily.

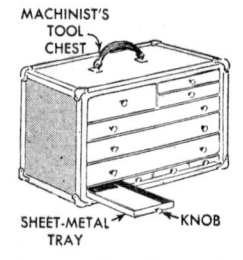

Assembling the Machinist's Toolbox

DESIGNED and built for carrying machinist's tools both compactly and in an arrangement that permits quick selection of any particular item, this sheet-metal toolbox also provides a handy repair kit for the homeowner. The box is divided into four separate sections, two of them being sliding drawers for such tools as wrenches and clamps. A roomy center compartment immediately above the drawers has space for hammers, pliers, snips, etc., while the fourth section consists of a removable tray which fits over the center compartment directly beneath the lid. The ends of the tray are triangular in shape so that a handgrip can be cut into a sheet-metal partition which divides the tray lengthwise. If the tray is to be used for bolts, screws and other small parts, U-shaped inserts may be fastened to the tray with small iron rivets, thus dividing each compartment into three smaller ones as in the pull-apart detail.

The bottom and sides of the toolbox are bent from one piece, as in the lower right-hand detail, except that the top edges at the corners are not crimped over tightly until the endpieces are soldered in place. After these parts have been assembled, the angle-iron drawer rails and mounting brackets for the bottom of the center compartment and top tray are soldered to the sides and ends. The drawers are cut and bent from sheet metal, the ends being crimped around the corners and soldered. A small brass pull is fitted to the front end of each drawer, and a hinged cover, which is locked with a turn button, holds the drawers in place. Note that these hinges are soldered in place. When bending the lid, or top, the two lower edges are crimped over to form $3/8$-in. seams. These seams and the ones along the top edges of the box sides are important, as they add considerable rigidity to the box and also eliminate the sharp edges. The wooden handle, which is reinforced with sheet metal, is fitted over the ridge of the toolbox lid. The flanges formed by the ends of the sheet metal provide brackets for riveting the handle in place. The lid is attached to the box with a length of piano hinge by riveting one leaf of the hinge to the lid and fastening the other leaf to the side of the box with sheet-metal screws or solder. The projecting tips of the screws are filed down a little on the inside of the box. A hasp and staple for a padlock are riveted to the lid and front of the box. If a bending brake for the sheet-metal work is not available, one can be improvised by sandwiching the metal between two hardwood strips and holding it in position with C-clamps. The sheet is located so that the bend is at the edge of the wooden clamp.

MACHINIST'S TOOL BOX

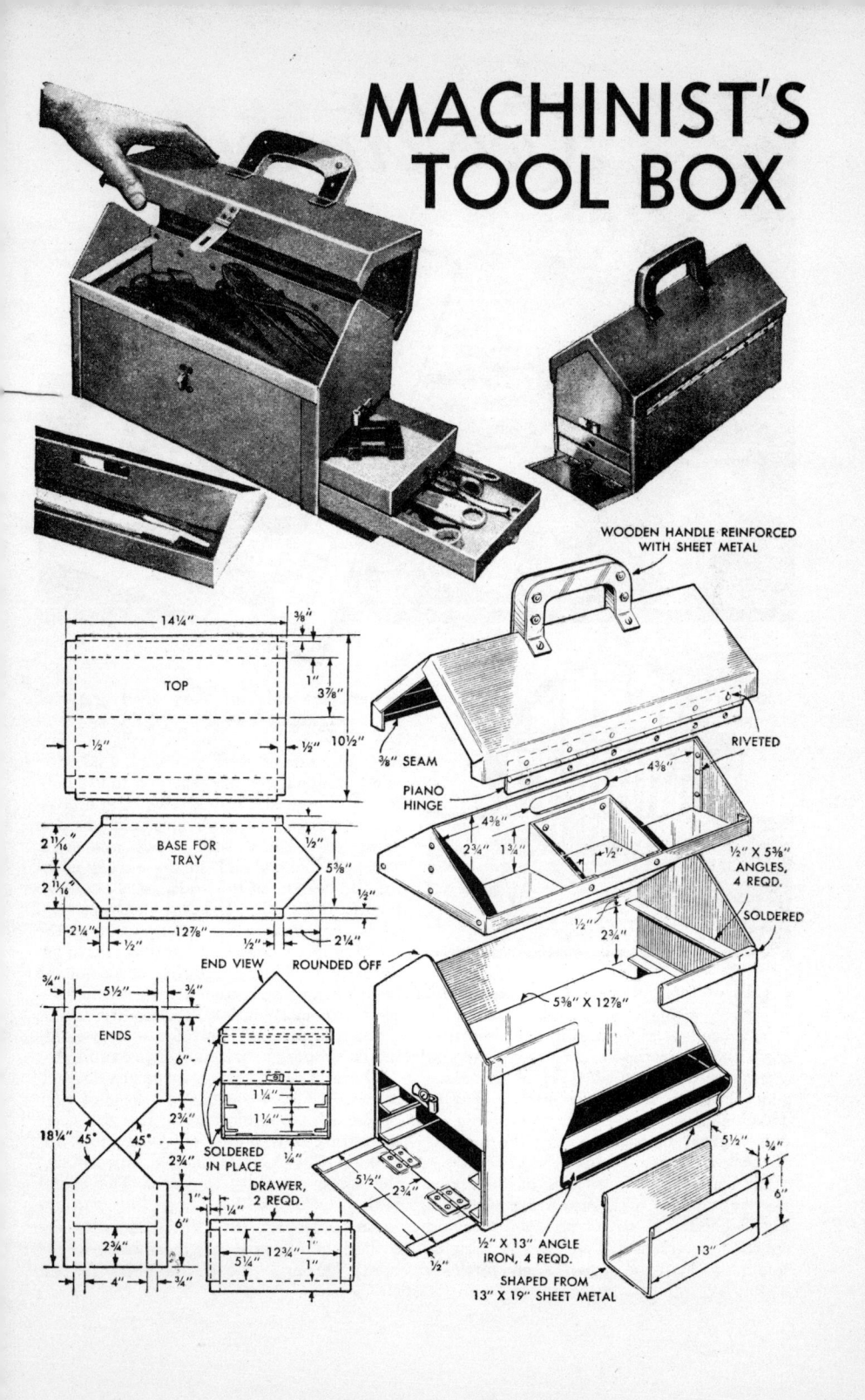

Two Unusual

ONE of these magazine baskets is entirely a jigsaw project, while the other is a product of your skill at woodturning, although in the absence of a lathe you can make a simplified type, of the same general design, from dowel rod. A motor-driven jigsaw is, of course, preferable, as it is a great time-saver over the method of using a fretsaw by hand.

Like most jigsaw articles of this type, the magazine basket, shown completed in Fig. 3, should be cut of plywood to avoid splitting, which would occur with single-thickness material when such intricate designs are cut out. Any good grade of plywood, about 6 sq. ft., will answer, and the thickness should be ¼ to ⅜ in. The feet are made of ¾-in. material, 4 in. wide and 9 in. long, and are fastened to the bottom of the basket with screws. Get a large sheet of paper or cardboard and line the sheet off with a soft pencil into 1-in. squares. Then sketch in the curves making the designs of the ends, sides and center members, Figs. 1, 4 and 5 respectively. These designs are then transferred to the wood with carbon paper, and sawing them out is next. When the different parts have been cut out and sanded they are assembled with small nails and glue, after which a finish to properly harmonize with existing or proposed surroundings is applied.

The second type of magazine basket, shown in Fig. 2, consists of a base of 1-in. plywood, or one built up to this thickness by gluing four pieces of ¼-in. stock together. The base is sanded smooth and the edges carefully rounded. The center is then located with a pencil line, and five ½-in. holes for the center spindles are drilled, taking care that these holes do not break through. Next, the holes for the

MAGAZINE BASKETS

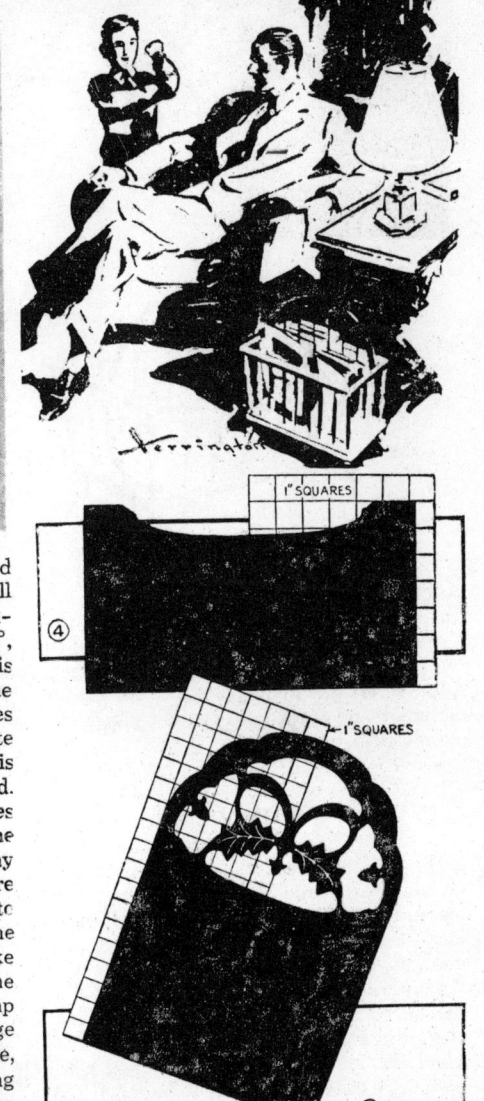

side and end spindles are laid out and drilled, centering ¾ in. from the edge all the way around the base. The corner spindles incline outward approximately 10°, measuring diagonally across the base. This position of the corners will incline the side spindles about 5°, and the holes should be drilled accordingly. An accurate guide for the bit is shown in Fig. 7. This may be made from a piece of hardwood. In use it is clamped to the base. The holes should not break through. Holes for the four legs which are set at a 15° angle may be drilled with a similar guide to assure uniformity. The top frame is made up to the dimensions shown in Fig. 10, and the necessary ½-in. holes are drilled to take the spindles. The mitered corners of the upper frame may be fastened with clamp nails and glue as in Fig. 8. At this stage the basket may be assembled on the base, using glue on all joining parts, assuming that you have already turned out the spindles, sizes of which are given in Fig. 9. The curved handle is built up of several segments as in Fig. 6. After the glue has dried, the wood is sawed roughly to the curvature shown in Fig. 9 and is finished with a sharp spokeshave and fine sandpaper. Holes are drilled for the five center spindles and the whole is then glued and screwed in place.

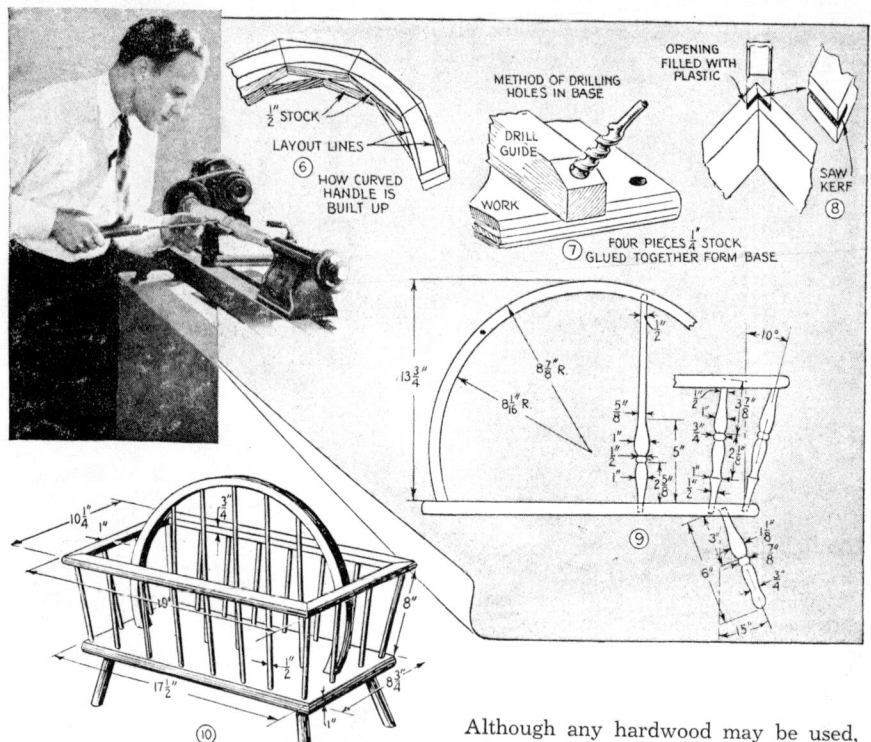

Although any hardwood may be used, maple or birch is perhaps best as the piece belongs to the Colonial period. It may be given an appropriate finish with a light oak oil stain, rubbed off to highlight the various parts and followed with two coats of white shellac lightly sanded and waxed. Fig. 10 shows the same type of basket in which plain pieces of dowel rod are substituted for the turned spindles.

Magazine Rack

Doubling as a desk that can be supported on your lap while sitting in a chair, this magazine rack permits you to lean back and relax while you write. Although almost any similar kind of rack will serve, the one shown was designed especially for the purpose. It has one vertical and one slanted side, and the bottom projects beyond the lower edge of the slanted side, forming a convenient ledge to hold pen and pencil. When the rack is on your lap, there is no projecting edge to cause discomfort. Light plywood or hard-pressed board is used for the sides and partition, but the ends should be ⅜ or ½-in. solid stock.

Novel Magazine Rack Is Made With Hand Tools

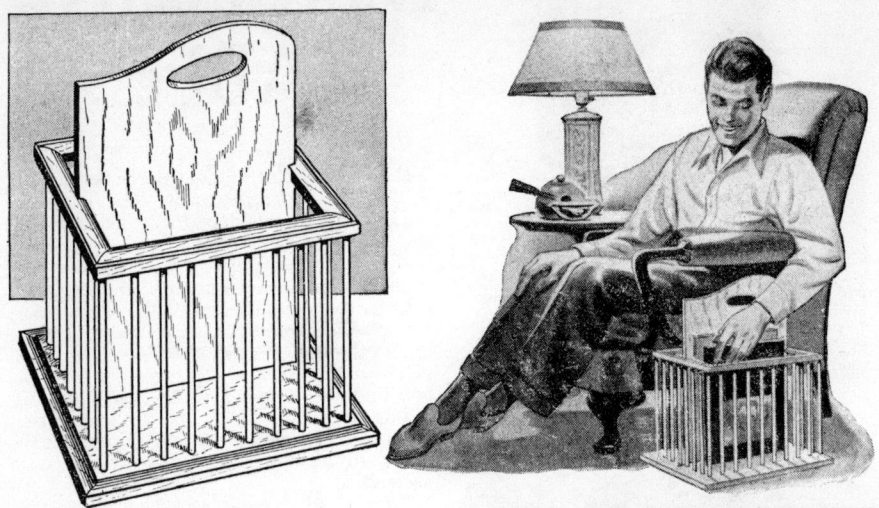

A BREADBOARD, a picture frame, a piece of ½-in. plywood, and a number of ¼-in. dowels are the principal parts of this unusual magazine rack. The design fits in nicely with the furnishings of a den or basement recreation room. Although a breadboard or kneading board from an old kitchen cabinet makes a good base, you can build up a base by gluing together several wooden strips to make the required width. The base is detailed with molded edges but if no shaper is available it can be simplified by merely rounding the edges evenly. Holes for the ¼-in. dowels are spaced and drilled to a uniform depth in the base as indicated. Groove the base as shown. Top frame of the rack is a hardwood picture frame of the dimensions given. Holes are drilled in the back of this frame to register with those already drilled in the base. Cut all the dowels to an equal length and insert in the holes in the picture frame. A spot of glue on the end of each dowel will hold it in place. Next, place a drop of glue in each hole in the base. Then invert the picture frame over the base and insert the dowels one by one into the holes. Finally, tap the top of the frame lightly with a mallet to seat the dowels in the holes. Cut the center panel to the shape and dimensions given in the squared pattern, sand the top edge and the edges of the handhole smooth, notch the top frame as indicated and glue the part in place. Finish in the natural color with shellac and varnish or clear lacquer, or with colored enamels to match other furniture.

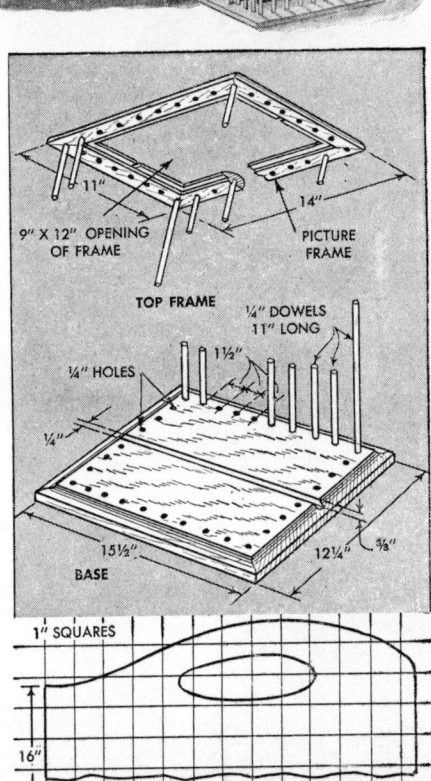

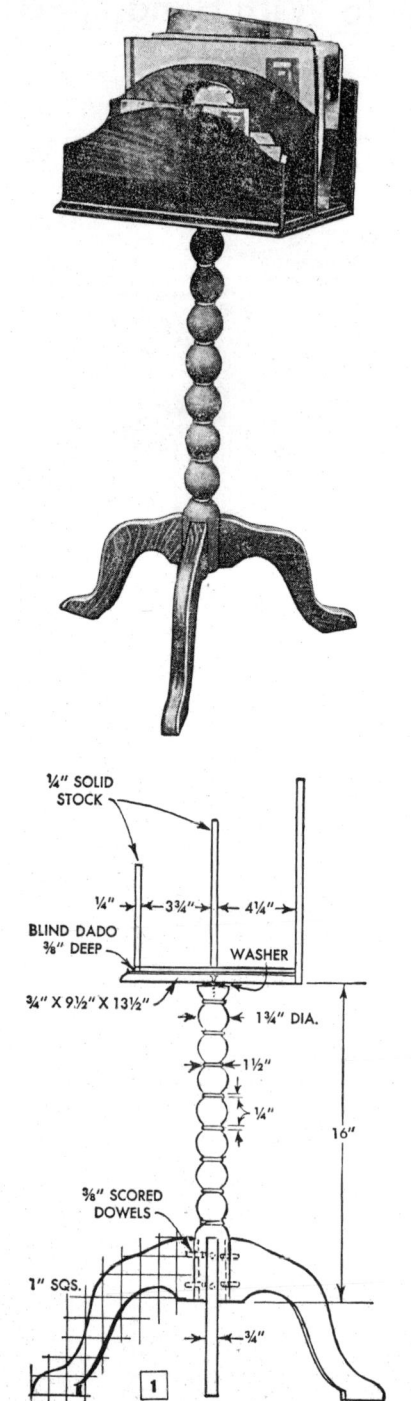

MAGAZINE TABLES

IF YOU have a flair for Early American furniture, you'll surely find a place among your furnishings for at least one of these charming little magazine tables. Embodying characteristic spool-turning in their design, both tables provide opportunity to try your hand at lathe work. Although the magazine compartment of only one of the tables is designed to rotate, both can be made to feature this point or the tops can be made stationary. In either case, it is important to true up the top end of the column in the lathe so that the compartment will be square with the column when assembled.

Walnut, birch, maple or cherry are appropriate woods to use. The method of doweling the three legs to the turned column is the same for both tables, the legs being spaced 120 deg. apart. Figs. 1, 2 and 3 detail the table with the rotating top. The bottom of the magazine rack is cut to size, the edge molded on three sides and the blind dadoes are cut for the scrolled frontpiece and the center divider. One simple way to form the blind dadoes on a small piece, such as the bottom of the rack, is to run the grooves clear through, then rip a strip from the same wood as the bottom to the sectional size of the groove. Cut off two short lengths and glue these into the ends of each groove. Sand flush after the glue is dry.

Fig. 3, left-hand sectional detail, shows how the pivot is assembled. A tap hole for a No. 5 wood screw is drilled in the top end of the column and filled half full of glue. Then a hole is drilled through the bottom of the rack, centered in the groove. After inserting the screw, a large 1/8-in.

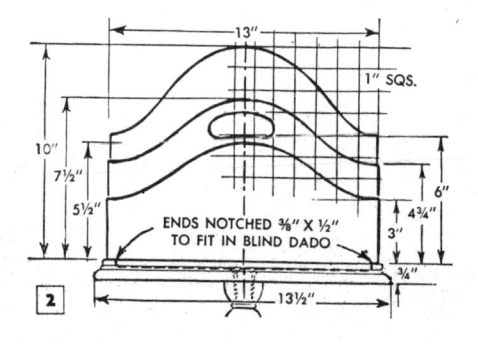

fiber washer is placed between the bottom and the top of the column. Draw the screw up fairly tight. Notch the lower corners of the center partition to fit in the blind dado and glue in place. In fitting the curved legs, the ends are sanded or filed to a concavity with a radius equal to that of the lower end of the column. This permits them to fit snugly against the curved surface. After bandsawing the legs, smooth the sawed edges and round the top edges to a radius by sanding. Do the same with the top edges of the front and back pieces as well as the center partition of the magazine rack. This done, the assembly is ready for finishing.

The hopper-type rack, pictured at the right and detailed in Fig. 4, is simply a tray with a handhole in each end. The sloping sides and ends are assembled with compound miters, or "hopper" joints, at the four corners. A special feature of the handholes is that they are bandsawed from the edge, that is, you simply saw in and saw out. After the miter cuts have been made on all the pieces, the corners are joined, one at a time, with screws and glue. The screw holes are countersunk and fitted with wooden plugs cut from the same wood as that from which the tray is made. Then the edges of the bottom are beveled and the sides joined with screws and glue. Sand all parts smooth and round the corners slightly. Then join to the column as in the right-hand detail in Fig. 3.

Methods of applying a finish depend on the kind of wood used and the type of finish desired. If an open-grained wood, such as walnut is used, then a filler must be applied. After staining and filling the grain, apply a sealer and follow this with shellac or two coats of varnish or lacquer. Maple and birch do not require a filler, but many craftsmen apply a filler to cherry as it helps to bring out the beauty of the grain.

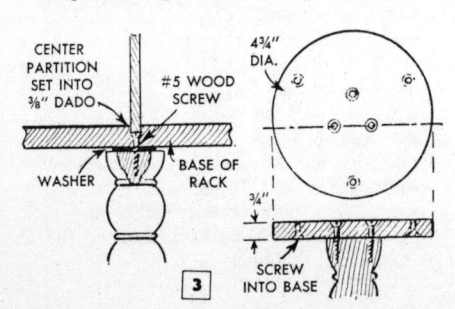

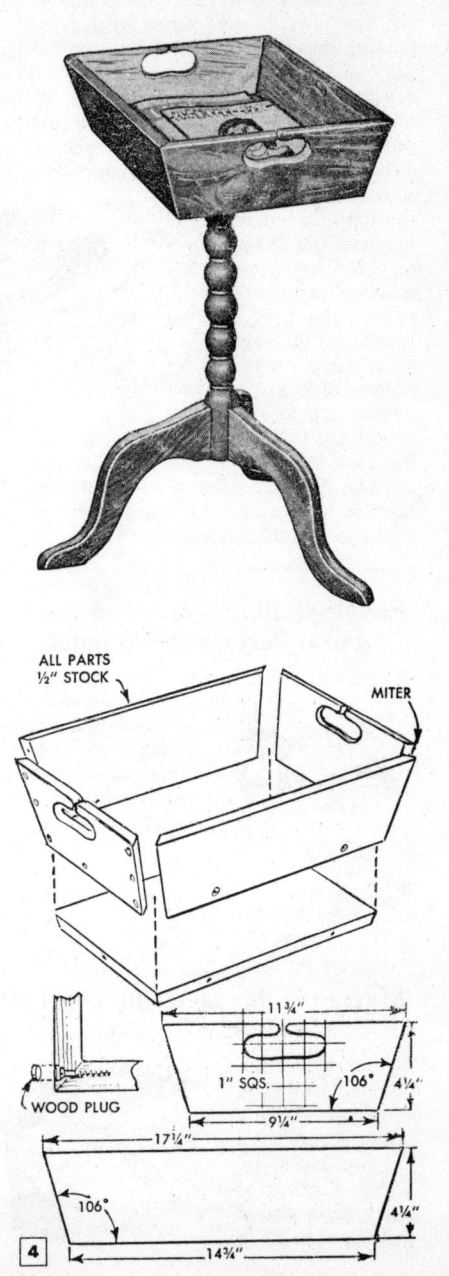

MAGNETS

The usual array of small tools that clutter a bench when you are working, can be stored efficiently on this magnetic rack. It will support pliers, screwdrivers, compasses, chisels, small wrenches, etc. Just place them against two iron strips which contact the faces of several magnets. Model-T Ford magnets are well suited for the purpose. They are placed in a rack made as indicated and the iron strips are screwed to the front of the rack at the top and bottom. It is important that like poles of the magnets be matched throughout, otherwise the pull will be lessened greatly. The poles can be matched by holding two magnets together as indicated in the center photo. If they show no attraction for each other when held together in this position, the like poles are together, but if they cling to each other, one of the magnets must be turned over.

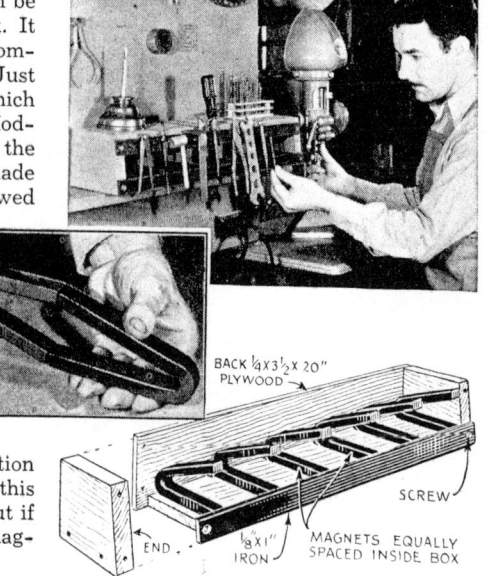

Envelope Removes Small Tacks From Permanent Magnet

Before using a permanent magnet to gather up small tacks or metal filings, slip it into a thin paper sack or envelope. The magnet will lose little of its power, and the paper can be removed to dislodge the metal.

Magnet Holds Lettering Brushes In Handy Position

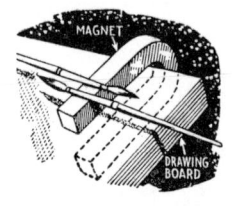

One showcard writer keeps a couple of lettering brushes handy on his workboard by using a magnet taken from an old magneto. The magnet hooks over the top of the drawing board in the manner shown in the illustration. The metal ferrules of the brushes adhere to the magnet.

Fishhooks Dispensed in Store With Small Magnet

To avoid injuring the fingers in removing small fishhooks from drawer compartments, when selling or invoicing, one merchant employs a small magnet. This lifts the hooks so that any number may be removed. The same idea can be used by fishermen who lose hooks or small flies in the grass while fishing. The magnet will locate and pick up the hooks that could seldom be found in the usual manner.

Tools Permanently Magnetized With Battery and Wire

If you have a screwdriver, tack hammer or other tool that you wish to magnetize, this may be done easily with a storage battery and a length of heavy-gauge, flexible wire. First, slot the face of the tool with a hacksaw to provide two poles, similar to those on a horseshoe magnet. Then, form the wire into a coil of about 10 turns and fasten each end to a storage-battery post. Insert the tool through the coil, allowing it to remain there approximately 10 min. Rub it lightly on a piece of iron or steel during this period. To make a demagnetizer, take a coil of bell wire and connect it in series with a lamp bulb on a.c. current. Pass the tool through this coil several times, testing it for magnetism each time until the tool is completely demagnetized.

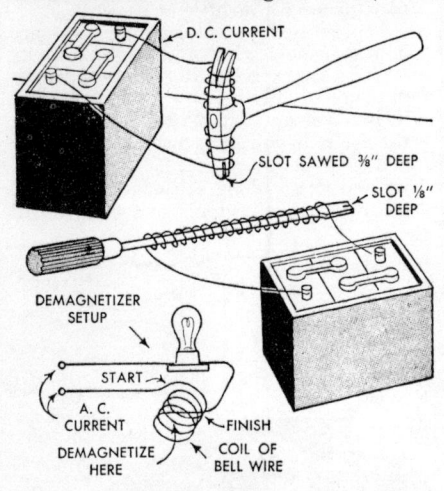

Swinging of Screen-Door Hook Avoided by Magnet

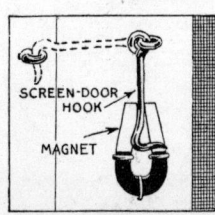

To prevent the hook on a screen door from swinging between the frame and door when the latter closes, you can employ a small permanent magnet as indicated. Wire staples hold the magnet under the hook in the position shown.

Tin-Can Tops Easily Removed With Magnet

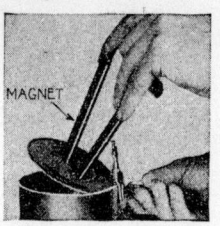

When using a can opener of the rotary type where the lid has a tendency to drop down inside the can after it has been cut away, you will find a small magnet handy to hold a lid. When the latter has been almost cut away, just place the magnet against it, as indicated in the photo, and then complete the cutting, after which you can easily lift it up.

Assembling of Small Ball Bearings Aided With Magnetized Nail

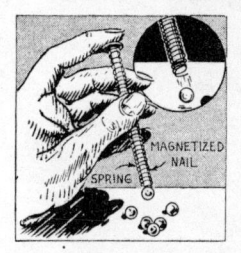

Small ball bearings or other steel parts may be handled easily for assembly work by means of a magnetized nail and a coil spring. The spring, which is of a size to just fit over the nail, is stretched so it is slightly longer than the nail. Held as shown, the nail is pushed out of the spring to pick up a bearing, which is then pushed off the nail end by releasing the spring.

Magnet Holds Lettering Brush In Cleaning Fluid

A fine-pointed lettering brush can be kept in a glass of cleaning fluid so that it will not rest on the bottom and destroy its shape if you use a toy magnet over the edge of the glass to hold the brush as indicated in the sketch at the right. The magnet will attract the steel ferrule of the brush and hold it in place securely so that the bristles remain at any desired depth in the liquid.

Make Your Own

MAGNIFIERS are fun to make and find dozens of practical uses in home, shop or office. Lenses with chipped edges are quite satisfactory and by purchasing these factory rejects you can make really nice glasses for as little as 20 cents.

Lenses used in making magnifiers are of the plano-convex and double-convex type, as shown in Fig. 10. The focal length of a lens usually is known when you buy it, but it is an easy matter to determine the focus if you get lenses mixed. This is done as shown in Fig. 1. Move the lens up and down along the edge of a ruler until the light rays focus to a sharp, small

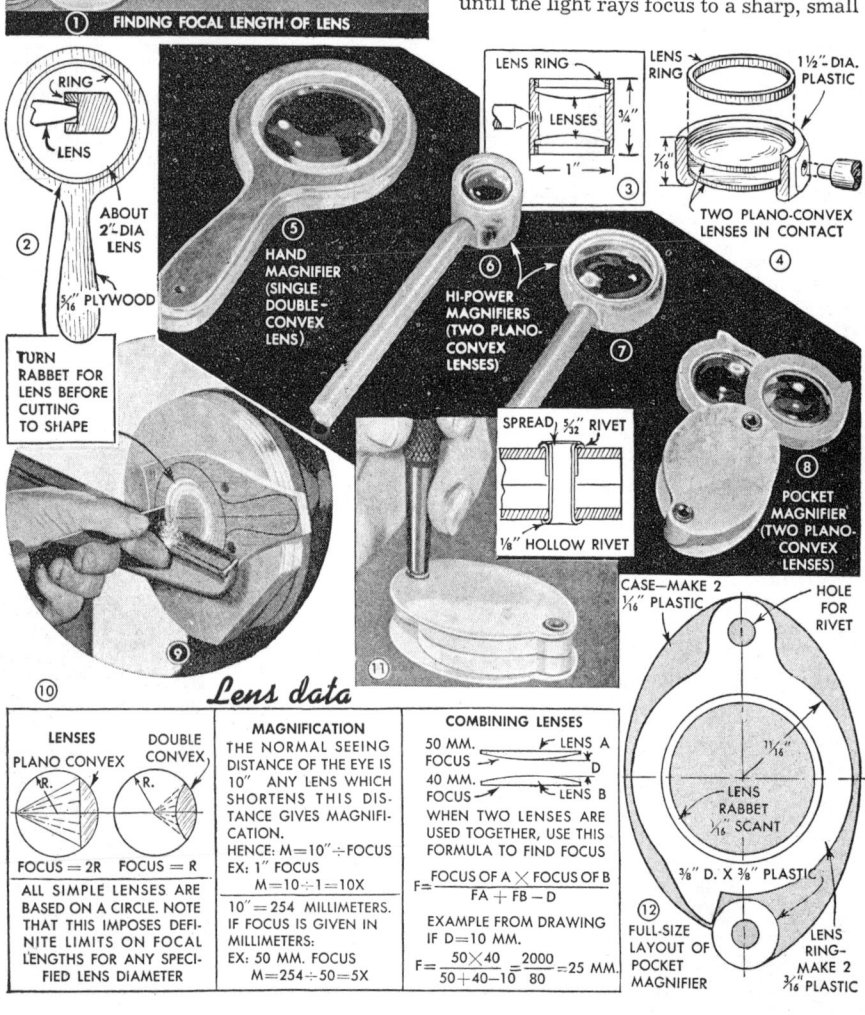

① FINDING FOCAL LENGTH OF LENS

② RING / LENS / ABOUT 2"-DIA. LENS / 5/16" PLYWOOD / TURN RABBET FOR LENS BEFORE CUTTING TO SHAPE

③ LENS RING / LENSES / 3/4" / 1"

④ LENS RING / 1½" DIA. PLASTIC / 7/16" / TWO PLANO-CONVEX LENSES IN CONTACT

⑤ HAND MAGNIFIER (SINGLE DOUBLE-CONVEX LENS)

⑥ HI-POWER MAGNIFIERS (TWO PLANO-CONVEX LENSES)

⑦

⑧ POCKET MAGNIFIER (TWO PLANO-CONVEX LENSES)

⑨

⑪ SPREAD 5/32" RIVET / 1/8" HOLLOW RIVET

⑫ CASE—MAKE 2 1/16" PLASTIC / HOLE FOR RIVET / 11/16" / LENS RABBET 1/16" SCANT / 3/8" D. X 3/8" PLASTIC / FULL-SIZE LAYOUT OF POCKET MAGNIFIER / LENS RING— MAKE 2 3/16" PLASTIC

⑩ Lens data

LENSES	
PLANO CONVEX	DOUBLE CONVEX
FOCUS = 2R	FOCUS = R

ALL SIMPLE LENSES ARE BASED ON A CIRCLE. NOTE THAT THIS IMPOSES DEFINITE LIMITS ON FOCAL LENGTHS FOR ANY SPECIFIED LENS DIAMETER

MAGNIFICATION
THE NORMAL SEEING DISTANCE OF THE EYE IS 10". ANY LENS WHICH SHORTENS THIS DISTANCE GIVES MAGNIFICATION.
HENCE: M = 10" ÷ FOCUS
EX: 1" FOCUS
M = 10 ÷ 1 = 10X
10" = 254 MILLIMETERS.
IF FOCUS IS GIVEN IN MILLIMETERS:
EX: 50 MM. FOCUS
M = 254 ÷ 50 = 5X

COMBINING LENSES
50 MM. FOCUS — LENS A
40 MM. FOCUS — LENS B
WHEN TWO LENSES ARE USED TOGETHER, USE THIS FORMULA TO FIND FOCUS

$$F = \frac{FOCUS\ OF\ A \times FOCUS\ OF\ B}{FA + FB - D}$$

EXAMPLE FROM DRAWING IF D = 10 MM.

$$F = \frac{50 \times 40}{50 + 40 - 10} = \frac{2000}{80} = 25\ MM.$$

MAGNIFIERS

VIEW THROUGH ILLUMINATED MAGNIFIER (ABOUT 3X) ⑬

⑭ ILLUMINATED MAGNIFIERS ARE VERY HELPFUL ON FINE DETAIL WORK. LENS SHOULD BE 1½" TO 2" DIAMETER FOR TWO-EYE VISION. MAGNIFICATION 2 TO 4X TO SUIT

⑯ SWITCH — TO 110-V. OUTLET WIRING

⑮

LENS—1½" TO 2" DIA.—3" TO 5" FOCUS

DRILL ½" HOLE FOR CANDELABRUM-SOCKET SHELL ⅞" X ⅞"

³⁄₁₆ HOLE FOR LENS

CANOPY SWITCH

6-WATT 110-VOLT LAMPS

¼"

DRILL ⅜"

³⁄₃₂"
³⁄₁₆

1⅜"

3½"   4⅞"   WOOD HANDLE

⑰ ⁵⁄₁₆" X ⅜" GROOVE   ⅛" PIPE NIPPLE

area, then read the focal length on the ruler. The best light to use for checking the focus is sunlight, but with small lenses you can get accurate results with an electric light, provided it is several feet away from the lens.

It is important to know the focal length of a lens because this determines its magnification. The shorter the focal length the greater the magnification. You can find the power of any single lens or any two lenses combined by applying the simple calculations given in Fig. 10. Combining lenses always gives a shorter focus. If two lenses of the same focus are placed in contact the combined focus will be one half of the individual focus.

Simple hand and pocket magnifiers are shown in Figs. 5 to 8. The frame for the hand magnifier, Fig. 5, can be wood or plastic. A turned ring cemented in place holds the lens securely as can be seen in Figs. 2, 3, 4 and 9. Magnifiers shown in Figs. 6 and 7 use two plano-convex lenses. The handy pocket magnifier, Fig. 8, is diagrammed full size in Fig. 12, while Fig. 11 shows how the assembly is made with hollow rivets. In this unit lenses are a snug, press fit in the plastic mounts and are simply cemented in place without using a retaining ring. The lenses are plano convex, arranged so that their curved surfaces face each other.

Plano-convex lenses should be used whenever possible in a combined lens system and always with the curved sides facing each other. This assembly gives good correction and a clear, flat field. Double-convex lenses also can be combined, but the correction is not so good. The double-convex lens is a logical choice for single-lens magnifiers as you can

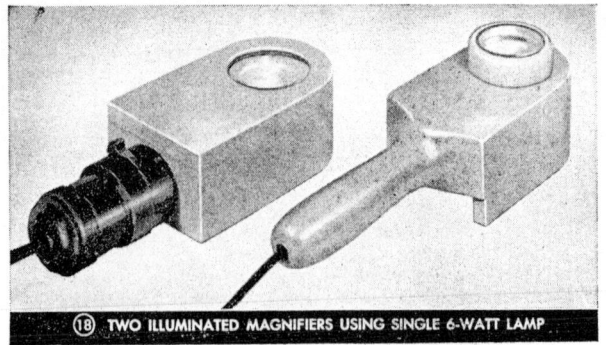

110-volt with candelabrum screw bases; sockets are candelabrum socket shells. If you have any trouble buying these you may be able to solve your problem by breaking up an old Christmas-tree string. Another way is to solder the wires directly to the lamp, but this means, of course, that you have to tear the box apart when a lamp goes bad.

look through either way and get the same result. If a single plano-convex lens is used, have the flat side next to your eye if you use the glass close to your eye; if you hold the magnifier a distance away from your eye have the curved side facing you. Check this yourself and you will note there is considerable distortion if these suggestions are not followed.

A magnifier is useless unless there is good light on the object being examined. Hence, it is logical to combine a light directly with the magnifier. A good unit of this kind is shown in Figs. 13, 14 and 15 and detailed in Fig. 17. Fig. 16 diagrams the wiring. The lamps used are standard 6-watt,

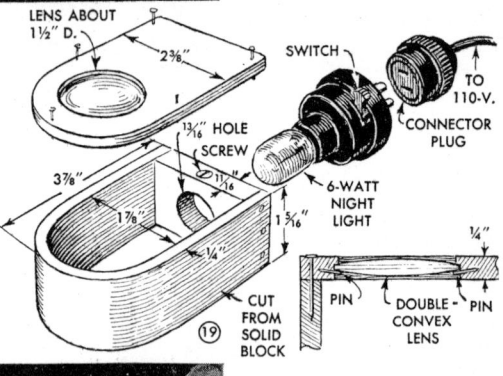

For certain kinds of work, you may want this magnifier on a permanent stand. The best mounting is a 10 or 12-in. flexible gooseneck, as can be seen in Fig. 14. The gooseneck has a 1/8-in. pipe coupling at either end, making it easy to attach a base and the magnifier by means of pipe nipples.

Two illuminated magnifiers each using a single lamp are shown in Fig. 18. The night-light socket shown in one of the units, Fig. 19, can be purchased at most electrical stores. Discard the small plastic shade that comes with it and hold the socket in a wooden frame by driving in a machine or wood screw to make contact with the groove in the socket. The top and handle for the style shown in Fig. 21 are worked in one piece, the handle being turned first, as shown in Fig. 20.

Lenses for these illuminated magnifiers should be at least 1½ in. in diameter if you want to use them for two-eye vision. It is best to keep the magnification rather low —two or three power will be plenty and at this magnification you will get a wide, flat field without distortion.

Other useful magnifiers are shown in

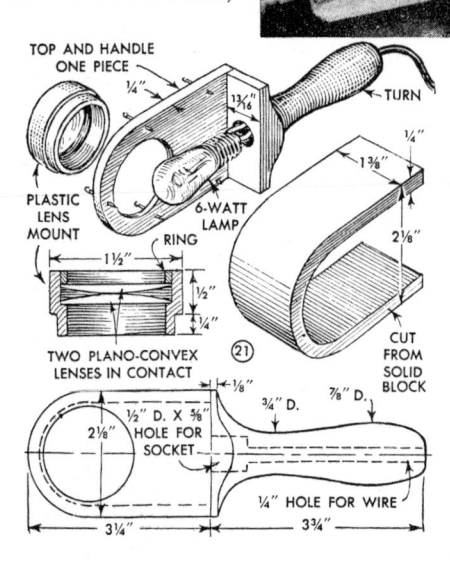

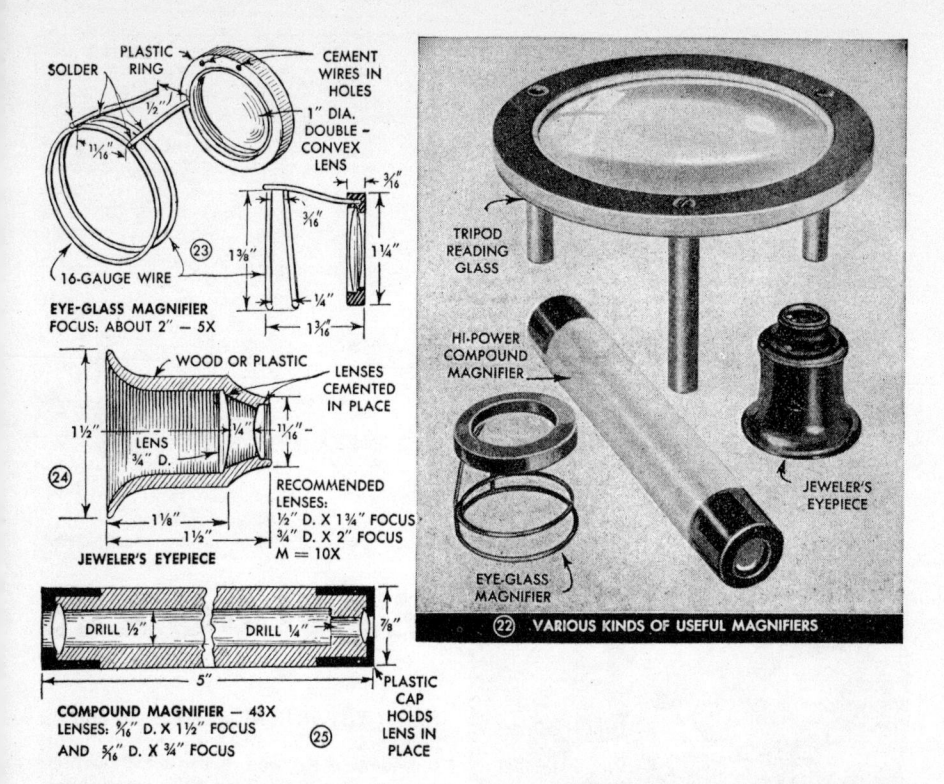

Fig. 22. The tripod style is excellent for retouching or other jobs where a fixed focus over the work is required. The lens should be 2 to 4 in. in diameter and of low magnification. If you wear glasses you will like the little eye-glass magnifier detailed in Fig. 23. It slips over one lens of your glasses and you can look through it or under it at will. The lens focus should be about 2 in. for 5X magnification. For critical examination of small objects the jeweler's eyepiece, Fig. 24, makes a nice glass. It can be turned in either wood or plastic. The recesses for lenses should be made for a snug fit, the lenses being cemented in place with transparent adhesive. This glass should be made 8 to 10X. It is used close up, being clamped directly to the eye in monocle fashion. If you want extremely high magnification the best simple instrument is the compound magnifier shown in Fig. 25. It is really a low-power microscope in which an image is picked up and magnified by one of the lenses and then further magnified by the second lens. The final image will be upside down. The big lens is commonly the eye lens, but you can use it either way with same magnification. In a system of this kind the eye lens is calculated for magnification the same as any other magnifier; that is, 10 in. (the eye's normal seeing distance) divided by the focal length of the lens. The object lens is calculated by dividing its focal length into 5 (the length of the tube). The two powers thus obtained are then multiplied to get the total magnification. Plastic caps shown can be bottle caps, while the tube itself can readily be turned and bored in wood with the ends recessed to take the lenses.

Foreign Particle in Eye Located with Aid of Magnifying Glass

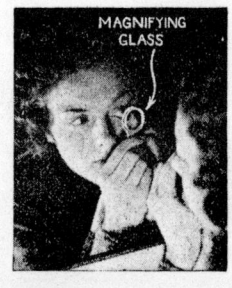

A foreign particle in the eye that is hard to locate often can be seen with the aid of a magnifying glass held in front of a mirror. Get as close to the mirror as possible. The hand that holds the glass can pull down the lower eyelid, thus leaving the other hand free to remove the particle.

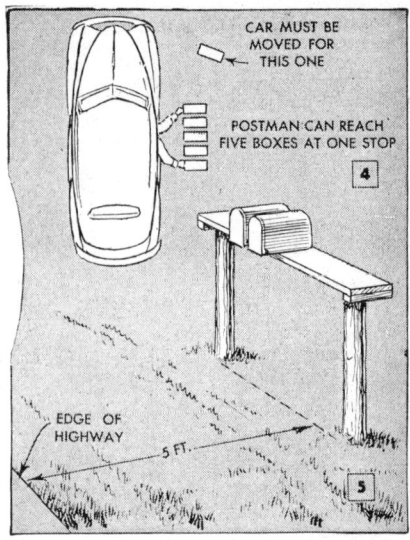

MAILBOXES

The three illustrations, above and below, should make it easier for any community to decide to do something about that leaning cluster of rural mailboxes. Almost everybody will be glad to pitch in and help line them up

BOTH THE RURAL mail carrier and the neighbors will thank you for promoting the idea of arranging the neighborhood mailboxes in a neat, orderly row. They'll all be glad to lend a hand in obtaining the material and constructing such attractive units as those shown in Figs. 1 and 3. Compare these with the usual unsightly group in Fig. 2 and you'll agree that either one is a much simpler arrangement than mounting the boxes on individual posts. Moreover, these structures are approved by the Post Office Department.

Where the boxes are at a uniform height of 4 ft., the postman can service five boxes with one car stop, Fig. 4, but he must move his car for any beyond this reach. Placed in two tiers, ten boxes can be handled at one stop.

A very simple arrangement suitable for a rustic setting is to mount the boxes on a horizontal log raised 4 ft. above ground, as in Fig. 6. The top is hewn flat and also cut in low steps for boxes. Another one, Fig. 7, consists of iron pipe and fittings assembled in a candelabra style. Weld or pin the pipe fittings so that they cannot turn.

Where wood wagon wheels are available, boxes can be mounted in a turntable style very convenient for the postman. A 36-in. wheel is recommended, its spindle being anchored solidly to the post in a vertical position, Fig. 9. These wheels

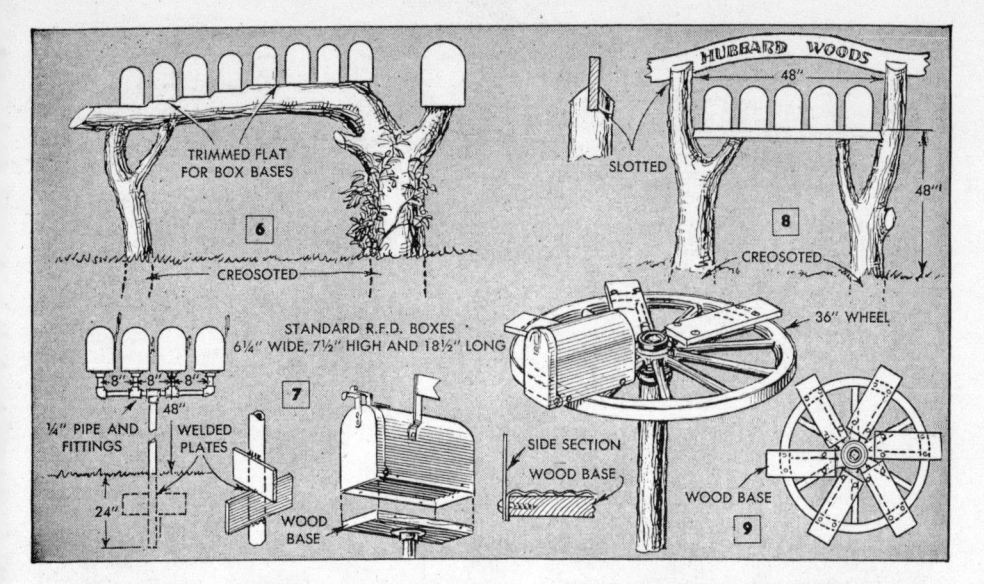

CAN BE MADE ATTRACTIVE

are large enough so that five boxes can be mounted radially, or six placed at an angle. This refers to the standard R.F.D. mailbox which is 6¼ by 7½ by 18½ in., inside dimensions. A wooden base is first screwed to the wheel rim and spokes, and the box is then screwed to the base.

An arbor effect is illustrated in Fig. 8, built along conventional lines and with the curved top members made in three pieces, glued and doweled together. Set two crotched logs in the ground to support the shelf. A curved branch adzed to a flat section and mounted in slots, with the community name burnt into the wood, makes a pleasing job. All rustic construction should be of thoroughly seasoned logs, with the bark removed.

The design of the unit in Figs. 1 and 10 is patterned after the English timbered style. It can be built to shelter about eight boxes of the parcel-post size. Brick piers are mounted on concrete slabs. Where a single parcel-post box is to be used, place it in the center of the row of boxes as in the V-type mounting, Fig. 13.

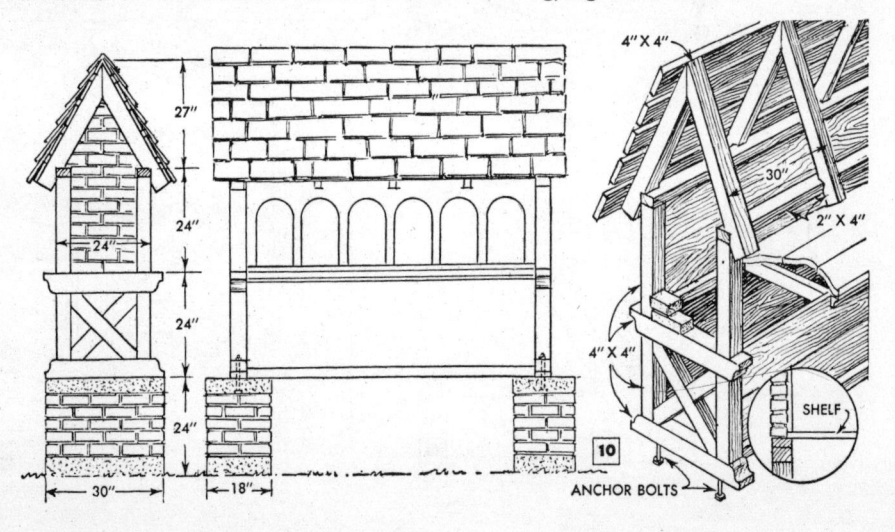

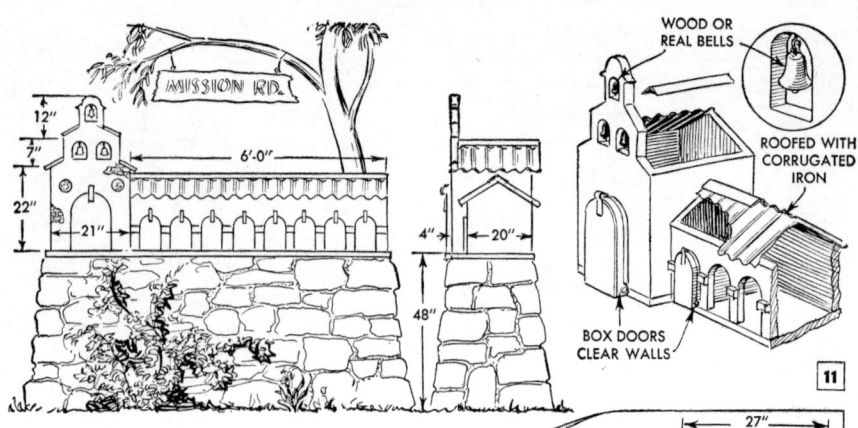

Another one, Fig. 11, is patterned after the mission style. It's mounted on a stone pier which can rest on a concrete slab. Use stock sizes of pine lumber for the box unit. The archways, typical of mission design, house the mailboxes. These should extend out far enough so the hinged fronts will swing out freely, Fig. 11, the right-hand detail. Corrugated iron roofing, or aluminum roofing painted brick-red resembles tile roofs quite closely. Wooden bells, or if you desire, real bells of suitable size can be mounted in the tower.

In Fig. 12, a simple canopy accommodates a number of boxes arranged in one or two tiers. Essentially the same design is pictured in Fig. 3. Allow 8 in. of space for standard boxes, and at least 12 in. for the parcel-post size. Another design, Fig. 14, is suitable for roadside installation in the country.

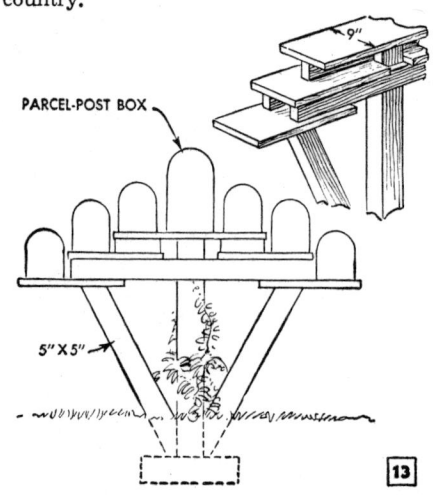

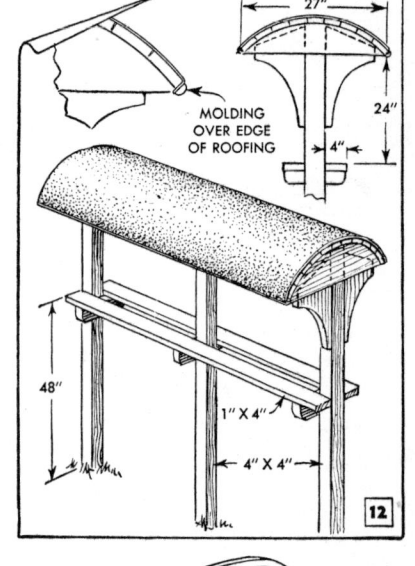

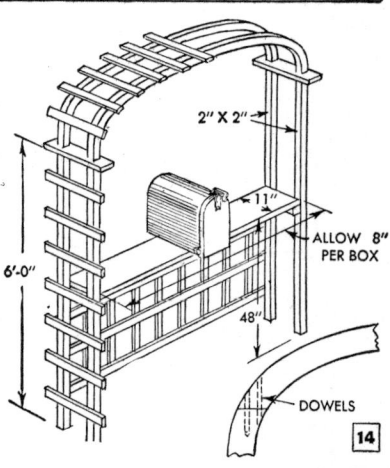

Wood Mailbox Has Hinged Lid for Magazines

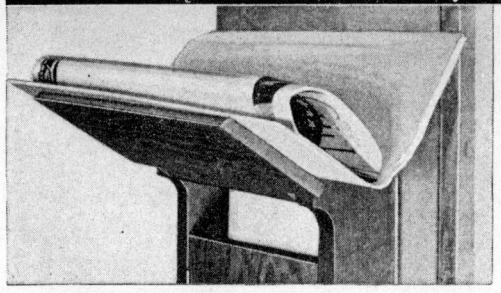

SIMPLE, neat appearance and a false, hinged lid that opens to provide a rack for holding small packages, newspapers and magazines, combine to make this wooden mailbox thoroughly practical as well as attractive. Wood used in its construction may be maple, white pine or waterproof plywood, but ordinary plywood will, of course, be unsuitable when exposed to rain or snow. All parts except the hinges are butted and nailed, with a little waterproof glue added here and there if desired for extra durability.

The various pieces are cut to the net sizes as shown in the cutting list. For ease in bandsawing and sanding, the side pieces can be nailed together and worked as one piece. Hinge gains are mortised by chisel, or they can be routed on a drill press with a 1/8-in. router bit. Notice that the lower edge of the lid is beveled inward to provide a stop for the false lid, which is hinged on it to swing down. A wooden turnbutton can be made, according to the dimensions in the circular detail, to lock the door.

If a bright finish is desired, a brown-orange stain should be used. This is followed by two coats of spar varnish or three coats of clear lacquer. Two screws driven through the back panel secure the box to a porch post or wall. Knobs, of course, may be fastened to the door and lid, but they are not necessary. Leave the false lid open the first day the box is installed to acquaint the mailman with its use.

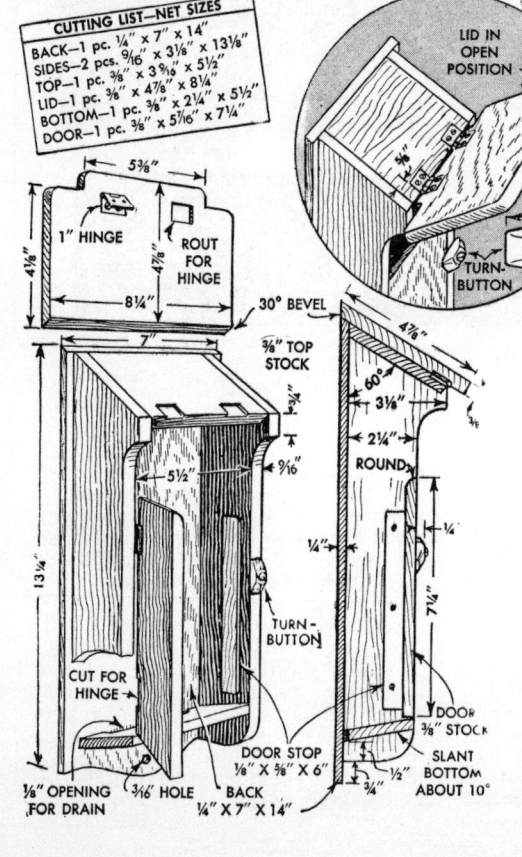

CUTTING LIST—NET SIZES

BACK—1 pc. 1/4" x 7" x 14"
SIDES—2 pcs. 9/16" x 3 1/8" x 13 1/8"
TOP—1 pc. 3/8" x 3 5/16" x 5 1/2"
LID—1 pc. 3/8" x 4 7/8" x 8 1/4"
BOTTOM—1 pc. 3/8" x 2 1/4" x 5 1/2"
DOOR—1 pc. 3/8" x 5 7/16" x 7 1/4"

Hammered-Copper Mailbox Has Antique Finish

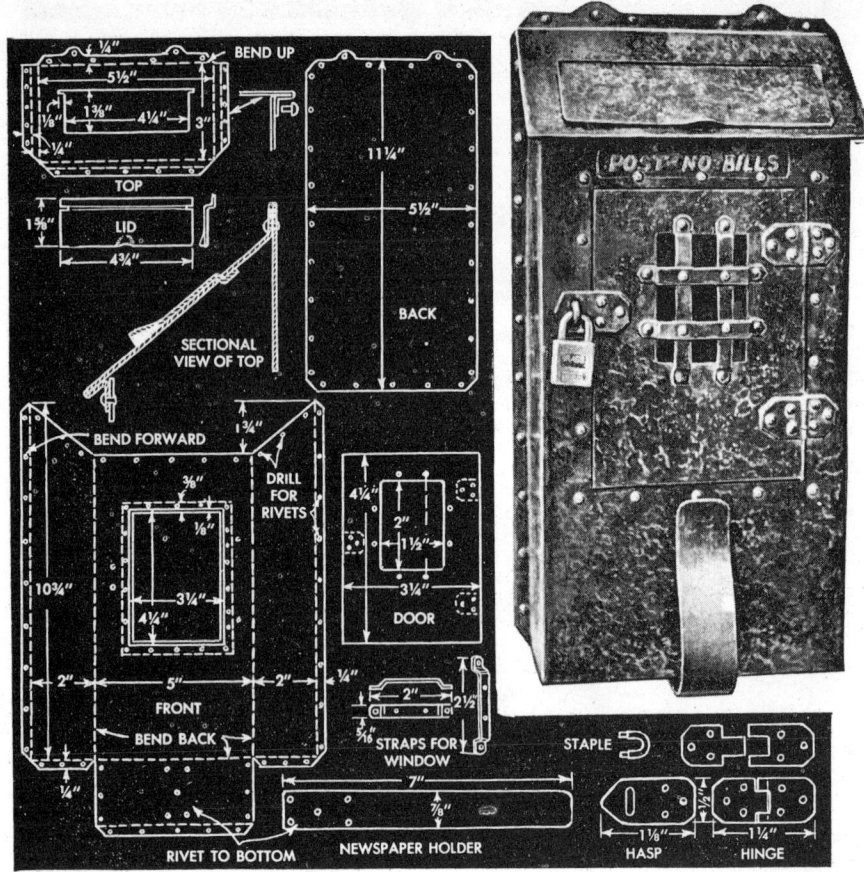

With the exception of the spring holder for newspapers and larger pieces of mail, this antiqued mailbox is made of 20-ga. soft sheet copper. The pieces are cut out with tin shears and a chisel, after which the edges are filed smooth. Then the face side of each piece is peened with a hammer. The lettering which appears over the door is made in the following manner: First, paint the letters on with asphaltum varnish; let dry completely, then outline the box around the letters with modeling clay. Now apply sufficient commercial nitric acid to cover the area, let it stand for few moments and wash off with water. Remove the asphaltum with lacquer thinner. Drill all holes and bend the parts to shape as shown on the drawing. Before assembling, antique all parts by the following method: Pour a small quantity of commercial nitric acid into a wide-mouth jar, being careful to keep the solution off hands and clothing, then drop pieces of scrap copper into the acid. When the acid stops boiling, it is ready for use. With a cloth swab, cover the surface of all parts. Then heat with a blowtorch until the surface is an even brown. Dust off when cool, burnish the highlights and rub to a high gloss with floor wax. The box is designed for use in private homes. If used in an apartment building, first check postal regulations for such buildings.

ADJUSTABLE CONE MANDREL

WHERE THE OUTSIDE diameter of a bushing, sleeve or other tubular piece of work must be machined concentric with the bore, or inside diameter, an adjustable cone is required to mount the work between lathe centers. As you will see in the drawing below, the mandrel consists of a fixed and movable cone which automatically center the work as they engage the ends. The movable tailstock cone is knurled for tightening by hand and is held in position on the threaded mandrel with a locknut. A flat ground at the opposite end of the mandrel permits gripping it with an end wrench while the locknut is adjusted. The mandrel detailed is dimensioned for work having an inside diameter of 9/16 to 1 in. However, by enlarging the size of the cones and the diameter of the mandrel, work having a larger inside diameter can be accommodated. To make the mandrel, cut the stock to length, center-drill each end and turn as indicated. To thread the mandrel accurately, it is best to use a newly sharpened threading tool, taking a number of light cuts and lubricating with plenty of oil. The adjustable cone is made by knurling, drilling and tapping a bar of tool steel and then turning the cone roughly to shape. After the work is cut off, the cone is screwed onto the threaded end of the mandrel where it is held with a locknut. The mandrel is mounted between centers and a finishing cut is taken to true the cone. This will assure its running concentric with the mandrel when the finished assembly is mounted in the lathe. When the work will permit, break the corners of the bore 1/64 in. before mounting on the mandrel.

A — Mandrel holds sleeve having straight bore, as above

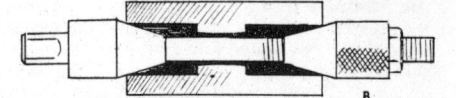

B — Cone mandrel will also handle counterbored sleeve

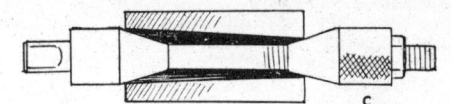

C — Tapered bore does not affect alignment of the work

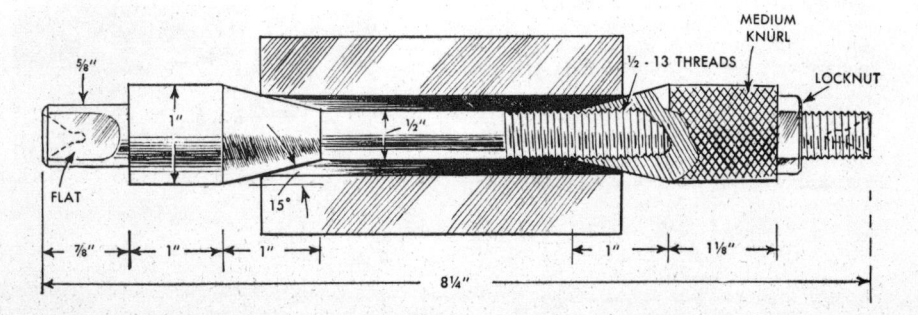

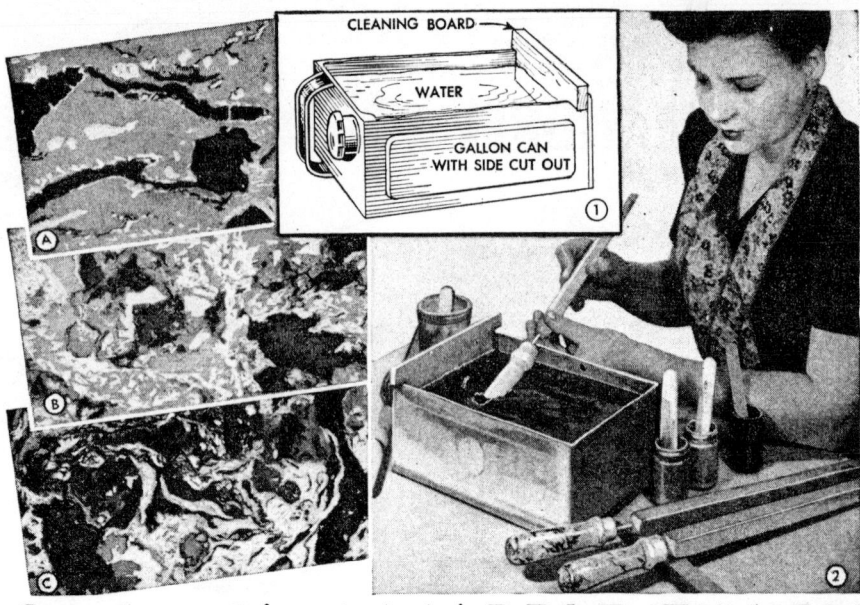

Water-Float MARBLE FINISH

PRODUCING imitation marble by means of colors floating on water is a novelty technique used in finishing turnings, handles, small boxes, etc. The work itself is child's play, but the production of a specific design in direct imitation of a certain kind of marble demands considerable experience and practice. You can use this finish on any material—glass, paper, metal or wood. Wood should be coated previously with lacquer of suitable color to obtain a smooth working surface.

On a small scale, the work can be done nicely in a 1-gal. paint can having one side removed as shown in Fig. 1. Fill this nearly to the top with cold water. You will need a film solution made up as indicated in Fig. 3. Also, you will need several colors in japan, which can be obtained in either tubes or cans, the latter being the best since only the pigment is needed and the fluid floating on top can be poured off. Reduce the paste with lacquer thinner, mixing it well, to a consistency a little thinner than you would use for brushwork.

The actual finishing schedule is started by putting a few drops of the film solution on the surface of the water. A lot of solution is not needed; all you want is a thin, almost invisible

Film Solution
ONE PART WHITE SHELLAC OR DAMMAR VARNISH, TWO PARTS LACQUER THINNER, ABOUT TEN DROPS LINSEED OIL PER PINT

Colors
THIN COLORS IN JAPAN WITH LACQUER THINNER TO AVERAGE BRUSHING CONSISTENCY

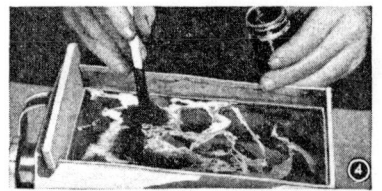

FLOAT COLORS ON WATER

DIP WORK FACE DOWN

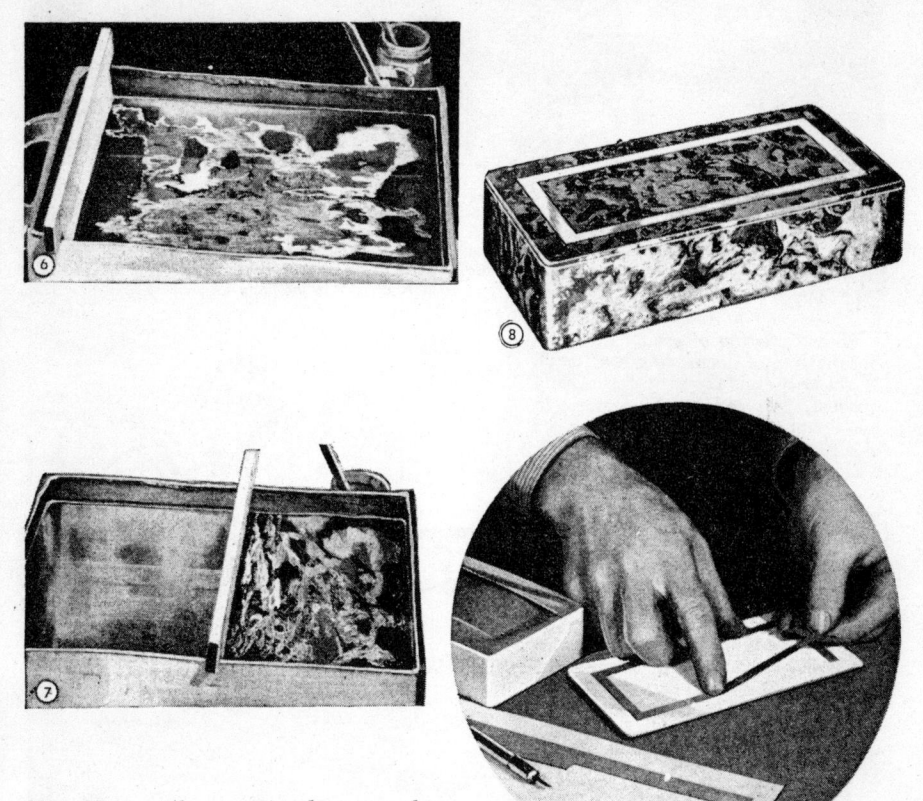

film. Next, pick up any color on a thin piece of wood and draw the paddle through the water as in Fig. 4. Repeat with other colors, mixing the whole pattern gently to get whatever effect you desire. Finally, dip the work. Good practice can be had by using small pieces of white cardboard. Flat work like this is always dipped face down, Fig. 5. The color design floating on water adheres instantly to the underside of the card, which then is withdrawn immediately. Typical designs are shown in illustrations A, B and C. If the work is a turning, it should be placed on the surface of the water and then rotated to pick up the design as in Fig. 2.

Any kind of color combination can be used in this work, ranging all the way from brilliant multicolor effects to soft brown and black tones in imitation of wood grain.

The cleaning board, Fig. 1, is pushed from one end of the tank to the other to skim off any surplus film after each operation. Also, it can be used to compress the design as shown in Figs. 6 and 7. In the dipping process, the paint film will adhere to all surfaces so that suitable protection must be given any area that is not to be treated. This is illustrated in Figs. 8 and 9, where the inside of the box and the inlay on the lid are masked. After dipping, the work should dry overnight.

First, mix powdered chalk and water to a consistency of thick cream and pour it onto a sheet of aluminum. Tap edge of metal to level mixture to uniform depth, slightly greater than the desired depth of the design. Use a blotter at the edges to draw off excess water and let the chalk mixture stand until thoroughly hard

When the chalk is hard, the surface is sanded lightly to remove irregularities and produce a flat surface. This is important. Use fine-grade sandpaper wrapped around a small wooden block. Plaster of paris can be used instead of chalk, although it is somewhat more difficult to tool, and air pockets form in mixture

MATRIX CASTING

Matrix casting offers an interesting method of decorating craftwork of sheet metal. By a direct-casting process, you can apply designs in bas-relief to any flat, metal surface to which stereotype metal or solder will adhere. The design can be a monogram or a simple carving, which is tooled in a chalk-coated aluminum plate and filled with bits of stereotype metal. The metal is heated in the design by placing the matrix on a hot plate. Then the craftwork is placed face down on the matrix. When the casting cools, the work is lifted from the matrix, and the casting will be affixed to it. It is important that the matrix surface be perfectly flat so that the molten metal will contact the work at all points.

The design is traced or drawn directly on the chalk surface. Remember that an initial must be reversed. Tooling is done by scraping the chalk down to the aluminum plate with the point of a pocket knife. The design should not be undercut. Tool it to give a slight draft so that the casting can be withdrawn

Stereotype metal and linotype metal are best for casting. Place matrix on a hot plate and fill design with bits of the metal. Add enough to bring it level with surface when molten. Craftwork to receive design is tinned and placed face down on matrix. Upon cooling, work is lifted from matrix with casting affixed

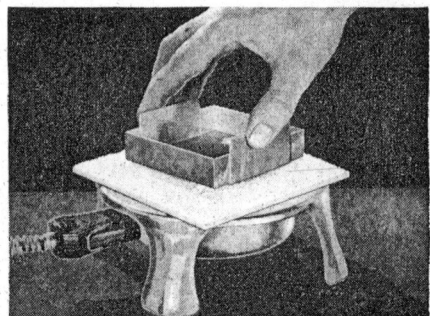

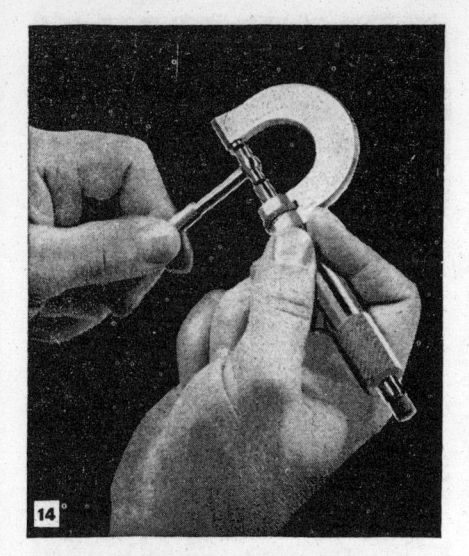

MEASURING TOOLS

IN ORDER to use precision measuring instruments efficiently it is necessary to keep in mind that there never is a perfect measurement. There are always tolerances and limits. Generally speaking, the term tolerance refers to the allowable error, while limit, or limits, refers to the specification of the large and small dimensions. As an example, a hole is to be bored 1 in. in diameter. If the work is of such a nature that the diameter must be held within fairly close limits, large and small dimensions will

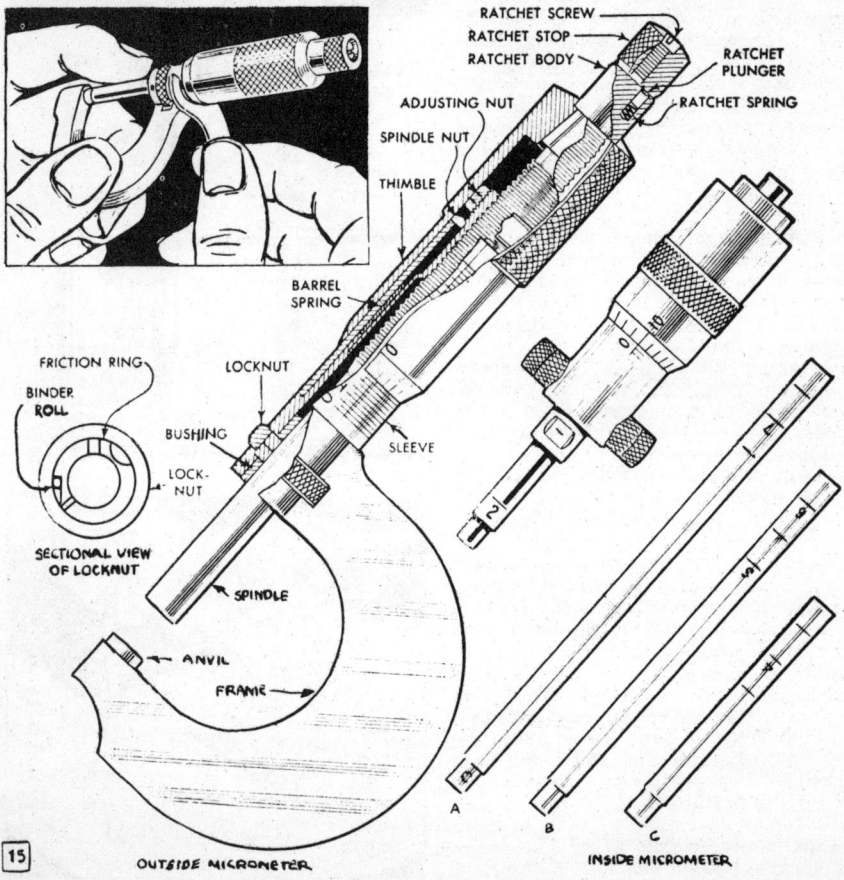

OUTSIDE MICROMETER

INSIDE MICROMETER

be specified as plus and minus, that is, the hole can be bored slightly over 1 in. or slightly under and still meet the requirements. But the measurements of the finished work cannot exceed the limits specified as otherwise the job will not serve its purpose.

As pointed out previously, the important thing is the ability to make measurements of consistent accuracy and to follow through with the production of work which checks within the specified limits with an equal consistency. Specified limits define the necessary degree of accuracy. For ordinary machine work in production shops, dimensions nearly always are specified in thousandths and in some cases in "tenths," or ten thousandths, of an inch. Production shops usually work within thousandths while limits specified in ten thousandths are used by manufacturers of highly specialized products. Ten thousandths are used also in toolrooms and die shops and in experimental laboratories where mockups and scale models are made for reference or testing purposes and where, for various reasons, work must be produced within very close limits.

A graduation line on the average rule is several thousandths of an inch in width thus making fine measurements impracticable. The double contact of the ordinary caliper makes consistently accurate measurements within close limits dependent to a considerable extent upon the skill and experience of the individual. Although very accurate work still is done with these simple measuring tools, shops that must work to close limits, with a consistent accuracy not attainable with ordinary rules and calipers, require more precise measuring instruments, such as the micrometer caliper.

The micrometer caliper, or "mike," as it is commonly called, is simply a slide caliper with a fixed contact, or anvil, and a movable contact, or spindle, actuated by a precision screw adjustment provided with graduations which can be read in thousandths of an inch. The common types of outside and inside mikes shown in Fig. 15 have many uses in the shop and are made in a wide range of sizes. Inside mikes are supplied with extensions, A, B and C in Fig. 15, to increase their capacity. Many other types are available for special purposes. Reading an ordinary mike in thousandths is quite simple to do. The screw, Fig. 15, is accurately made with a pitch of

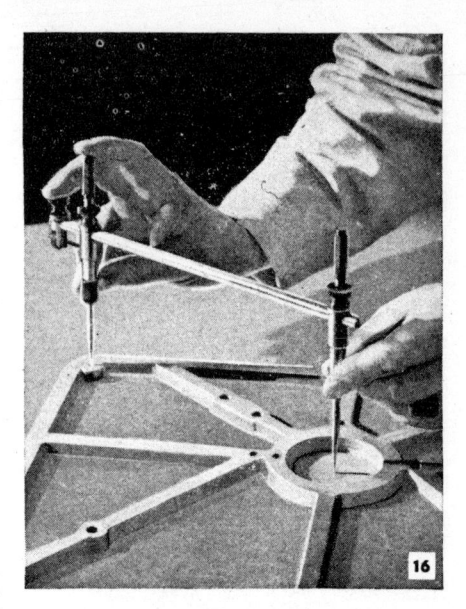

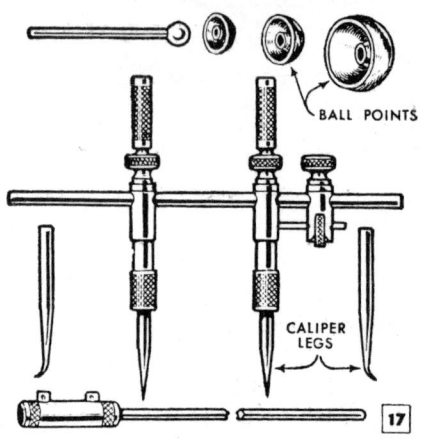

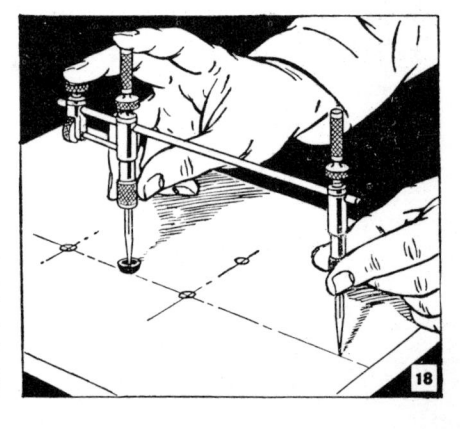

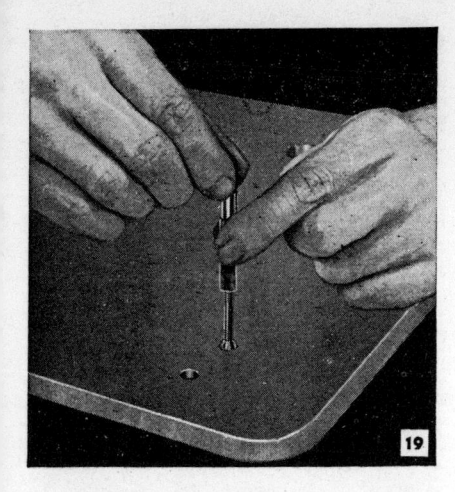

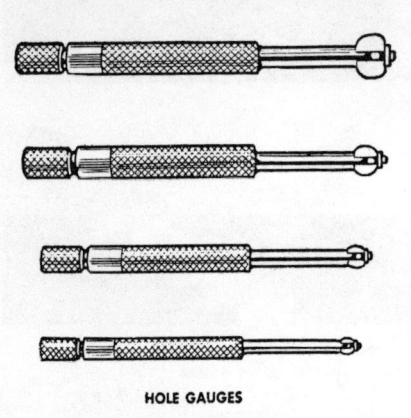

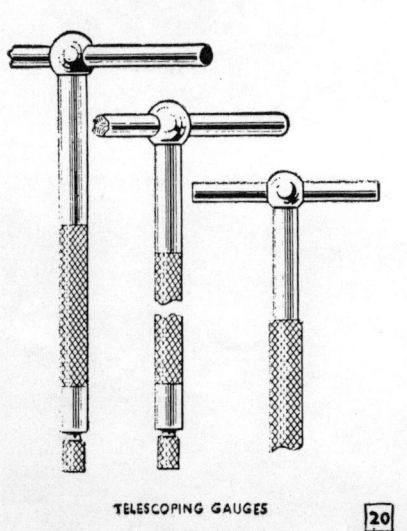

40 threads to the inch and advances one fortieth (.025) of an inch with each complete turn. On a mike of 1-in. capacity, Fig. 14, the sleeve is graduated longitudinally with 40 lines to the inch. Every fourth line is longer and is numbered 1, 2, 3 and so on, to indicate divisions in tenths of an inch. The beveled end of the thimble, Fig. 15, is graduated into 25 divisions numbered from 0 to 25. As one turn of the thimble, from 0 to 0, advances or retracts the spindle one fortieth (.025) of an inch, rotation of the thimble from 0 to the first graduation will move the spindle one twenty-fifth of one fortieth, or one twenty-fifth of twenty-five thousandths, which is .001 in.

On certain types of mikes, such as the micrometer depth gauge, the sleeve usually is numbered from 10 to 0 and it should be kept in mind that in this case, the total reading will be a value less than the lowest graduation visible on the sleeve. The inside mike is shown in use in Fig. 22. Ordinary outside mikes are available with a ratchet stop built into the end of the thimble, as in Fig. 15, or into the thimble itself. This feature is helpful when a number of measurements must be made quickly and accurately or when the same instrument is used by more than one person. The ratchet stop reduces the possibility of error to the minimum as it is so designed that it will slip and thus prevent the spindle from turning after a given amount of pressure is applied when taking the measurement. Nearly all mikes are provided with an adjustment to compensate for wear in the threads. This adjustment is made with a spanner wrench as in the inset, Fig. 15.

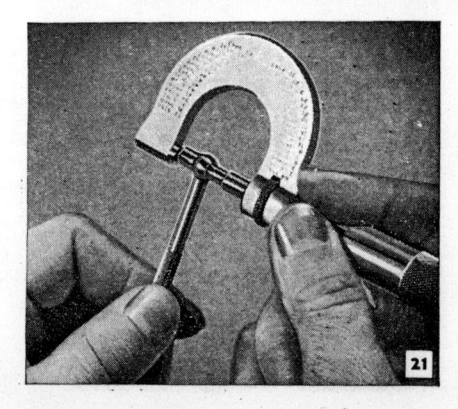

When it is necessary to scribe, lay out, or transfer dimensions of greater lengths than can be spanned with dividers, machinists often use a trammel, Fig. 17. The instrument consists of a polished bar, or beam, on which the points are adjustable by means of special clamps, or carriers. One clamp is provided with a slow-motion screw to permit fine adjustments. Various types of points, or legs, are furnished to adapt the unit for scribing, Fig. 16, measuring from holes, Fig. 18, and for measuring distances across or inside which are too great for ordinary calipers or mikes. For the latter use, the trammel is fitted with caliper legs, Fig. 17. Trammel beams and couplings are available for increasing the span of the trammel to 36 in. or more.

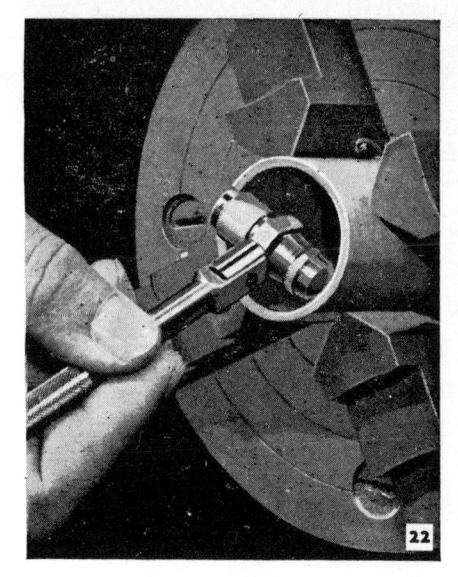

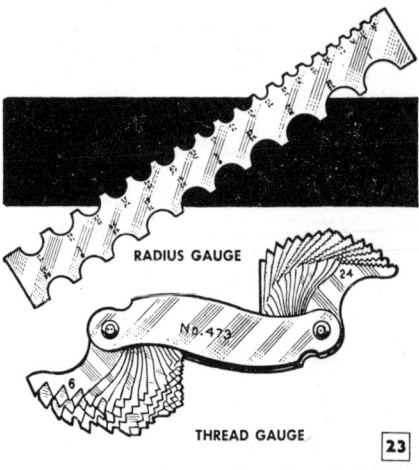

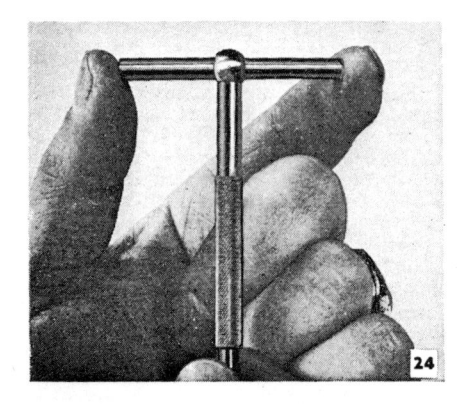

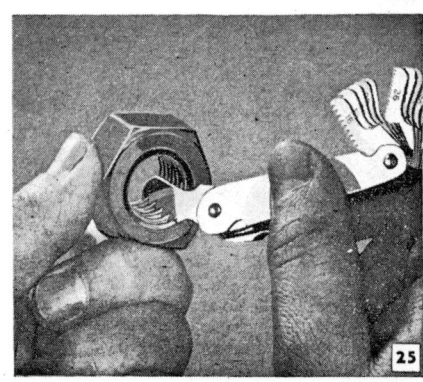

Telescoping gauges, Fig. 20, are often used when it is necessary to obtain precise measurements of internal diameters, Fig. 26. The head of the telescoping gauge is made with one fixed and one telescoping member, and in ordinary use the head is telescoped by pressing on the ends of the members as in Fig. 24. When the distance across the head is slightly less than the diameter of the work to be measured, the head is locked by turning the knurled screw at the end of the handle. The gauge is inserted in the work and the head is released by slacking the screw. The telescoping head then expands across the opening, the spring tension keeping it in contact with the walls until the binding screw is tightened to hold the adjustment. Then the distance across the head is miked, as in Fig. 21, to determine the precise diameter of the hole. Before taking a measurement with a telescoping gauge, be sure there are no metal chips or other small particles on the contacts. When taking internal measurements with the telescoping gauge it should be held with the handle in line with the axis of the opening and the handle should be rocked slightly to make sure the contacts are seated firmly against the walls. Small-hole gauges, Figs. 19 and 20, serve much the same purpose as the telescoping gauges for measuring holes ranging from ⅛ to ½ in. in diameter. Hole gauges are made with a split ball at the contact end which is expanded by means of a tapered sleeve, Fig. 20. The sleeve is actuated by

and, therefore, it is necessary that they be pushed into the opening far enough to assure that the contacts touch the walls at the highest point of the radius. Before taking a measurement with any of these units, clean the work thoroughly to remove chips and grime which might cause inaccuracies of several thousandths of an inch.

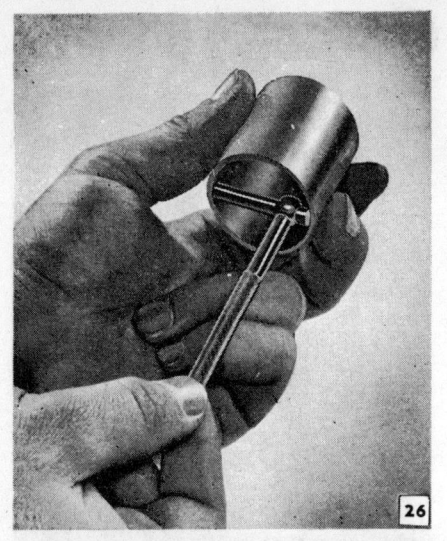

Two other gauges widely used in the shops are the radius, or ball, gauge and the thread gauge, Fig. 23. The former is especially made for diesinker's use but it also is used in some shops for making a quick check of the diameters of rod stock. As will be seen from the detail, the gauge is simply a length of polished steel with 180-deg. radii milled into each edge. Sizes are stamped on the body of the gauge and generally range from $\frac{1}{8}$ to 1 in. in 32nds. Similar types of gauges are made for measuring wire, drills and screws. Likewise, the thread gauge, Fig. 23, is one variation of many similar types of gauges made for checking fillets and angles. Although commonly called a thread gauge, it is perhaps more properly referred to as a screw-pitch gauge. It consists of a number of thin leaves of polished steel pivoted in each end of a short frame. When not in use all the separate leaves fold inside the frame to prevent damage. On one edge of each leaf there are teeth corresponding to standard thread sections. When the pitch of a thread is not known, the leaves of similar size are opened and placed successively over the thread until one is found which meshes with that particular thread. The pitch of the thread is then read from the stamping on that individual leaf. The free end of each leaf is made narrow so that it can be inserted in a nut as in Fig. 25.

The depth mike, Fig. 27, or micrometer depth gauge, is simply an ordinary depth gauge fitted with a micrometer head instead of a graduated bar, or spindle. It is especially designed for accurate measurement of holes, grooves, recesses, projections and offsets in work of an irregular shape. The base is ground flat and the sides are finished perpendicular to the bottom so that the unit can be placed either on top of a finished surface or against a machined shoulder to obtain accurate readings. The micrometer head usually is furnished with a ratchet stop, and extra rods, or plungers, can be obtained in various lengths up to 6 in. or more for measurement of deep holes. Also the bases are furnished in different widths, or spans, up to 6 in. The rods are available in the round or flat type with flat ends, or ends turned and lapped convex. These features extend the instrument's usefulness and versatility to include a greater variety of work.

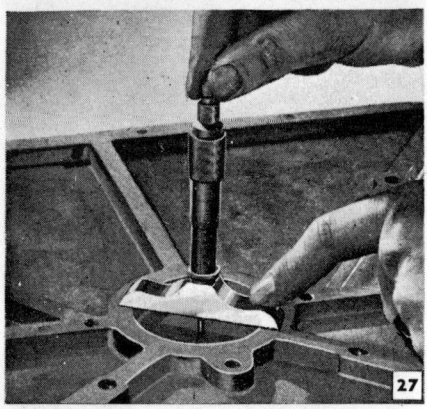

turning the knurled tip of the handle. This advances or retracts the sleeve, thus changing the diameter of the split ball. The gauge is used as in Fig. 19, the ball end being placed on the hole and the knurled tip turned until the ball will slide into the hole. Then the knurled tip is turned in the opposite direction until the ball contacts the sides of the hole firmly, but not so tightly that it cannot be withdrawn easily. The trick in taking consistently accurate measurements with the hole gauges, and also with the telescoping gauges, is in judging the degree of frictional contact between the measuring instrument and the work. Another thing to be careful about when taking measurements with hole gauges, telescoping gauges and inside micrometer calipers is to be sure that the contacts are in full engagement with the walls of the work. All such instruments have rounded contacts

[Certain technical information courtesy L. S. Starrett Co.]

MEDICINE CABINETS

Before

After

An old medicine cabinet often can be modernized quite simply by fitting it with a hinged plywood panel to support a full-length mirror. In the one shown above, the door, inner frame and trim are removed

Photo below shows the plywood subframe fitted after removing the original mirror door and trim. In some types of built-in cabinets a subframe may not be necessary. Be sure subframe is exact size of new mirror

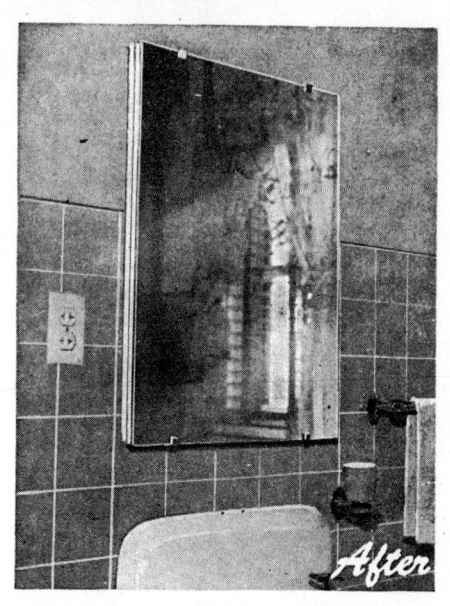

Here's the cabinet at the left after fitting subframe, mirror panel and new full-size mirror. Exposed edges of plywood frames are filled with paste white lead and enameled. Mirror is held securely by metal clips

Where a subframe is required in addition to the mirror-backing panel it must be cut carefully to assure an accurate fit over the built-in portion of the original cabinet. Attach mirror to backing as in detail

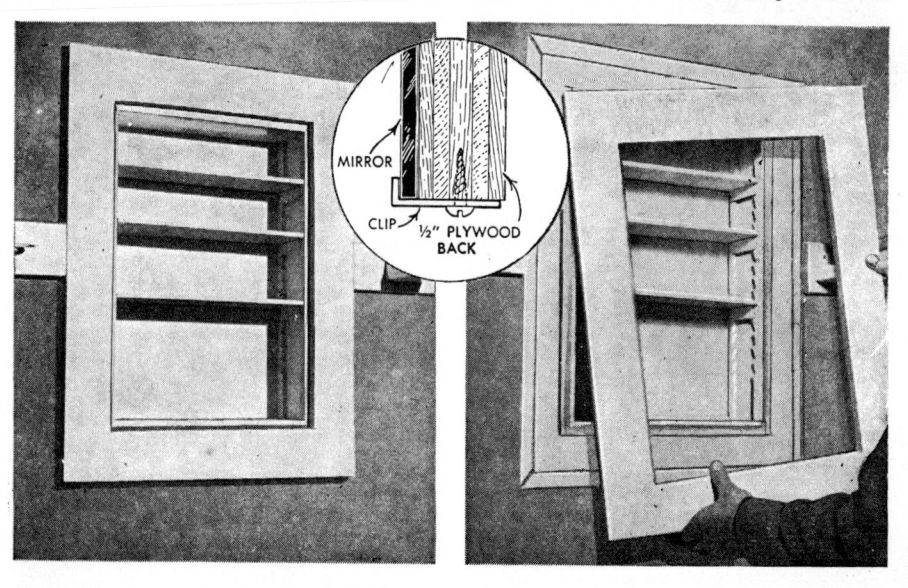

MIRROR
CLIP ½" PLYWOOD BACK

MERRY-GO-ROUND
is Hand Propelled

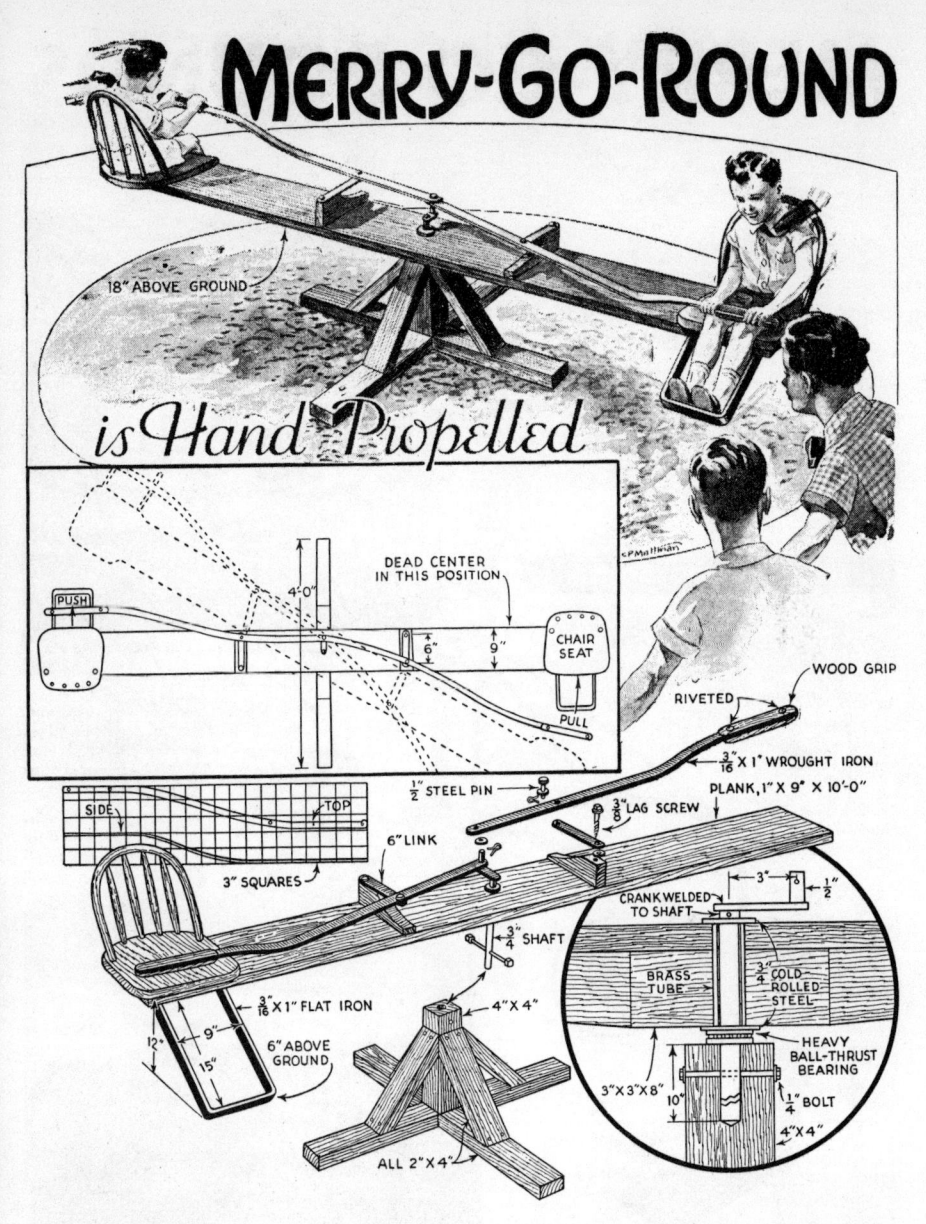

Operated something like a railroad hand car, this merry-go-round not only offers beneficial exercise, but it is inexpensive and is easy on shoes. It consists of a well-braced standard, a strong plank and two seats. A steel shaft, pinned rigidly to the standard, passes through a bushed hole in the center of the plank, and a crank, having a 3-in. throw, is welded to the upper end. Or, a longer shaft can be used and a crank bent on the end. Friction between the plank and the standard is reduced to a minimum with a ball thrust bearing. For the seats, old chairs are used, the leg assemblies having been removed. U-shaped footrests of flat iron are bolted under the seats to project downward at an angle. Push handles or bars, with which the merry-go-round is driven, are lengths of heavy flat iron to which wood handgrips are riveted. A 6-in. link connects each handle to a block screwed to the plank, and permits lateral movement of the bars

MERRY-GO-ROUNDS

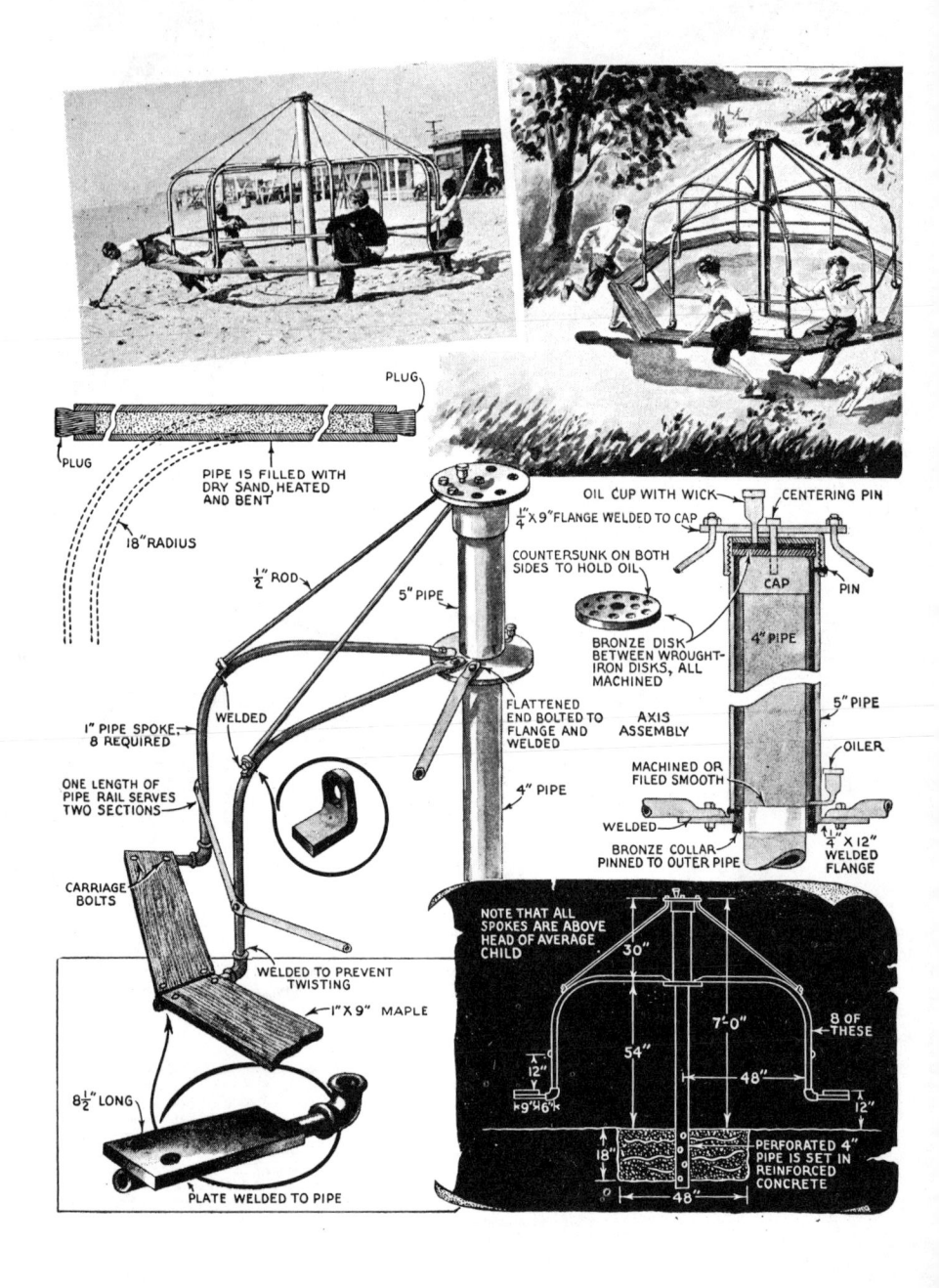

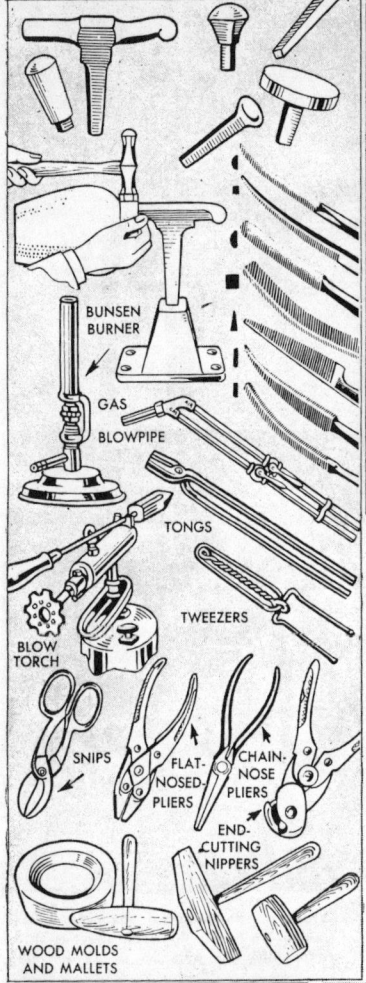

ART METAL HAND TOOLS

ART METAL for the beginner runs to cutout work in sheet metal with chased or hammered decoration. Cutting is most conveniently done on the jigsaw. The No. 1 hammer is the chasing hammer; this has a wide, flat head and is used for striking chasing tools, decorative stamps and gravers. More advanced work gets into raising and forming the sheet metal to make bowls and trays. A simple way of doing this is to

At the top of the panel at left are tools peculiar to the art-metal shop—the variously shaped forming "stakes" over which the projects are shaped on the stake holder, somewhat as a cobbler's last is used. Different kinds of forming files also are shown. In the photo above, an art-metal craftsman uses his forge, an essential for working with heavy metals

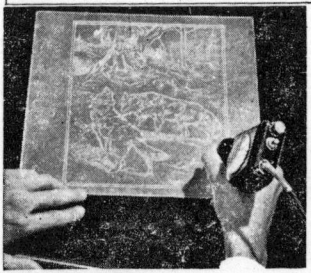

BUNSEN BURNER
GAS BLOWPIPE
TONGS
TWEEZERS
BLOW TORCH
SNIPS
FLAT-NOSED-PLIERS
CHAIN-NOSE PLIERS
END-CUTTING NIPPERS
WOOD MOLDS AND MALLETS

VIBRATING TOOL 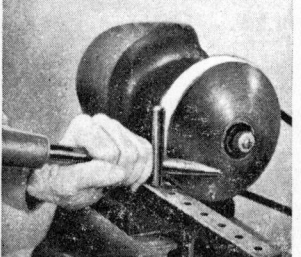   SPINNING LATHE    ELECTRIC HAMMER

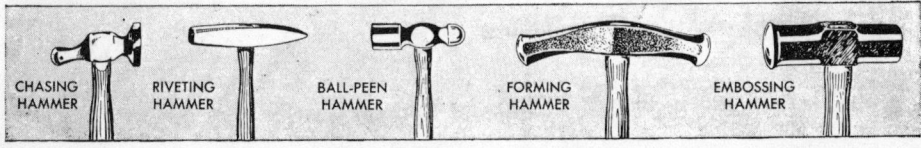

CHASING HAMMER   RIVETING HAMMER   BALL-PEEN HAMMER   FORMING HAMMER   EMBOSSING HAMMER

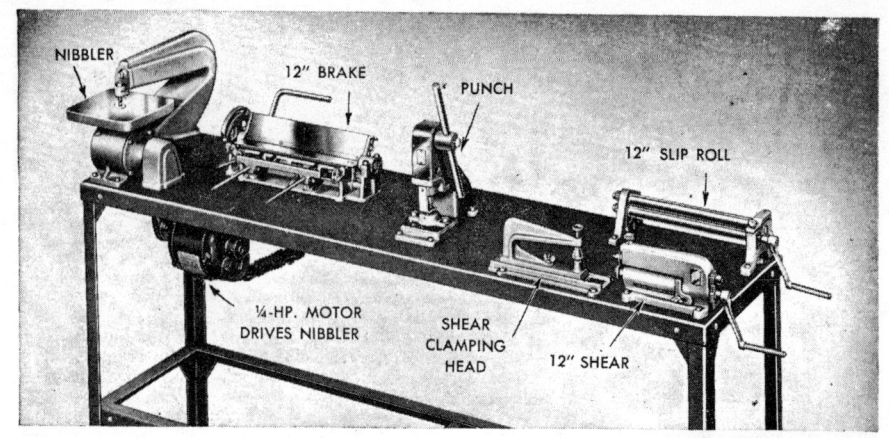

A sheet-metal shop mounted on a single bench. This outfit will bend, cut and punch out metal up to 20-gauge thickness

purchase a wooden mold; then use a mallet to drive the sheet metal into the mold. In the expert class, the metal is formed externally over various shapes of stakes or anvils. You can buy a 30-piece art-metal kit including chasing hammer, several pliers, files, snips, gas blowpipe, gravers, clamps, tweezers, soldering iron, hand scrollsaw, etc., for about $25.

The **brake** is the most important piece of sheet-metal equipment, and is used for angular bending; the **slip roll** forms cylinders and cones; the **shear** does straight or curved cutting and, equipped with a clamping (pivot) head, it will cut perfect circles; the **punch** makes holes from 1/8 to 1/2 in. and, with suitable dies, will also do vee and square notching. The only power tool in the group is the **nibbler**. The nibbler resembles the hand punch except it makes a series of holes instead of a single hole, and in this way does cutout work. Drilling and cutting jobs may require a drill press and metal-cutting bandsaw.

The brake is the principal sheet-metal tool. The unit shown here will handle sheet metal of 20-gauge thickness and up to 12 in. wide

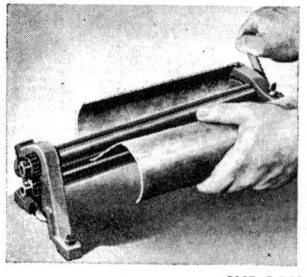

SLIP ROLL 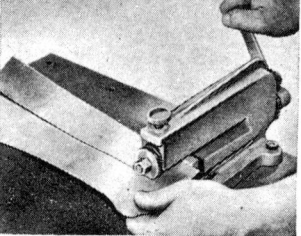   SHEAR 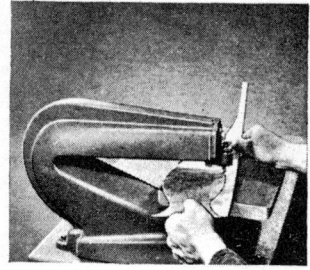   NIBBLER

RAISING HAMMER   BOTTOMING HAMMER   CHASING TOOLS AND STAMPS

ART METAL HAMMERING

Plant Holder

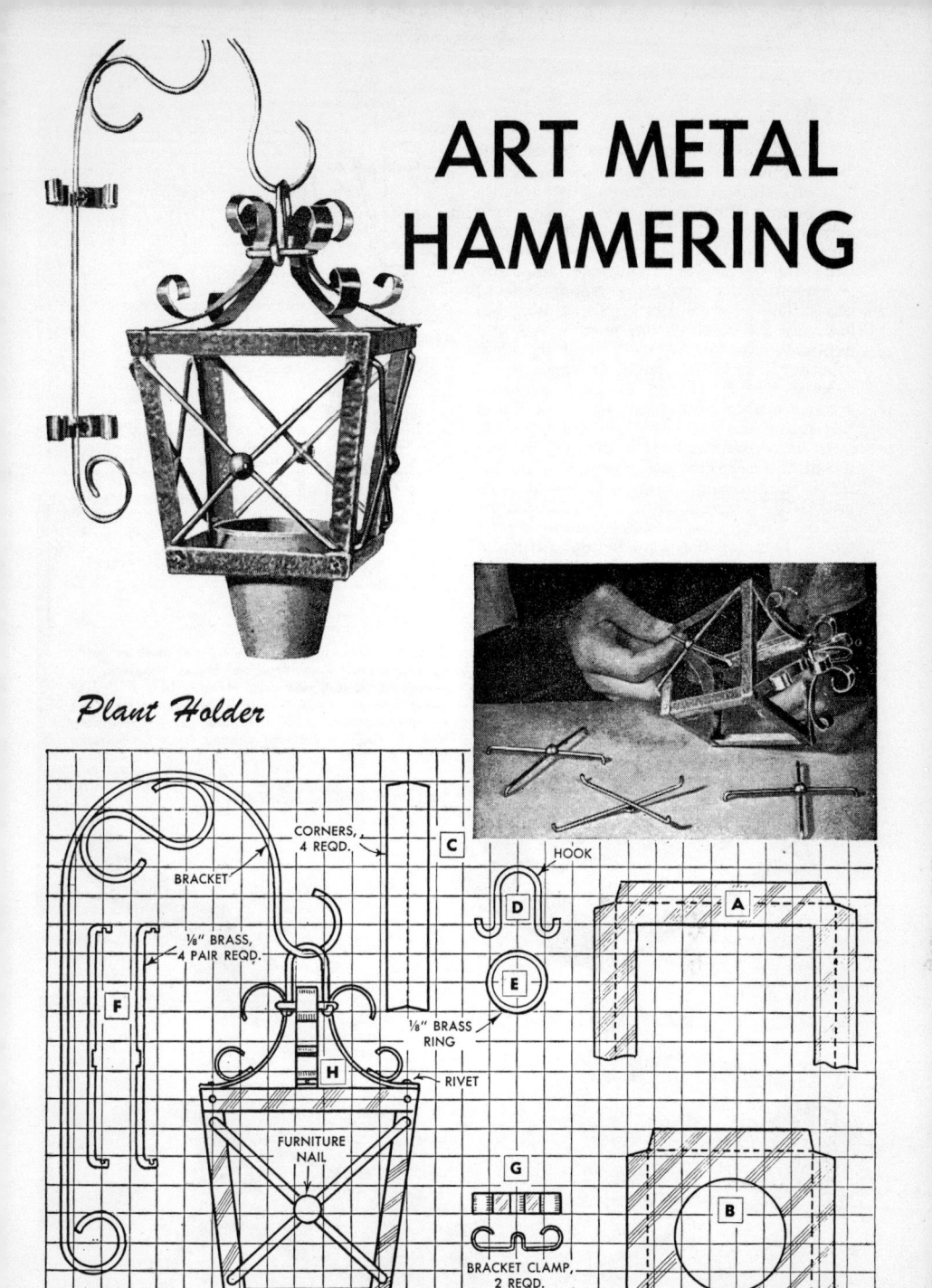

CORNERS, 4 REQD.

BRACKET

1/8" BRASS, 4 PAIR REQD.

HOOK

1/8" BRASS RING

RIVET

FURNITURE NAIL

BRACKET CLAMP, 2 REQD.

1/2" SQUARES

CONSIDERING the original investment in time and money, there are few crafts that offer finished products of more value and beauty than items made from hammered sheet metal. The plant holder pictured on the opposite page takes its lines from the old-fashioned carriage lamp, and is exceptionally attractive when made of brass. The silent-butler cigarette box, below, and the combination coaster rack and napkin holder on the next page are ideal in either aluminum, copper or brass.

None of the three projects requires a considerable amount of sheet metal, and the gauge used is not important—just so it is heavy enough to withstand normal usage. Chances are you'll have enough material in the shop scrap pile for at least one of the projects. But, if you have none of the softer metals and are unable to purchase them for the time being, substitute light-gauge sheet steel and paint the completed project in lively colors.

Plant holder: All parts of the plant holder are riveted together and should be hammered before they are cut out to prevent

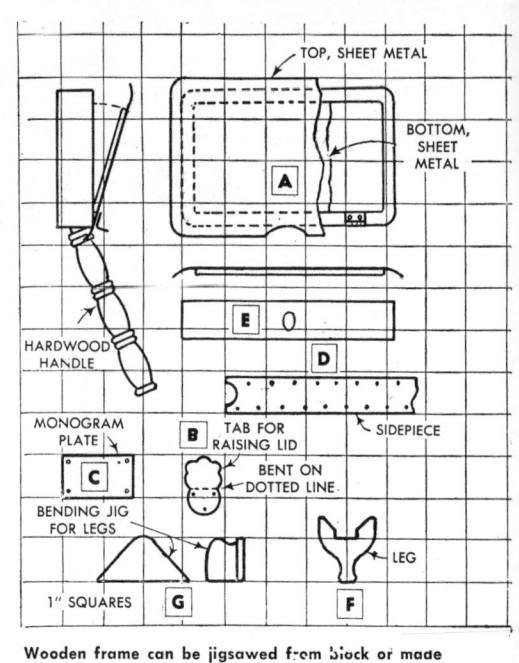

Wooden frame can be jigsawed from block or made by gluing four pieces together. End of handle is tenoned and slotted, and then glued in hole drilled at angle through frame. Wooden wedge is driven in slot to spread tenon and after glue dries, projecting end of tenon is cut and sanded flush with frame

Cigarette Box

Coaster Rack

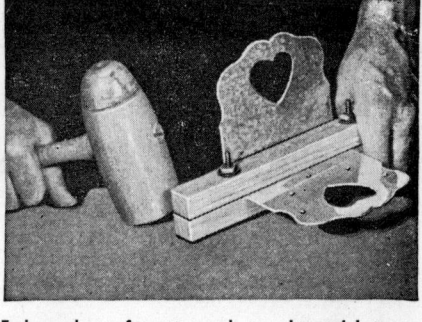

End members of coaster rack are clamped between two pieces of hardwood for bending with wooden mallet. Second right-angle bend is not made until ends of ¼-in. rods have been fastened to sheet metal

⅛" HOLES IN INNER SECTION

¼" X 5" ALUMINUM ROD, 5 REQD.

½" SQUARES

BALL FOOT, RIVETED

distortion. The hole in the bottom piece, B, is cut just large enough so the rim of a flowerpot will rest on the edge. At the top of the holder, the ornamental parts, H, are inserted through the brass ring, E, during assembly. The crossed pieces of brass wire, F, which form the sides of the holder, are half-lapped and joined at the center with a furniture-upholstery nail. The latter is driven through both wires and peened over. The notched ends of the wires are inserted through holes drilled in the corner pieces, A, so the notches engage the edges of the metal. Sheet-metal clamps, G, hold the graceful wire bracket to the wall, the vertical length of wire being bent slightly to keep it from slipping. A U-shaped hook, D, is hung from the upper end of the bracket and engages the brass ring.

Cigarette box: In this project, hammered sheet metal covers a wooden frame which is fitted with a turned hardwood handle. The squared pattern details the necessary parts. The sheet-metal sidepiece, D, is fastened to the sides of the frame, E, with escutcheon pins, and a sheet-metal bottom piece is nailed to the underside of the frame. Four sheet-metal legs, F, are formed over a wooden bending jig made as in detail G. A piece of ⅛-in. plywood is riveted to the underside of the top, A, to serve as a stiffener, and the top is hinged to the frame. The monogram plate, C, the finger tab, B, and the hinges are riveted to the top.

Coaster rack: The ends of the rack, which serve as napkin holders, are jigsawed from sheet metal and joined with five lengths of ¼-in. rod to form a trough for storing coasters or ash trays. The end members are drilled to take ⅛-in. tenons filed on the ends of the rods. The rods can be tenoned quickly by chucking them, one at a time, in a drill press and supporting the lower end in a vertical half-round groove cut in a hardwood block, the groove having a ⅛-in. radius. The block is nailed to a board which is clamped to the drill-press table. After insertion through the holes in the end members, the tenons are peened over.

ART METAL MODELING

FASHIONING decorative animals and novelties in gleaming metal is an interesting pastime, which can be turned into a profitable one after experience enables the hobbyist to turn out pieces people will buy eagerly. The suggestions given are only a few of the many figures which can be reproduced. All the tools you will need to work the paper-thin metal to shape are an ice pick for punching holes, a pair of long-nose pliers, a dressmaker's tracing wheel and a pair of household shears. For material, No. 30-ga. non-tarnishable monel metal is preferred, although sheet aluminum, while it does not have such a bright finish, can be used if you are unable to obtain the former. For fastening the pieces together, you will need a spool of No. 30-ga. bright-finish wire, and a few feet of 1/8-in. aluminum tubing will also be needed to form a framework required by the larger figures.

To model the horse shown in Figs. 1 and 3, first enlarge the pattern of the body given in Fig. 4. As both sides are identical only half of the pattern need be enlarged full size. In cutting the metal, you'll find it much easier to follow the outlines if the paper pattern is fastened directly to the metal as in Fig. 5. Ordinary household shears will do to cut the metal, and by cutting the complete pattern from one side, a slightly rounded edge is obtained. Where indicated on the patterns, lines of small dots, which are made on the underside with a dressmaker's tracing wheel, are used to represent hair on a figure and to give a shaggy effect. Bend the body on the dotted lines, studying Figs. 1 and 3 to determine the approximate shape. The body is wired to a supporting framework of aluminum wire, Fig. 2, with short lengths of silver-finish wire. The head is made in three pieces, the patterns for these being given in Fig. 6. These are bent and wired together as shown in Fig. 7 and the photo at the right. The two neck pieces, Fig. 10, are shaped as in Fig. 9. Piece (b) fits over piece (a) and they are wired together at the points indicated on the patterns. The head is joined to the neck by wiring the neck tab in the position shown in Fig. 7. Finally, the neck is wired to the body at the shoulders, after which the slits in the mane are curled as in Fig. 8. The feet are formed by simply turning the metal under at right angles.

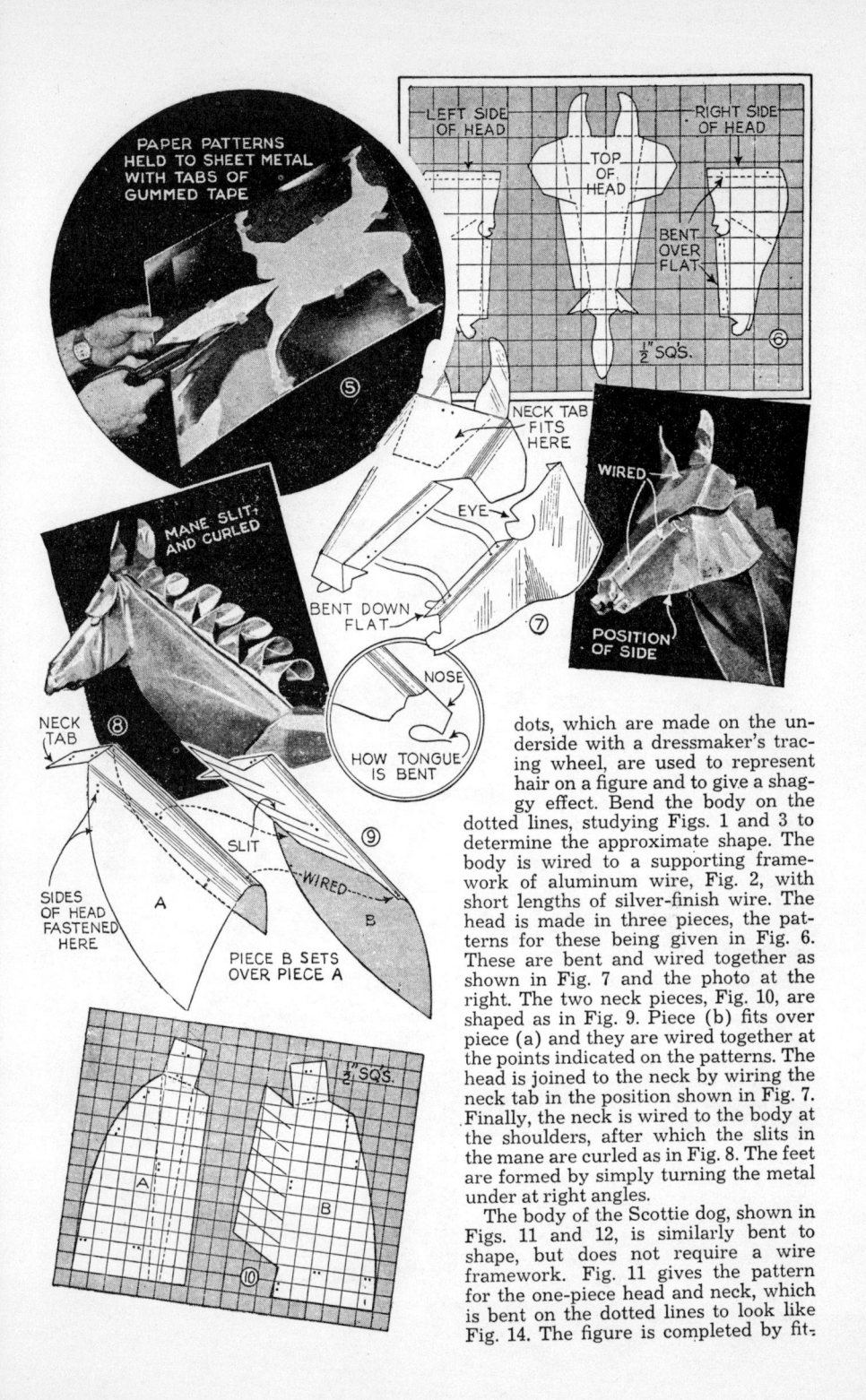

The body of the Scottie dog, shown in Figs. 11 and 12, is similarly bent to shape, but does not require a wire framework. Fig. 11 gives the pattern for the one-piece head and neck, which is bent on the dotted lines to look like Fig. 14. The figure is completed by fitting the neck between the V-cut in the body and fastening with wire at each side. Figs. 13 to 18 inclusive, with the exception of Fig. 14, show other examples which can be reproduced. Remove finger marks with metal polish and then lacquer to protect the sheen.

Bluing Brass Ware

Articles of brass may be given a blue or black color by immersing in a solution made by dissolving sodium thiosulphate, ½ lb., and lead acetate, 2 oz., in one gallon of water. More lead acetate may be added to increase the depth of color and the speed of the action. To use the solution, heat it almost to the boiling point and immerse the work. Watch the color carefully and remove when of a suitable shade. Brass wire should be used to suspend the work. It is important that the work be well cleaned before treating. The colored work should be protected with clear lacquer.

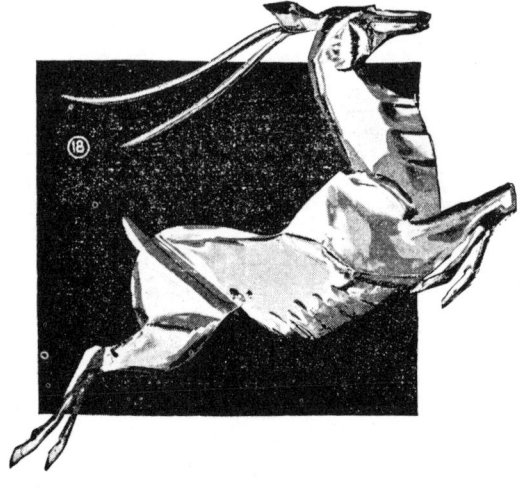

ART METAL MODELING

MANY PROJECTS in sheet metal require cylindrical and conical shapes, which often present a difficult problem to the beginner in sheet-metal work. The slip roll, Fig. 2, simplifies forming these shapes from flat sheets and, as it also will handle strap stock and wire, the project possibilities of the tool are almost unlimited. The table lamp pictured in Fig. 1 is a typical example.

A study of the drawing below and the various work pictures makes it apparent how the tool is used. There are two feed rolls which grip and feed the flat work against a single forming roll. If the forming roll is close to the feed rolls, the sheet, strap or wire work will be rolled into a cylindrical shape with a minimum diameter of about 1⅛ in. Larger cylinders are formed with the forming roll positioned a greater distance from the feed rolls. In ordinary practice, the work first is rolled to a greater diameter than required and then gradually reduced by advancing the forming roll for successive passes of the work.

Figs. 3 to 9 show standard slip-roll shapes. The cone shape, Fig. 4, requires side-slipping the work to keep it square with the feed rolls. This is done most readily when the feed tension is light so as to allow the work to be pushed in a sidewise direction as the feed is made. If the work cannot be side-slipped with one hand, the feed can be

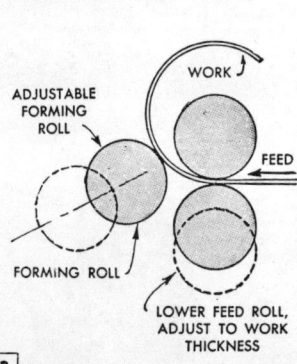

TYPICAL SLIP-ROLL PROJECTS are lamp, cigarette server and wall planters. Machine is shown at right

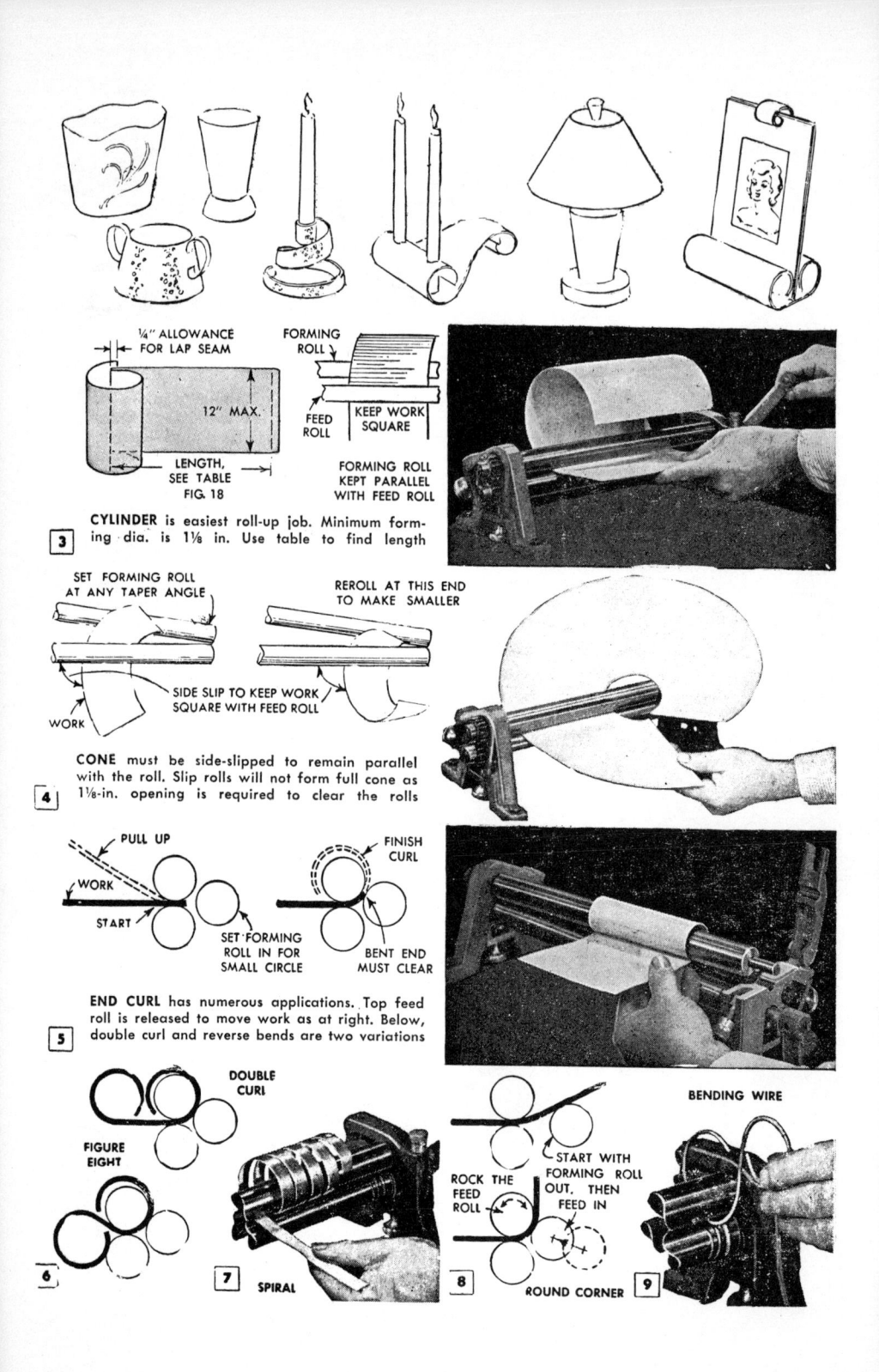

3 **CYLINDER** is easiest roll-up job. Minimum forming dia. is 1⅛ in. Use table to find length

4 **CONE** must be side-slipped to remain parallel with the roll. Slip rolls will not form full cone as 1⅛-in. opening is required to clear the rolls

5 **END CURL** has numerous applications. Top feed roll is released to move work as at right. Below, double curl and reverse bends are two variations

6 FIGURE EIGHT / DOUBLE CURL

7 SPIRAL

8 ROUND CORNER — ROCK THE FEED ROLL — START WITH FORMING ROLL OUT, THEN FEED IN

9 BENDING WIRE

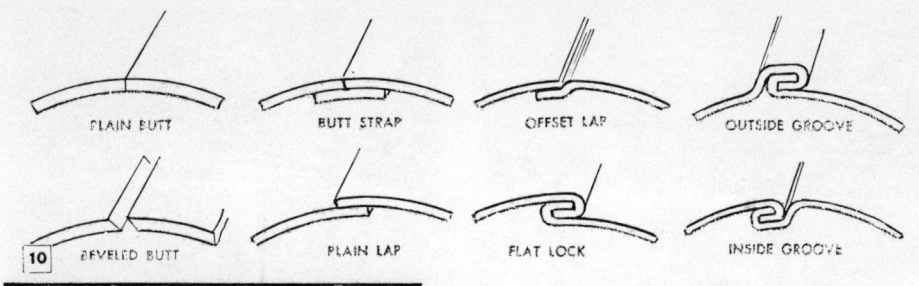

PLAIN BUTT · BUTT STRAP · OFFSET LAP · OUTSIDE GROOVE
BEVELED BUTT · PLAIN LAP · FLAT LOCK · INSIDE GROOVE

10

Eight Ways to Make Seams

made in short steps of about 1 in. each, squaring the work with two hands after each advance. The useful round corner, Fig. 8, is made without advancing the work. It is a plain pressure operation done by advancing the forming roll as the work is rocked backward and forward slightly.

After rolling a cylinder or cone, you have the job of securing the seam. Fig. 10 shows eight ways of doing this. The simple lap joint is most popular with beginners, using either rivets or small bolts for fastening. Holes for rivets or bolts are easy to drill if the seam is clamped with two small C-clamps and then drilled, as shown in Fig. 11. Fig. 12 is a simple, practical setup for riveting. If you have a brake and can make the necessary bends for a lock seam, you can do a neater job by "grooving" the lock to the inside, as shown in Figs. 13 and 14. For occasional work, a hardwood grooving block is satisfactory for soft metals to 22 ga.

An attractive beginning project to try on the slip roll is the teapot lamp shown in Figs. 17 and 19. The bottom alone makes a neat planter, Fig. 15. A variation of the lamp is pictured in Fig. 16, using a standard socket with clip-on shade. The planter, Fig. 15, is of 22-ga. stainless steel with ball-peen texture hammered on the sheet. ★ ★ ★

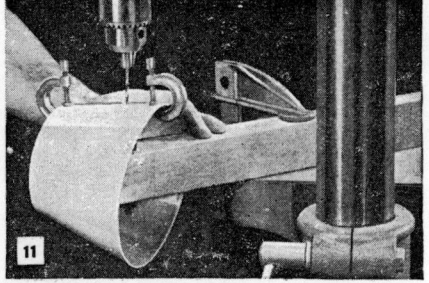

11 
DRILLING holes for seams is done easily with work supported on bolt fitted in end of wooden block clamped to the drill-press table

12
RIVETING IS DONE ON SIMPLE BAR ANVIL

13
GROOVING MAKES A NEAT LOCKED SEAM

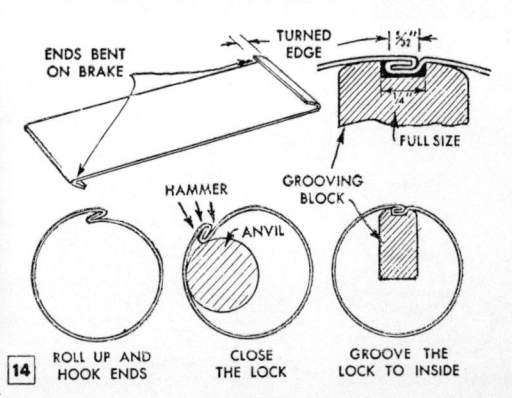

14
ENDS BENT ON BRAKE — TURNED EDGE — FULL SIZE
HAMMER — GROOVING BLOCK — ANVIL
ROLL UP AND HOOK ENDS · CLOSE THE LOCK · GROOVE THE LOCK TO INSIDE

18 CIRCUMFERENCE OF CIRCLES

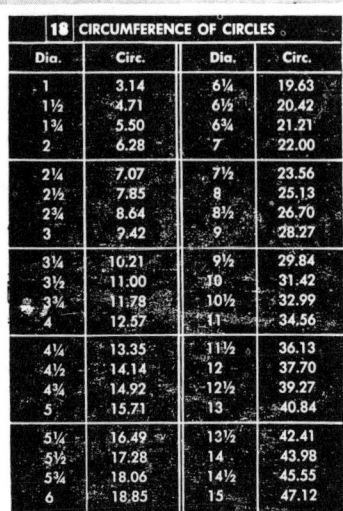

Dia.	Circ.	Dia.	Circ.
1	3.14	6¼	19.63
1½	4.71	6½	20.42
1¾	5.50	6¾	21.21
2	6.28	7	22.00
2¼	7.07	7½	23.56
2½	7.85	8	25.13
2¾	8.64	8½	26.70
3	9.42	9	28.27
3¼	10.21	9½	29.84
3½	11.00	10	31.42
3¾	11.78	10½	32.99
4	12.57	11	34.56
4¼	13.35	11½	36.13
4½	14.14	12	37.70
4¾	14.92	12½	39.27
5	15.71	13	40.84
5¼	16.49	13½	42.41
5½	17.28	14	43.98
5¾	18.06	14½	45.55
6	18.85	15	47.12

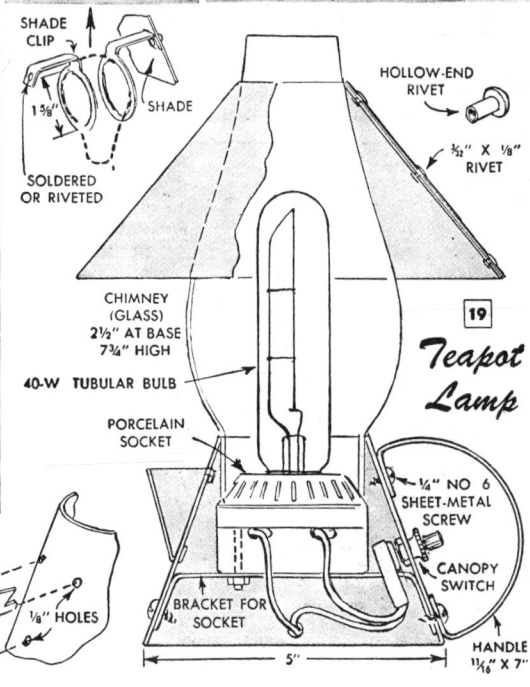

19 Teapot Lamp

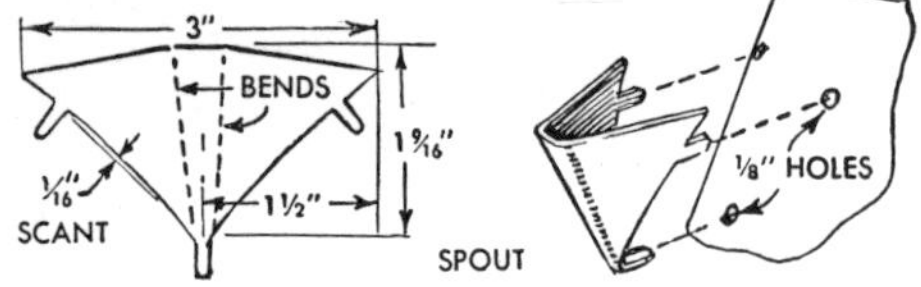

SHADE CLIP — SHADE — SOLDERED OR RIVETED — 1⅝" — HOLLOW-END RIVET — 3/32" × 1/8" RIVET — CHIMNEY (GLASS) 2½" AT BASE 7¾" HIGH — 40-W TUBULAR BULB — PORCELAIN SOCKET — BRACKET FOR SOCKET — ¼" NO 6 SHEET-METAL SCREW — CANOPY SWITCH — HANDLE 1 1/16" × 7" — 5"

20 3" — BENDS — 1 3/16" — 1½" — 1/16" SCANT — ⅛" HOLES — SPOUT

How to Make Stretchout of Cone

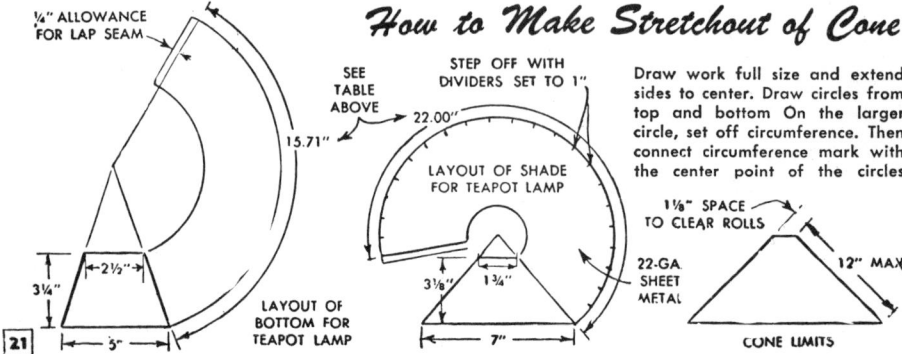

Draw work full size and extend sides to center. Draw circles from top and bottom On the larger circle, set off circumference. Then connect circumference mark with the center point of the circles

21 ¼" ALLOWANCE FOR LAP SEAM — SEE TABLE ABOVE — 15.71" — 2½" — 3¼" — 5" — LAYOUT OF BOTTOM FOR TEAPOT LAMP

STEP OFF WITH DIVIDERS SET TO 1" — 22.00" — LAYOUT OF SHADE FOR TEAPOT LAMP — 3⅛" — 1¾" — 7"

1⅛" SPACE TO CLEAR ROLLS — 22-GA SHEET METAL — 12" MAX — CONE LIMITS

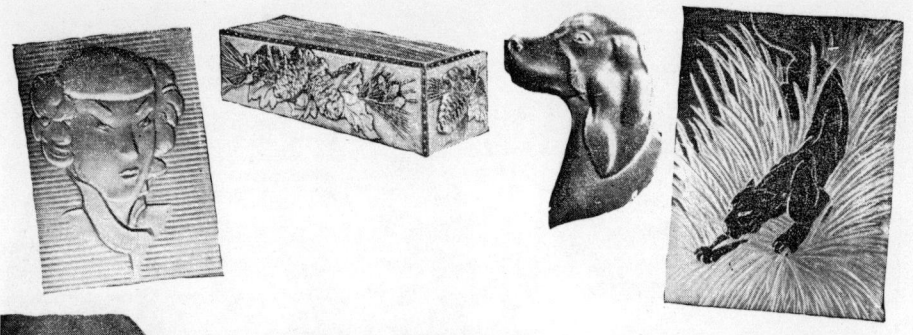

ART METAL PROJECTS

IF YOU CAN trace around a design with a pencil, you can tool metal — it's that easy! What's more, if you are the least bit artistic by nature, you'll enjoy this means of producing attractive bas-relief plaques and trays as well as decorative coverings for wastebaskets, jewel and cigarette boxes similar to those illustrated above and on the opposite page. Currently the most

Here's What You Need!

MATERIALS: TOOLING METAL, OXIDIZING CHEMICAL

TOOLS: 2/0 STEEL, ¼" MAPLE DOWEL, POINTED, FLAT CHISEL POINT, FLATTENED, PENCIL ERASERS FOR MODELING

WORK SURFACES: FIRM (MAGAZINE OR CARDBOARD), HARD (GLASS), SOFT (FELT, LEATHER, ETC.)

This Is How It's Done!

1 DRAW THE DESIGN, USING 3H PENCIL OR POINTED DOWEL

2 RETRACE DESIGN WITH POINTED TOOL TO DEEPEN OUTLINE — ON MAGAZINE

3 TURN WORK OVER AND TRACE INSIDE OUTLINE — ON MAGAZINE OR THIN, SOFT SURFACE

popular of the metal crafts, metal tooling has gained much of its widespread acceptance through offering an end product that can be produced at comparatively low cost without need for special skills.

Simple tools used: The designs may be drawn freehand or traced from a drawing or photograph. Of the few simple tools needed, the basic one is a short length of 1/4-in. maple dowel flattened at one end and pointed at the other. Also necessary is a narrow block of wood tapered at one end, (previous pg.). The only other "tools" required for the more simple designs are two pencils, one of which is fitted with a slip-on eraser.

Three kinds of surfaces for supporting the work will be necessary. Felt or rubber can be used for a soft surface, while a magazine or piece of 1/8-in. mounting board will provide a firm surface. The hard surface can be glass, metal or marble.

Materials needed: The tooling metal can be either copper, aluminum, bronze or brass. Copper is by far the most suitable, inasmuch as the soft, annealed condition in which it is sold has just the right amount of "give" for easy tooling. Other advantages favoring the use of copper are its golden-orange color and the ease with which it may be oxidized or antiqued. From a mechanical standpoint, aluminum is easiest to tool. Though not antiqued as readily as copper,

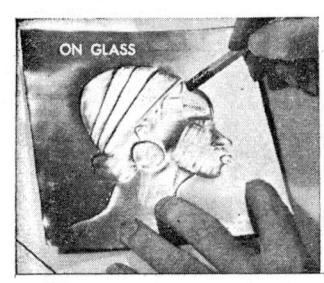

4	WITH WORK ON GLASS, PRESS OUT FROM BACK WITH ERASER
5	TURN WORK FACE UP AND FLATTEN THE BACKGROUND
6	POLISH WITH STEEL WOOL AND APPLY CLEAR LACQUER

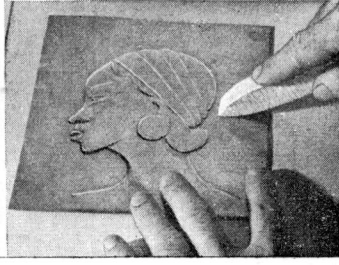

7 STIPPLED BACKGROUND IS POPULAR. BESIDES BEING ATTRACTIVE, IT CONCEALS TOOL MARKS AND DEFECTS

CARDBOARD

1/8" BALL TOOL

8

9 POINTED STEEL TOOL CAN BE USED TO WORK CLOSE

10

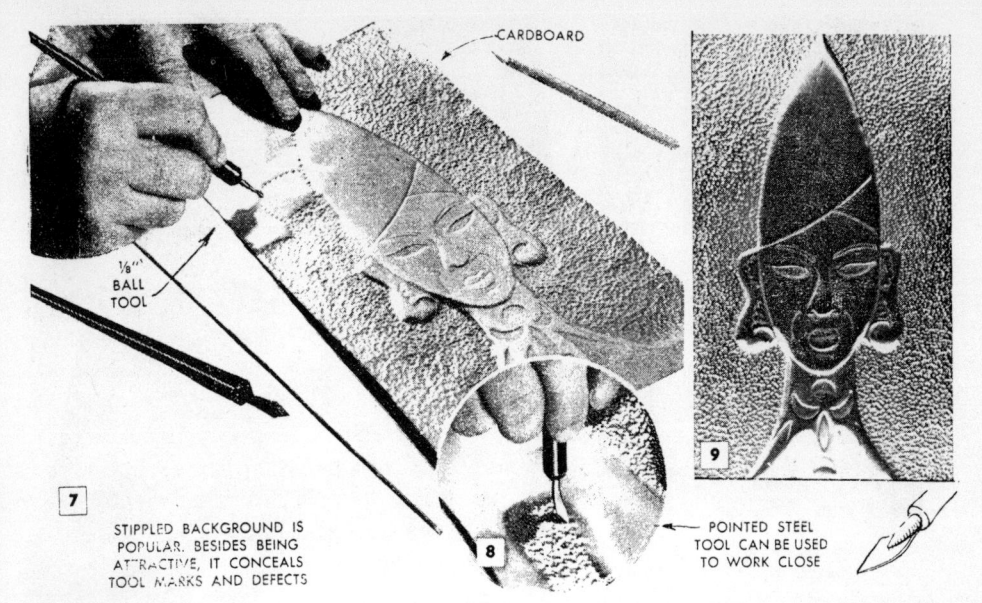

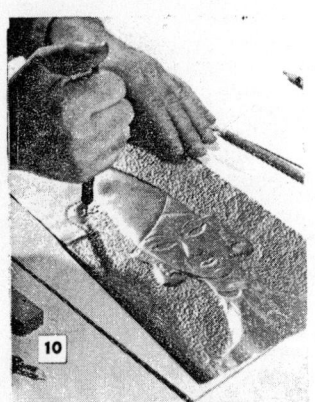

aluminum can be painted with transparent lacquers to produce a variety of beautiful effects. All metals are available in roll form 12 and 18 in. wide, and are polished and ready for tooling. For additional data, see Fig. 11.

How tooling is done: Besides hundreds of design patterns that are available, your own can be made from photographs and illustrations. Start by taping the design to the metal and tracing the outline with the dowel or a hard-lead pencil, Fig. 1. The pencil is most suitable because it leaves a light mark on the pattern which indicates how much has been traced. The work is placed on the firm surface for this operation, keeping the convex side of the slightly curved metal up. This is preferable for the face or front of the work. Pressure when tracing should be slightly heavier than when writing. Ease up at the beginning and end of each stroke to avoid making undesirable dots in the metal. After

11 *Quick Data on Metals for Tooling*

Metal	Best Weight	Tooling Pressure	Chemical Coloring	Remarks
COPPER	35 OR 36 GA.	TOOLS EASILY WITH MEDIUM PRESSURE	BLACK AND BROWN SHADES ARE OBTAINED WITH LIVER OF SULPHUR IN WATER SOLUTION	BEST ALL-AROUND METAL FOR TOOLING. WORKS EASILY WITH WOOD TOOLS. POLISHES NICELY WITH 2/0 OR 3/0 STEEL WOOL. PRICE ABOUT 50c SQ. FT. (1952).
ALUMINUM	34 OR 35 GA.	VERY SOFT; EASY TO TOOL	NO SIMPLE METHOD OF COLORING; IS ATTRACTIVE PAINTED WITH TRANSPARENT LACQUERS	THE EASIEST METAL TO TOOL—IS ACTUALLY TOO SOFT FOR FINE DETAIL, HENCE THE HEAVIER WEIGHT RECOMMENDATION. EXCELLENT FOR DEEP RAISING. PRICE ABOUT 25c SQ. FT.
BRONZE	35 OR 36 GA.	ALMOST AS SOFT AS ALUMINUM; EASY TO TOOL	NO SIMPLE METHOD OF COLORING; CAN BE BLACKENED BY HEATING	TOOLS EASILY; A BIT SOFT FOR FINE DETAIL BUT EXCELLENT FOR CONTOUR WORK. BRIGHT GOLD COLOR POLISHES NICELY. IDEAL FOR HEADS. PRICE ABOUT SAME AS COPPER.
BRASS	36 GA.	FAIRLY HARD; REQUIRES FIRM PRESSURE	VARIOUS SHADES OF BLACK ARE OBTAINED WITH COPPER-SULPHATE SOLUTION	MUST BE DEAD SOFT; NOT A BEGINNER'S METAL SINCE IT DOES NOT YIELD READILY TO TOOLING PRESSURE. EXCELLENT FOR FINE DETAIL. GOOD FOR ORIENTAL MOTIF. SAME PRICE AS COPPER.

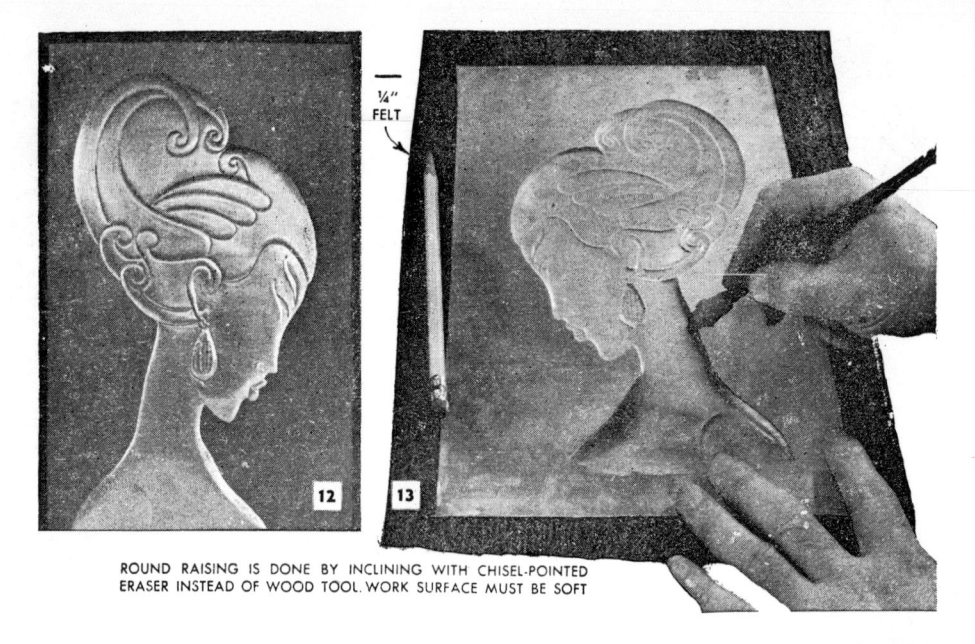

ROUND RAISING IS DONE BY INCLINING WITH CHISEL-POINTED ERASER INSTEAD OF WOOD TOOL. WORK SURFACE MUST BE SOFT

tracing the design, remove the pattern and go over the work once again to deepen the lines, Fig. 2.

Next, turn the work over and trace a line just inside the raised outline, Fig. 3. This operation also is done on the firm surface. After you have gained more experience, this operation may be done on leather or thin rubber to permit deeper impressions, the one disadvantage being a slight increase in distortion around the edges of the design. The "in" line should be made with the tool held at an angle over the outline, with the latter serving as a guide. In determining which side of the outline should be inlined, always keep in mind that you inline on the side that is to be raised.

Now place the work face down on the hard surface and flatten all large areas with the pencil eraser, Fig. 4, rubbing the work out with light, even strokes. Then turn the work face up and flatten the background areas with the large flattening tool, Fig. 5. Use the flat end of the dowel for areas adjacent to the lines. Make all strokes in the same direction to flatten the metal without undue buckling. The strokes should be made with the "grain" which, in tooling metal, consists of fine, parallel lines on the surface. When the entire design has been pressed out, polish it with 2/0 or 3/0 steel wool and brush or spray with clear lacquer, Fig. 6.

Backgrounds: One of the most common troubles when tooling are the marks usually left on plain backgrounds. A simple remedy for this is the textured background, Fig. 9. Of the many which may be produced easily, the plain stipple is one of the most attractive. On average work this is done with a 1/8-in. ball tool, or an escutcheon or clinch nail driven in the end of a dowel can be used. First dot in a circular area about 1 in. in diameter, Fig. 7, and then use a pounding stroke, Fig. 10, to stipple inside this area, repeating the operation until the entire background is completed. For working close to the design, any kind of pointed tool may be used, Fig. 8, but it must be used with care to keep the metal inside the line free of distortion.

Parallel lines provide an excellent background and also help level buckled metal. Stippled lines can be used to depict clouds. Crisscross lines, Figs. 23 and 24, provide an attractive background and are made with the point of the dowel guided along a straightedge. Painted backgrounds, usually painted black, are the simplest of all, Fig. 12.

Round raising: After doing work in the basic flat-raising technique, you undoubtedly will want to try round raising, Fig. 12. The only difference between the two is that instead of inlining with the dowel point, the design is raised with pencil erasers when round raising, Fig. 13. The work is placed on a soft surface and a round eraser used to raise all large areas inside the design. Use light, fast strokes but avoid excessive pressure which might leave ridges in the metal. This technique is ideal for heads and other designs having large open

14 2/0 STEEL WOOL

F-GRIT SILICON CARBIDE

OXIDIZED WORK IS HIGH LIGHTED WITH 2/0 STEEL WOOL OR F-GRIT SILICON-CARBIDE ABRASIVE WITH WATER

15 *Oxidized*

16

Right, spachtling compound, to which varnish has been added for flexibility, is troweled into depressions on back of work to prevent crushing. Plywood frame having design cut out of center supports work during operation. A damp cloth is used afterward to clean metal around the design

areas of varying depths. The two techniques very often are combined, since most designs will have one or more large areas for which round raising is more suitable, together with details requiring the clear definition possible only with flat raising.

Oxidizing: If the work is to be oxidized which, along with antiquing, is done to make various lines stand out in greater relief, copper should be used because it is the easiest and most economical of the metals to treat. All you need is an ounce of liver of sulphur to supply you for several jobs. After dissolving a couple of pea-size lumps in a glass of water, use a wad of steel wool to swab the work with the solution. The metal will begin to darken almost immediately and eventually turn black, Fig. 15. When the metal is the shade desired, rinse under running water to stop the action, then allow the work to dry naturally—do not wipe. When the work is dry, rub it with steel wool to high light, Figs. 14 and 16. If this seems too mild, use F or FF-grit silicon-carbide grains with water and

18 OILED STICKS

17 PLYWOOD SUPPORT FRAME

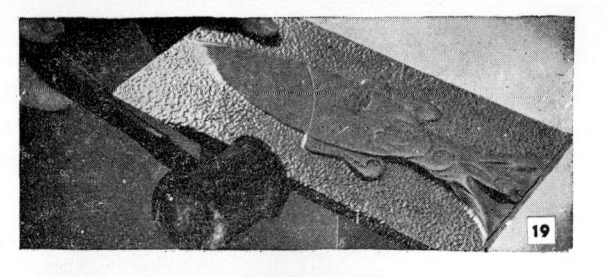

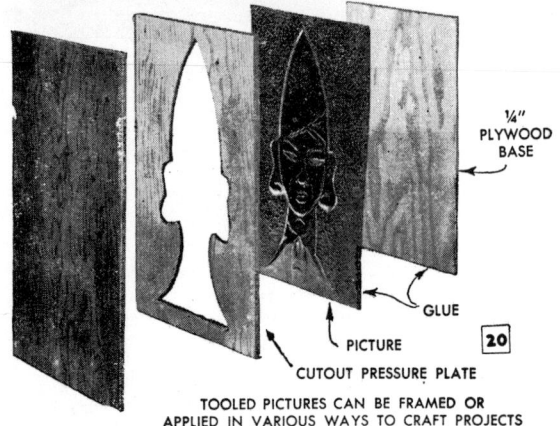

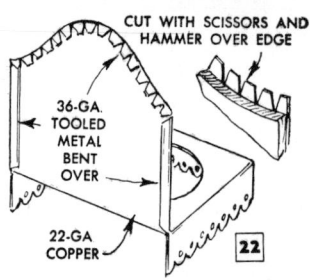

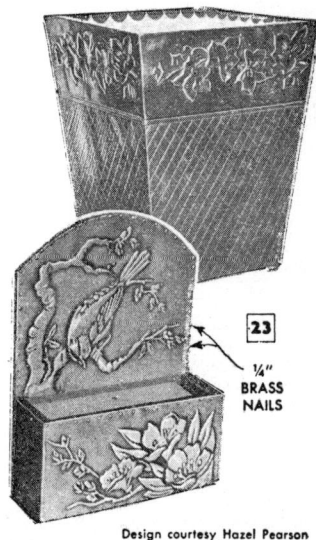

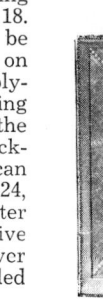

rub with steel wool. For best results, oxidize the copper to a deep brown rather than black, controlling the intensity by the strength of the solution.

Backing: All large, deeply formed work should be backed to prevent possible crushing. The backing can either be wax, which is flowed on hot, or plaster which is troweled into the depressions, Fig. 17. Plaster or spachtling compound, to which varnish, 1 teaspoon per cup, has been added for flexibility, is preferable for this purpose. Mix the plaster with your fingers until there are no lumps left which might make a mark when drawn across the work. The work should be supported closely on wooden blocks. After leveling it with a trowel or wooden straightedge, use a wet cloth to clean the metal around the design. Hold the work flat for drying by using oiled sticks under a weight, Fig. 18.

Mounting: Properly done, the work can be laid flat for framing as easily as a design on paper. For simple outlines such as heads, a plywood cutout may be used, Fig. 20, for pressing solidly against the background when gluing the work to a plywood backing. This type of backing may be trimmed by sawing or the edges can be bent over, Fig. 19. If used for a tray, Fig. 24, the work should be placed under glass to better withstand wear. When used for a decorative covering, the light metal is bent easily over heavier metal objects, Figs. 21 and 22, or nailed to wooden objects as in Fig. 23.

METAL CASTINGS

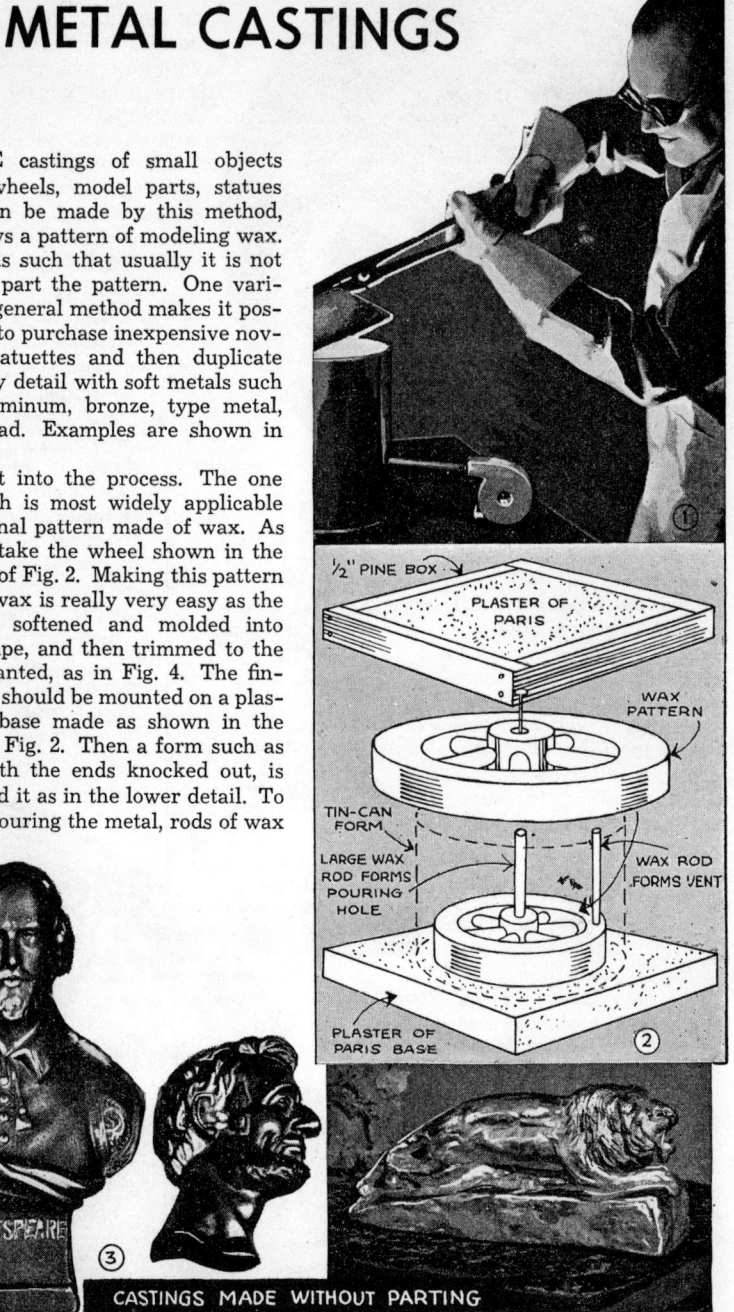

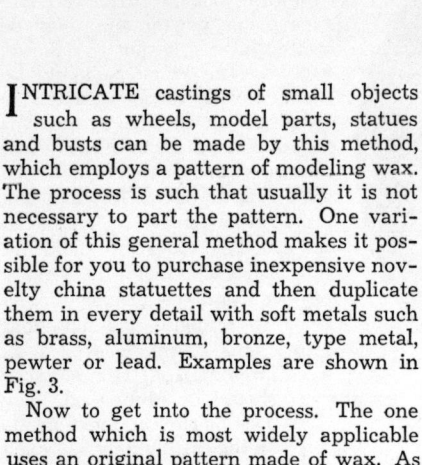

INTRICATE castings of small objects such as wheels, model parts, statues and busts can be made by this method, which employs a pattern of modeling wax. The process is such that usually it is not necessary to part the pattern. One variation of this general method makes it possible for you to purchase inexpensive novelty china statuettes and then duplicate them in every detail with soft metals such as brass, aluminum, bronze, type metal, pewter or lead. Examples are shown in Fig. 3.

Now to get into the process. The one method which is most widely applicable uses an original pattern made of wax. As an example, take the wheel shown in the center detail of Fig. 2. Making this pattern of modeling wax is really very easy as the wax can be softened and molded into most any shape, and then trimmed to the exact size wanted, as in Fig. 4. The finished pattern should be mounted on a plaster-of-paris base made as shown in the upper detail, Fig. 2. Then a form such as a tin can with the ends knocked out, is placed around it as in the lower detail. To provide for pouring the metal, rods of wax

CASTINGS MADE WITHOUT PARTING

FINISHING A WAX PATTERN

are softened and stuck on the highest part of the model, Figs. 2 and 6. These should extend higher than the metal form and one should be large enough to provide a passage for pouring the molten metal; the others serve only to let air out of the form. The wax pattern must be fastened down so that it can not move when the plaster is poured over it.

A thin mixture of plaster of paris is made by taking enough cold water in a large pail to fill the form, and slowly dropping in plaster of paris until some of the latter comes to the surface and floats, Fig. 5. Then the mixture is stirred well and poured into the form, as shown in Fig. 9. When the plaster is dry, the wax pattern is melted out as shown in Fig. 8, by standing the mold over a gas flame, with the vents down, and with a pan provided to catch the melted wax. Care must be taken that the wax does not catch fire. The mouth of the pouring hole should be enlarged so that it is funnel shaped.

ADD PLASTER UNTIL IT FLOATS

The mold must be thoroughly heated to drive out the last trace of moisture. The metal should be melted down in a crucible placed in a forge or a gas furnace. Lead, type metal, pewter, etc., may be melted in an iron pot in a forge, Fig. 16. The crucible should be handled with tongs similar to those shown in Fig. 8, and the metal poured as in Fig. 7. All pouring of molten metal should be done over dry sand. Excess metal must not be allowed to "freeze" in the crucible. After the metal has solidified in the mold, the form is removed and the plaster broken away by plunging the hot mold into a tank of cold water. With large molds, the plaster should be broken off. The pouring riser and vent prints on the casting may be sawed away and the surface dressed with a file or grinder.

Another method must be used where there is already a model to work from and where a single duplicate of any article is needed, or when a broken casting must be replaced. The broken

WAX RODS FORM POURING HOLES

parts can be fastened together temporarily and used as a model for a new casting. Also duplicates of antique or unusual parts can be made from the original without in any way injuring it. The model is first set in position on a flat board, as shown in Fig. 10. In order to keep it in this position and to return it to the same place, spotting pins or small brads with their heads cut off can be driven into the base and pressed down into the model. The model next should be covered with two or three layers of wet newspaper, and then a layer of modeling clay, at least ¼ in. thick as in Fig. 11. The outside of this should be coated with paraffin oil or grease. A wood flask, spotted with pins, Fig. 12, is placed around the model to retain the plaster. The inside of the flask should be greased thoroughly. The plaster mold can be in one piece if the draft or taper of the model runs all in one direction from the base as in Fig. 14. If not, the mold will have to be parted approximately along the center line of the model by imbedding a light length of piano wire in the clay along the desired line and drawing it out when the plaster has partly set.

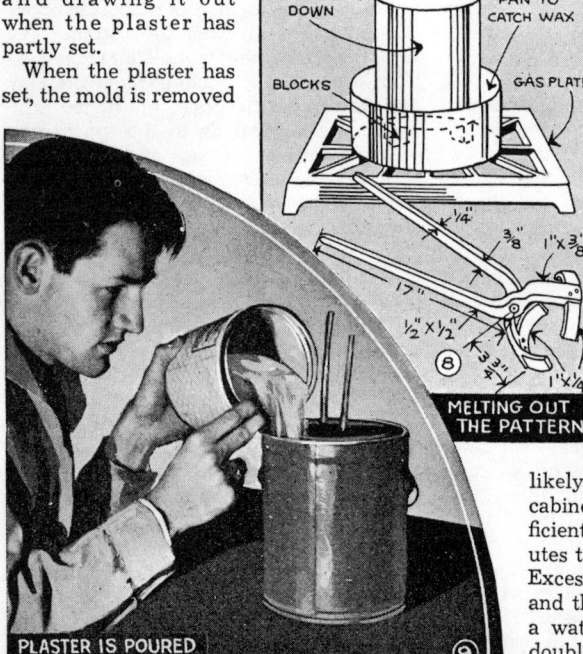

When the plaster has set, the mold is removed carefully from the clay. All the clay and the wet newspaper are now removed from the model and plaster mold. A pouring hole is drilled through the mold at the top, using a knife and working from the inside. The mold is then set back in position around the model as shown in Fig. 15. Both the model and the plaster mold should be coated with oil. Small vent holes must be drilled any point where the air is likely to be trapped. Next, soak cabinetmaker's white glue in sufficient cold water for fifteen minutes to swell the glue completely. Excess water is then poured off and the swollen glue is placed in a water-jacketed glue pot or a double boiler, and melted. When melted, it is poured into the mold until the space between the model and the plaster is filled. Give this plenty of time to set. The plaster case is then removed and the glue mold carefully split and skinned off. The model is no longer needed. Now cut a pouring hole at the top of the glue mold, as was done with the plaster case, and put the glue mold and the plaster case together and in place on the board. Vents should be provided if necessary. If the mold is in two parts it must be clamped together. Melt modeling wax and fill the glue mold. When the wax has set, you have a wax duplicate of the original model, and the casting is made from the wax pattern as previously described.

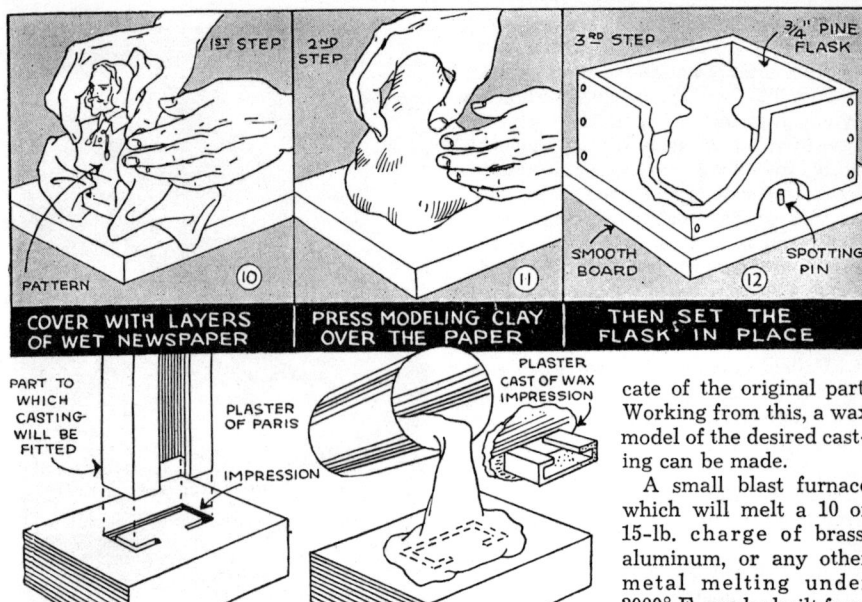

Frequently it is desired to make a casting, one side of which shall accurately fit a contoured surface, which may be cast, stamped, or formed in some other way. Fig. 13 shows one way of doing this. This method is particularly useful when it is required to cast a mounting bracket or ornament to fit a formed section, beam, or machine part exactly. A wax impression of the contour which is to be matched is made first and then a plaster cast of this impression which gives a positive dupli- cate of the original part. Working from this, a wax model of the desired casting can be made.

A small blast furnace which will melt a 10 or 15-lb. charge of brass, aluminum, or any other metal melting under 2000° F. can be built from a 5-gal. oil can with the top cut away, Figs. 17 and 18. Two holes are drilled and reamed for $3/8$-in. pipe nipples in opposite sides, $3\frac{5}{16}$ in. from the bottom. The heat-resistant lining is built up from ganister, a mixture of equal parts of fire clay and pulverized firebrick, moistened with water and worked up to the consistency of heavy plaster. Scraps of firebrick are added as filler and to increase the strength. The bottom of the can is covered with 1 in. of ganister, which is packed down by ramming with a stick. Four pieces of firebrick, each $4\frac{1}{2}$ in. long and $3\frac{1}{2}$ in. wide, are pressed down into the ganister, and another piece is placed in the center to support the crucible, Fig. 18. The space between these bricks is packed with ganister to within 1 in. of the top of the bricks. Two pieces of $3/8$-in. gas pipe are inserted through the burner holes to form openings for the burner pipes. A smooth sheet-metal form 8 in. in diameter and free from dents is now set in the can, centered and propped in place. The space between the outside can and the center form is now filled with ganister and pieces of firebrick. When the wall has set sufficiently to support its own weight, which will be in about 12 hours, the center form and the $3/8$-in.

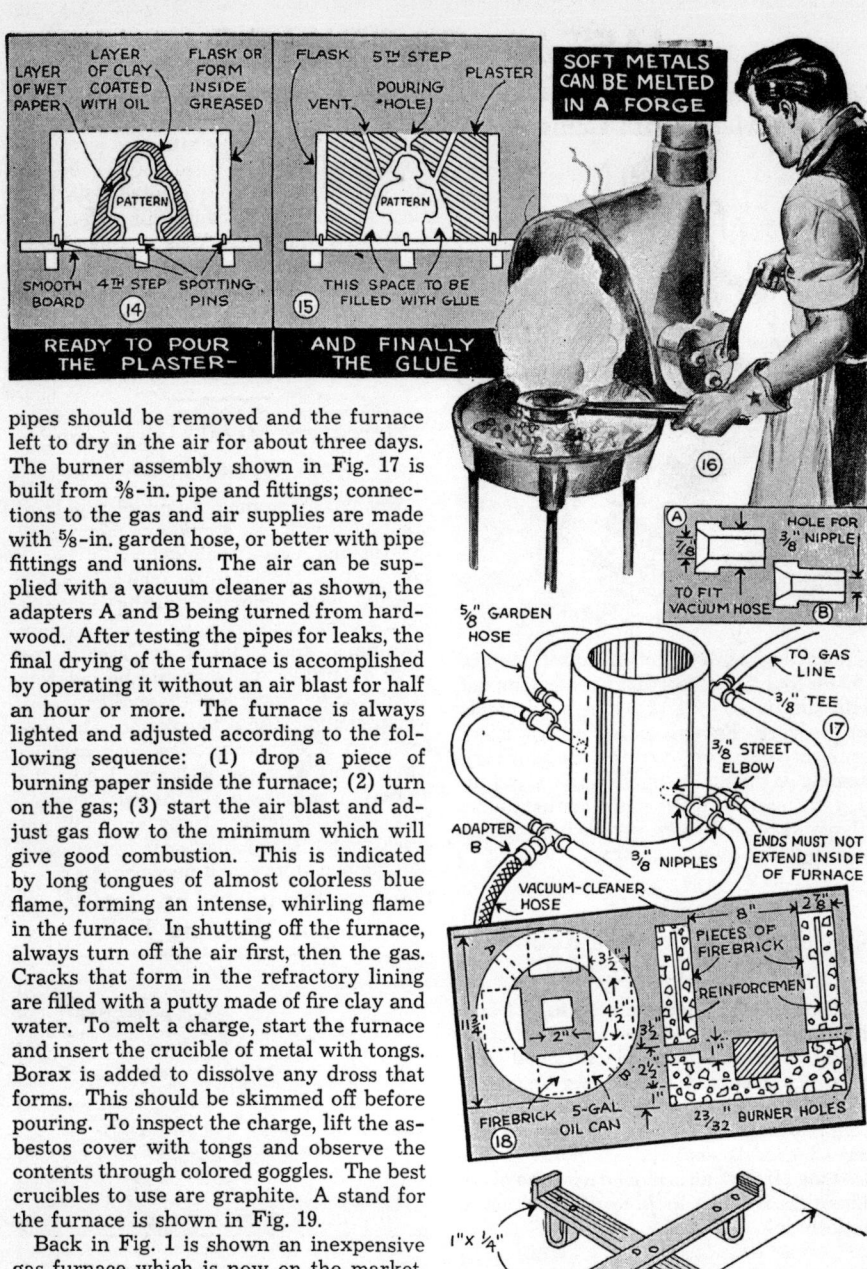

pipes should be removed and the furnace left to dry in the air for about three days. The burner assembly shown in Fig. 17 is built from ⅜-in. pipe and fittings; connections to the gas and air supplies are made with ⅝-in. garden hose, or better with pipe fittings and unions. The air can be supplied with a vacuum cleaner as shown, the adapters A and B being turned from hardwood. After testing the pipes for leaks, the final drying of the furnace is accomplished by operating it without an air blast for half an hour or more. The furnace is always lighted and adjusted according to the following sequence: (1) drop a piece of burning paper inside the furnace; (2) turn on the gas; (3) start the air blast and adjust gas flow to the minimum which will give good combustion. This is indicated by long tongues of almost colorless blue flame, forming an intense, whirling flame in the furnace. In shutting off the furnace, always turn off the air first, then the gas. Cracks that form in the refractory lining are filled with a putty made of fire clay and water. To melt a charge, start the furnace and insert the crucible of metal with tongs. Borax is added to dissolve any dross that forms. This should be skimmed off before pouring. To inspect the charge, lift the asbestos cover with tongs and observe the contents through colored goggles. The best crucibles to use are graphite. A stand for the furnace is shown in Fig. 19.

Back in Fig. 1 is shown an inexpensive gas furnace which is now on the market. In general construction it is similar to the furnace just described, but is fitted with a special mixing burner and an electric blower. It is supplied in several sizes, all suitable for home shop use.

METAL CASTINGS

Molds to Cast Small Metal Objects Are Made With Cement

SILICATE OF SODA AND CEMENT

In casting small objects from soft metals, I have found that portland cement mixed with silicate of soda provides good molds for the work. Enough soda is used to make a mixture of the right consistency for easy shaping of the mold. Such molds stand up well for casting toys, small statuary, etc. I have also used them for casting aluminum and zinc as well as brass.

Casting-Mold Corners Rounded With Ball-End Tool

When making aluminum castings, I find that a very neat job can be done by rounding the corners of the mold with wood putty, which I spread evenly and smoothly with a ball-end tool. This is made by soldering or brazing a 3/8-in. steel ball to one end of a small steel rod as shown in the drawing. If the tool is moved over the plastic with a twirling motion, you will get a uniform job.

Eliminating Pinholes in Castings

In making steel dies with a steel core in order to mold solid-babbitt bearings, there was only .015 in. allowance inside and outside for finishing. When finishing the castings, we found they were full of small pinholes just under the surface. To get rid of them both the steel die and core were first heated and then the inside of the die and the outside of the core painted with blue clay, pulverized very fine and mixed with water to a thin paste. This is applied with a small brush to the hot die and core quickly, as it will dry as soon as it touches the hot steel. This paint allows the gases to escape from the metal and eliminates all the holes.

Stereotype Metal Makes Good Casting Molds

Amateurs who have a limited supply of equipment will find stereotype metal good for making casting molds, as it has a low melting temperature, which is easily reached with ordinary heating methods. Wooden patterns can be used without burning, if the metal is not cast in too large quantities. Care should be taken to avoid getting water into the molten metal, which would cause it to splatter. The correct casting temperature has been reached when a piece of paper inserted into the molten mass turns a rich brown. Stereotype metal is fairly strong and will not crack under sharp blows. It can usually be obtained from small newspaper offices.

Castings Supported on Steel Wool While Welding Them

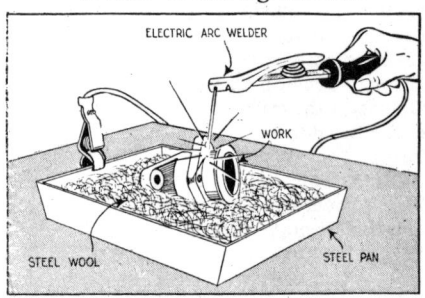

Sometimes when using an electric welding outfit on work that is irregular in shape it is difficult to attach the ground clamp to it. Try making a nest of steel wool in an iron or steel pan to hold the work and fasten the ground clamp to one edge of the pan. The steel wool assures a good electrical contact with the work

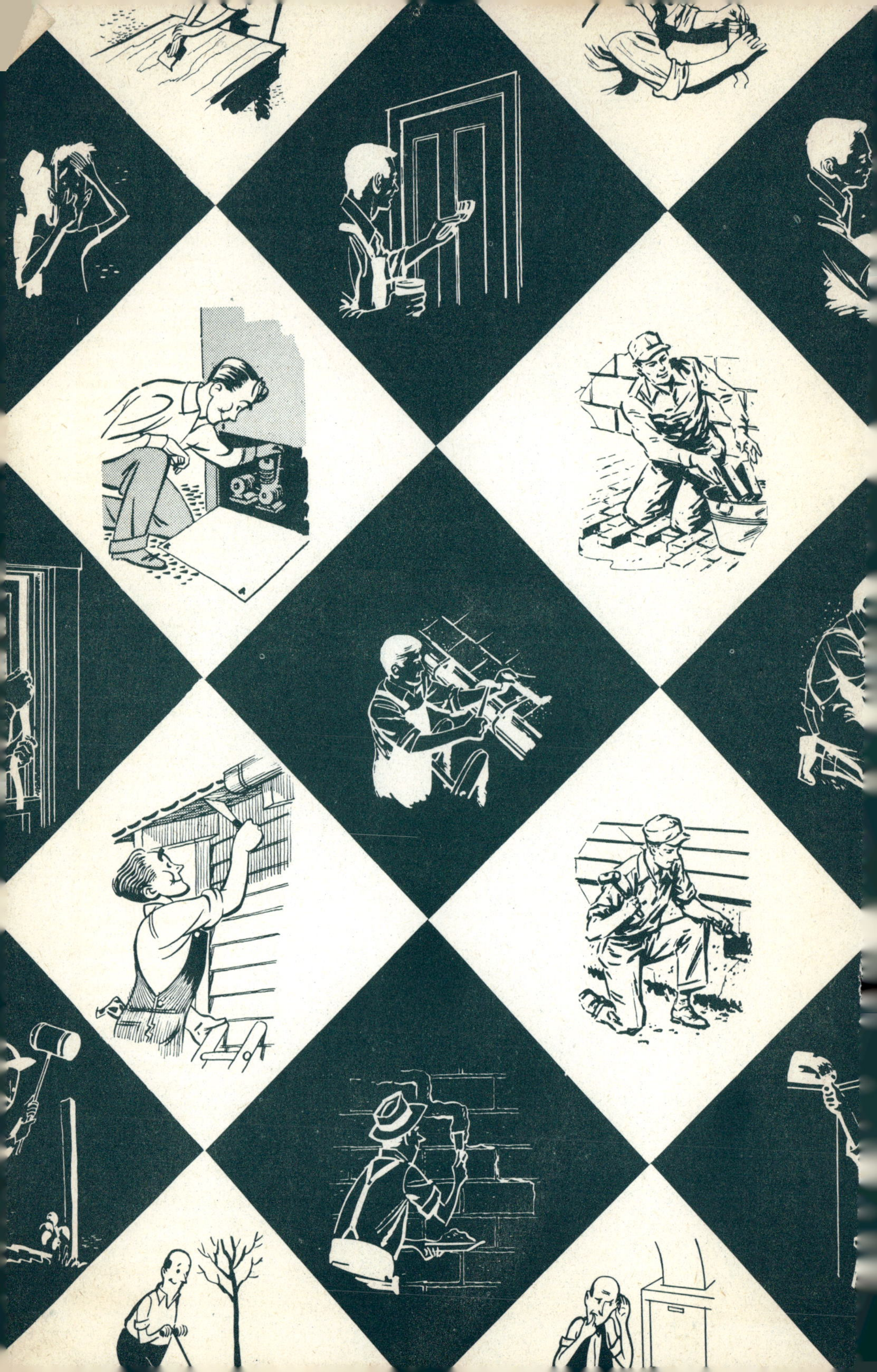